STUDIES ON MEXICAN PALEONTOLOGY

TOPICS IN GEOBIOLOGY

For detailed information on our books and series please visit: **www.springer.com**

Series Editors:
Neil H. Landman, American Museum of Natural History, New York, New York, landman@amnh.org
Douglas S. Jones, University of Florida, Gainesville, Florida, dsjones@flmnh.ufl.edu

---

Current volumes in this series

Volume 24: **Studies on Mexican Paleontology**
Francisco J. Vega, Torrey G. Nyborg, María del Carmen Perrilliat, Marison Montellano-Ballesteros, Sergio R.S. Cevallos-Ferriz and Sara A. Quiróz-Barroso
Hardbound, ISBN, 1-4020-3882-8, December 2005

Volume 23: **Applied Stratigraphy**
Eduardo A. M. Koutsoukos
Hardbound, ISBN, 1-4020-2632-3, January 2005

Volume 22: **The Geobiology and Ecology of *Metasequoia***
Ben A. LePage, Christopher J. Williams and Hong Yang
Hardbound, ISBN, 1-4020-2631-5, 2005

Volume 21: **High-Resolution Approaches in Stratigraphic Paleontology**
Peter J. Harries
Hardbound, ISBN 1-4020-1443-0, September 2003

Volume 20: **Predator-Prey Interactions in the Fossil Record**
Patricia H. Kelley, Michal Kowalewski, Thor A. Hansen
Hardbound, ISBN 0-306-47489-1, January 2003

Volume 19: **Fossils, Phylogeny, and Form**
Jonathan M. Adrain, Gregory D. Edgecombe, Bruce S. Lieberman
Hardbound, ISBN 0-306-46721-6, January 2002

Volume 18: **Eocene Biodiversity**
Gregg F. Gunnell
Hardbound, ISBN 0-306-46528-0, September 2001

Volume 17: **The History and Sedimentology of Ancient Reef Systems**
George D. Stanley Jr.
Hardbound, ISBN 0-306-46467-5, November 2001

Volume 16: **Paleobiogeography**
Bruce S. Lieberman
Hardbound, ISBN 0-306-46277-X, May 2000

Volume 15: **Environmental Micropaleontology**
Ronald E. Martin
Hardbound, ISBN 0-306-46232-X, July 2000

Volume 14: **Neogene, Paleontology of the Manonga Valley, Tanzania**
Terry Harrison
Hardbound, ISBN 0-306-45471-8, May 1997

Volume 13: **Ammonoid Paleobiology**
Neil H. Landman, Kazushige Tanabe, Richard Arnold Davis
Hardbound, ISBN 0-306-45222-7, May 1996

Studies on Mexican Paleontology

Edited by

FRANCISCO J. VEGA
Universidad Nacional Autónoma de México, Mexico

TORREY G. NYBORG
Department of Natural Sciences, Loma Linda University, CA, USA

MARÍA DEL CARMEN PERRILLIAT
Universidad Nacional Autónoma de México, Mexico

MARISOL MONTELLANO-BALLESTEROS
Universidad Nacional Autónoma de México, Mexico

SERGIO R.S. CEVALLOS-FERRIZ
Universidad Nacional Autónoma de México, Mexico

and

SARA A. QUIROZ-BARROSO
Universidad Nacional Autónoma de México, Mexico

Springer

A C.I.P. Catalogue record for this book is available from the Library of Congress.

ISBN-10 1-4020-3882-8 (HB)
ISBN-13 978-1-4020-3882-2 (HB)
ISBN-10 1-4020-3985-9 (e-book)
ISBN-13 978-1-4020-3985-0 (e-book)

Published by Springer,
P.O. Box 17, 3300 AA Dordrecht, The Netherlands.

www.springer.com

Cover illustrations:
Reconstruction of *Lambeosaurus laticaudus* with the skull of *L. lambei.* Campanian of Baja California (illustration by Rubén Rodríguez-de la Rosa).
Late Cretaceous ophiuroid from Chihuahua (photograph by Leonora Martín-Medrano).
Bauhcis moranii from the Oligocene of Puebla (photograph by Sergio R. S. Cevallos-Ferriz).

Top cover illustrations:
Tselfatia sp. from Turonian outcrops of Vallecillo, Nuevo León (photograph by Alberto Blanco-Piñón). *Megaxantho zoque* from Maastrichian outcrops of Ocozocoautla, Chiapas (photograph by Francisco Vega).

Printed on acid-free paper

Printed in the Netherlands.

Aims & Scope *Topics in Geobiology* Book Series

Topics in Geobiology series treats geobiology – the broad discipline that covers the history of life on Earth. The series aims for high quality, scholarly volumes of original research as well as broad reviews. Recent volumes have showcased a variety of organisms including cephalopods, corals, and rodents. They discuss the biology of these organisms-their ecology, phylogeny, and mode of life – and in addition, their fossil record – their distribution in time and space.

Other volumes are more theme based such as predator-prey relationships, skeletal mineralization, paleobiogeography, and approaches to high resolution stratigraphy, that cover a broad range of organisms. One theme that is at the heart of the series is the interplay between the history of life and the changing environment. This is treated in skeletal mineralization and how such skeletons record environmental signals and animal-sediment relationships in the marine environment.

The series editors also welcome any comments or suggestions for future volumes;

Series Editors:
Douglas S. Jones dsjones@flmnh.ufl.edu
Neil H. Landman landman@amnh.org

In Memoriam

While this book was in press, our dear and respected colleague Shelton P. Applegate (1928-2005), unexpectedly passed away. For many of us, more than a fellow researcher and a friend, he was an enthusiastic and esteemed teacher. Evidence for this can be found in the chapter by Alvarado-Ortega *et al.*, contained in this volume, where "Shelly" is considered the founder of "Mexican Paleoichtyology". We feel extremely fortunate to have had him share his passion and controversial ideas on Paleontology and Earth Sciences. For all his contributions and genuine love for Mexico, we will sorely miss him. We, editors and authors, would like to dedicate this book to his memory.

Foreword

During the last few years, the number of contributions to the Paleontology of Mexico has increased considerably. Paleontological work in Mexico has been focused on providing important information for petroleum exploration and specific studies dealing with pollen, foraminifera, radiolaria, dinoflagellates, rudists, and ammonites. Often these reports were published only in local or regional journals and therefore not available to the scientific community at large. The purpose of this book is to offer an updated review of the fossil groups from Mexico, providing their significance to the stratigraphy, tectonics, sedimentology, evolution and paleontology of Mexico whose study has proved to be relevant in stratigraphy, tectonics, sedimentology, and evolution. The fossil record of Mexico ranges from Precambrian to Pleistocene. Almost every Mexican State has reported fossil localities with ongoing studies and potential for the discovery of new localities. Even those localities that have been studied since the eighteen-century, such as the early Cretaceous San Juan Raya, have recently reported new fossil groups. Unfortunately, much of the fossil reports from Mexico have been published in Spanish from local journals, which represent a language barrier to the international community. There is little doubt that the paleontological history of Mexico deserves to be known in other countries. By making this book available to the international scientific community we hope that interest in the fossil record of Mexico will grow. Important topics, such as the Cretaceous/Tertiary controversy, were not included in this book because the topic has received a lot of attention and many papers in English have already been published. The main value of this book is the compilation, updated information and critical review of research on the diverse taxonomic groups, which include plants, pollen, corals, rudists, bivalves, crustaceans, echinoids, ophiuroids, brachiopods, fishes, amphibians, reptiles and mammals, as well as a review of one of the most famous localities of Mexico: Tepexi de Rodríguez, Puebla. The first chapters deal with the study of plant macrofossils and pollen, with emphasis on paleoenvironmental interpretations, stratigraphic implications, and paleogeographic considerations. The invertebrate chapters include reviews of taxa, biostratigraphy, and paleobiogeographic implications. The vertebrate chapters emphasize the importance of taxa recently found in different locations of Mexico, including evolutionary implications and correlation with other localities around the world. Chapter authors present this work not in Spanish but in English and we are grateful to all of them for this. There are excellent books on diverse topics, dealing with fossils of Mexico, but this is the first book to be published that represents the most active research groups in the country including interpretative reviews of many taxa. This book will be useful for teaching, reference, research and for the enthusiast to the relicts of the past.

Preface

Mexico, as a world paradigm of diversity and prolonged natural history, could not leave aside a fundamental aspect of her cultural inheritance, namely, the nature and evolution of life, as contained in the fossil register of animals and plants left in rocks of all epochs in the Mexican territory. This book, because of its timeliness and precious content, undoubtly will remain for years to come as a classical work in the scientific literature related to the subject. Throughout its 13 chapters it can be found either treated for the first time, or otherwise aptly reviewed, the results of many decades of research on fundamental fossil groups that tell us about the geologic, paleogeographic, and paleobiologic evolution of the past 350 million years in this part of the planet.

From the revision of the late Paleozoic fossil register of invertebrates (Chapter 8), represented in Mexico in formations that extend from Chiapas in the southeast, to Sonora and Baja California in the northwest, to the analysis of recent advances on the paleontology of the Tlayua Quarry in the State of Puebla (Chapter 13), a fossiliferous locality that is justly considered one of the most important paleontologic discoveries of the past century, this book also has the virtue to show the quality of life forms as a central element for the full understanding of how the Mexican territory was tectonically integrated, and also how these geologic changes of the earthly environment might have influenced the evolution of its inhabitants to the present time (Chapter 1). Particularly welcome is research material that deals with essential aspects of the plant fossil record (Chapter 2), Mammals (Chapter 12), or Dinosaurs (Chapter 11) that inhabited the Mesozoic or Cenozoic continental basins of Mexico, as well as others that bring to the attention of the international community animal groups for which the Mesoamerican and particularly the Mexican region was the theater of rapid changes and radiation, and yet little or nothing had been published before in Mexico (Chapters 5, 7, 9 and 10). Indeed, classical and internationally well known Mexican fossil groups such as Rudists (Chapter 4), Echinoids (Chapter 6), and Corals (Chapter 3), are more timely treated in the book to the benefit of readers interested in major episodes of natural history, as recorded in the Mexican geology and paleontology, and how this history integrated with the ever changing earth of those times.

Considering the various aspects handled by both original research, as well as in revision chapters, one very important added value should be the rather extensive bibliography of almost 1000 references related to the central topic contained in the book, an invaluable information which by itself constitutes a tribute to the memory of pioneer contributions by researchers that have worked in Mexico for the past two centuries.

Finally, as a student of the Mexican physical geology for several decades, and therefore may be as a relative outsider to the main subject of the book, I wish to congratulate sincerely the builders of this great work, and wish for it the best success it undoubtly deserves, and I am sure it will achieve.

Fernando Ortega-Gutiérrez
Instituto de Geología
UNAM

Contributors

Jesús Alvarado-Ortega Instituto de Geología, Universidad Nacional Autónoma de México, Ciudad Universitaria, 04510, D. F., México, brujaynerón@hotmail.com

Shelton P. Applegate Instituto de Geología, Universidad Nacional Autónoma de México, Ciudad Universitaria, 04510, D. F., México

Mouloud Benammi Instituto de Geofísica, Universidad Nacional Autónoma de México, Ciudad Universitaria, 04510, D. F., México

Alberto Blanco-Piñón Centro de Investigaciones en Ciencias de la Tierra, Universidad Autónoma del Estado de Hidalgo, Km. 4.5 Carretera Pachuca-Tulancingo, 42184 Pachuca, Hidalgo, México

Sergio R. S. Cevallos-Ferriz Instituto de Instituto de Geología, Universidad Nacional Autónoma de México, Ciudad Universitaria, Delegación Coyoacán, 04510, México, D.F., scrscfpb@servidor.unam.mx

Luis Espinosa-Arrubarrena Instituto de Geología, Universidad Nacional Autónoma de México, Ciudad Universitaria, 04510, D. F., México

Harry F. Filkorn Department of Invertebrate Paleontology, Natural History Museum of Los Angeles County, 900 Exposition Boulevard, Los Angeles, California 90007 USA, hfilkorn@nhm.org

Pedro García-Barrera Museo de Paleontología, Departamento de Biología Evolutiva, Facultad de Ciencias, UNAM. Ciudad Universitaria 04510, Coyoacán, Mexico, D.F. pgb@hp.fciencias.unam.mx

Katia A. González-Rodríguez Centro de Investigaciones Biológicas, Universidad Autónoma del Estado de Hidalgo, Km. 4.5 carretera Pachuca-Tulancingo, 42184 Pachuca, Hidalgo, México, katiag@uaeh.edu.mx

Enrique A. González-Torres Instituto de Instituto de Geología, Universidad Nacional Autónoma de México, Ciudad Universitaria, Delegación Coyoacán, 04510, México, D.F., eagtgaia@geologia.unam.mx

Eduardo Jiménez-Hidalgo Universidad del Mar - Campus Puerto Escondido, Km 3.5 Carretera Puerto Escondido-Oaxaca, Puerto Escondido, 71980 Oaxaca, México. eduardojh@zicatela.umar.mx

Leonora Martín-Medrano Museo de Paleontología, Departamento de Biología Evolutiva, Facultad de Ciencias, UNAM. Ciudad Universitaria 04510, Coyoacán, México, D.F.

Enrique Martínez-Hernández Instituto de Geología, UNAM. Departamento de Paleontología. Laboratorio de Palinología. Ciudad Universitaria 04510, Coyoacán, México, D.F. emar@servidor.unam.mx

Marisol Montellano-Ballesteros Instituto de Geología, UNAM. Circuito Exterior, Ciudad Universitaria, Delegación Coyoacán, 04510, DF, México. marmont@servidor.unam.mx

Itzia Nieto-López Museo de Paleontología, Departamento de Biología Evolutiva, Facultad de Ciencias, UNAM. Ciudad Universitaria 04510, Coyoacán, Mexico D.F. nlie@minervaux2.fciencias.unam.mx

Torrey G. Nyborg Department of Natural Sciences, Loma Linda University, Loma Linda, California, 92350 USA, tnyborg06g@ns.llu.edu

Jorge A. Ortíz-Mendieta Departamento de Paleontología, Museo del Desierto, A.P. 307, C.P. 25000, Saltillo, Coahuila, México

Ernesto Ovalles-Damián Museo Eliseo Palacios Aguilera, Calz. de los Hombres Ilustres s/n Parque Madero, Colonia Centro, CP 29000, Tuxtla Gutiérrez, Chiapas, México

María del Carmen Perrilliat Instituto de Geología, Universidad Nacional Autónoma de México, Ciudad Universitaria, Delegación Coyoacán, 04510 – México, D.F., mariacp@servidor.unam.mx

Sara A. Quiroz-Barroso Museo de Paleontología, Departamento de Biología Evolutiva, Facultad de Ciencias, UNAM. Ciudad Universitaria 04510, Coyoacán, México, D. F. saqb@hp.fciencias.unam.mx

Elia Ramírez-Arriaga Instituto de Geología, UNAM. Departamento de Paleontología. Laboratorio de Palinología. Ciudad Universitaria 04510, Coyoacán, México, D.F.

Víctor Hugo Reynoso Instituto de Biología, UNAM. Circuito Exterior. Ciudad Universitaria 04510, Coyoacán, México, D.F. vreynoso@ibunam.ibiologia.unam.mx

Héctor E. Rivera-Sylva Departamento de Paleontología, Museo del Desierto, A.P. 307, C.P. 25000, Saltillo, Coahuila, México, hriveras@yahoo.com.mx

Rubén Rodríguez de La Rosa Departamento de Paleontología, Museo del Desierto, A.P. 307, C.P. 25000, Saltillo, Coahuila, México

Francisco Sour-Tovar Museo de Paleontología, Departamento de Biología Evolutiva, Facultad de Ciencias, UNAM. Ciudad Universitaria 04510, Coyoacán, México, D. F. fst@hp.fciencias.unam.mx

Francisco J. Vega Instituto de Geología, Universidad Nacional Autónoma de México, Ciudad Universitaria, Delegación Coyoacán, 04510 – México, D. F., vegver@servidor.unam.mx

Contents

Chapter 3 • Mesozoic Corals of Mexico

Harry F. Filkorn

Chapter 4 • Rudists of Mexico: Historical Aspects and Selected Areas of Research

Pedro García-Barrera

Chapter 5 • Mesozoic and Tertiary Decapod Crustacea from Mexico

Francisco J. Vega, Torrey G. Nyborg and María del Carmen Perrilliat

Chapter 6 • Cretaceous Echinoids of Mexico

Itzia Nieto-López and Pedro García-Barrera

Chapter 7 • Fossil Ophiuroids of Mexico

Leonora Martín-Medrano and Pedro García-Barrera

Chapter 8 • Fossil Record of Upper Paleozoic Marine Invertebrates from Mexico

Sara A. Quiróz-Barroso and Francisco Sour-Tovar

Chapter 9 • Mesozoic Osteichthyans of Mexico

Jesús Alvarado-Ortega, Katia A. González-Rodríguez, Alberto Blanco-Piñón, Luis Espinosa-Arrubarrena, and Ernesto Ovalles-Damián

Chapter 10 • Research on Fossil Amphibians and Reptiles in Mexico, from 1869 to Early 2004 (Including Marine Forms but Excluding Pterosaurs, Dinosaurs, and Obviously Birds)

Víctor Hugo Reynoso

Chapter 11 • A Review of the Dinosaurian Record from Mexico

Héctor E. Rivera-Sylva, Rubén Rodríguez-de la Rosa and Jorge A. Ortiz-Mendieta

Chapter 12 • Mexican Fossil Mammals, Who, Where and When?

Marisol Montellano-Ballesteros and Eduardo Jiménez-Hidalgo

Chapter 13 • Revision of Recent Investigations in the Tlayúa Quarry

Shelton P. Applegate, Luis Espinosa Arrubarrena, Jesús Alvarado-Ortega and Moloud Benammi

Chapter 1

Geological setting and phytodiversity in Mexico

SERGIO R. S. CEVALLOS-FERRIZ and ENRIQUE A. GONZÁLEZ-TORRES

Instituto de Geología, Universidad Nacional Autónoma de México, Ciudad Universitaria, Delegación Coyoacán, 04510 – México, DF, scrscfpb@servidor.unam.mx, eagtgaia@geologia.unam.mx

1. Abstract

The joint discussion of biotic and abiotic factors allows proposal of a hypothesis in which the dynamic nature of the geological processes and of the organisms facilitate the understanding of the biodiversity of Mexico. In this respect, we highlight that since the Cretaceous, Mexico had a configuration similar to that of a peninsula that increased in size through time, but that with the establishment of the Isthmus of Panama approximately 3.5 my ago, it no longer had such a configuration. During this peninsular configuration, the main mountain belts of Mexico formed in different episodes of time attaining variable sizes through dynamic and fluctuant processes until they reached their extant physiognomy. At the same time the retreat of the seas was significant in exposing positive zones of the continent, where life was established. Finally, after occupying their current positions, the Baja California Peninsula and the Yucatan Peninsula started providing cover to the central continental mass of Mexico. The fossil record suggests that while all of this occurred, the newly generated environments were occupied by plants that were formerly part of higher latitude communities of North America. That through the land bridges that connected

F. Vega et al. (eds.), Studies on Mexican Paleontology, 1–18.

North America with Europe and Asia, plants with different geographical affinities arrived in Mexico. Also with the establishment of the Chortis Block as the northern part of Central America, at least part of the plants already growing on the Chortis Block arrived into the rest of Central America eventually expanding their distribution into South America. The interaction of the biological processes, as suggested by the fossil record with the environment, coupled with the geological setting, promoted the development and establishment of a distinctive biota among which, the endemic plants that characterize today the vegetation of Mexico stand out.

Some of the extant Mexican plants were already present in the Oligocene and possibly since the Eocene. Finally, the current landscape and the elements that conform it began their "modernization" some 3.5 million years ago, but they may have a longer history that started in the Cretaceous.

2. Introduction

The diversity and taxonomic relationships of the extant flora and vegetation of Mexico, compared with that of other geographical regions, has been reported since, at least, the beginning of the XIX century (e.g., Hemsley, 1879-1888; Engler, 1882, 1905; 1914; Standley, 1936; Johnston, 1940; Sharp, 1951, 1953; Matuda, 1953; Miranda, 1958, 1960; Toledo, 1982; Rzedowski, 1965, 1991). Recognizing that diversity and taxonomy support the natural history of the participating lineages, two hypotheses have been proposed to explain the conformation of the extant vegetation of Mexico and Central America. The first idea explains the wealth of vegetation in Mexico as a function of the pressures that the environment and/or the climate exercised upon the plant communities during the establishment of the Isthmus of Panama and through the Ice ages of the Plio-Pleistocene. This idea is possibly the most widespread one. This proposal implies that the biological rearrangement of plants took place only in the last 3.5 million years when the current landscape conformed (Axelrod, 1950; Raven, 1963; Sarukhán, 1968; Graham, 1973, Burnham and Graham, 1999). In contrast, a second hypothesis that has acquired greater importance in recent years suggests that the diversity and the geographical relationships of the vegetation that today develop in Mexico and Central America have a considerably longer history in the region, possibly going back to the Cretaceous (ca. 135 mybp), but at least to the Paleocene (65 mybp), or slightly earlier (Bray, 1898, 1900; Krueger, 1934; Johnston, 1940; Rzedowski, 1991; Wendt, 1998). This second hypothesis also acknowledges the influence of the environment and/or climate as prominent factors that guided the development of populations and communities, but emphasizes the age in which this interaction occurred. Although both hypotheses highlight the importance of physical parameters as dynamic factors influencing the biological track of plants, their discussion from a historic perspective has been reported, in the best of the cases, with little detail.

Communities composed of a particular mixture of organisms that coexist and interact in given areas produce the characteristic biodiversity of a region. The establishment of these communities is linked to a natural assemblage of variable factors such as air, water, soil, temperature, altitude and latitude, that due to their own interactions, offer the biota different settings in which life may develop. Defining these settings and the communities they contain has been an arduous task; to the point that today the topic continues to be the center of important debates (e.g., Hemsley, 1879-1888; Johnston, 1924; Wiggins, 1960; Miranda, 1952; Rzedowski, 1966, 1972; Trejo-Vázquez and González-Medrano, 2001). The form in which these factors had interacted through time in the area occupied today by Mexico and Central America surely favored the establishment of various settings that are not necessarily similar to the extant ones. Understanding how this region was geologically built up and identifying the plants that were able to establish in different settings at the same time, allows the proposal of this new hypothesis to explain the biodiversity that characterizes the region and introduces an historical biogeographic component to the lineages that conform it.

3. Organization of this Chapter

In the following sections, some prominent aspects of the integration of the geological setting of Mexico and the northern portion of Central America during the Upper Cretaceous and Cenozoic will be presented. These scenarios generate a general vision (not widely discussed here) of the orography, hydrography and distribution and extension of the seas during different time periods. These factors, along with others, influenced the distribution of air, marine currents and humidity; together they imposed restrictions to the distribution of the plants according to their ecological capabilities. Immediately following, some aspects of the paleobotanical record of the area will be presented, beginning in the Upper Cretaceous. Finally a mechanism to explain how the vegetation that today develops in the region is proposed.

4. Geological Setting

The geological evolution of Mexico and Central America during the Cretaceous and Cenozoic had significant influence on the processes that controlled the integration of their vegetation (Fig. 1). To better understand the main geological events that took place in Mexico during this time interval, the following general processes can been recognized: i) development of large magmatic provinces (Sierra Madre Occidental, Faja Volcánica Mexicana, Sierra Madre del Sur); ii) fragmentation and displacement of continental segments including the opening of the Gulf of California and the displacement of the Chortis Block; iii) marine regression that outlined the current contour of Mexico; and iv) formation and lifting of the eastern orogenic belt (Sierra Madre Oriental).

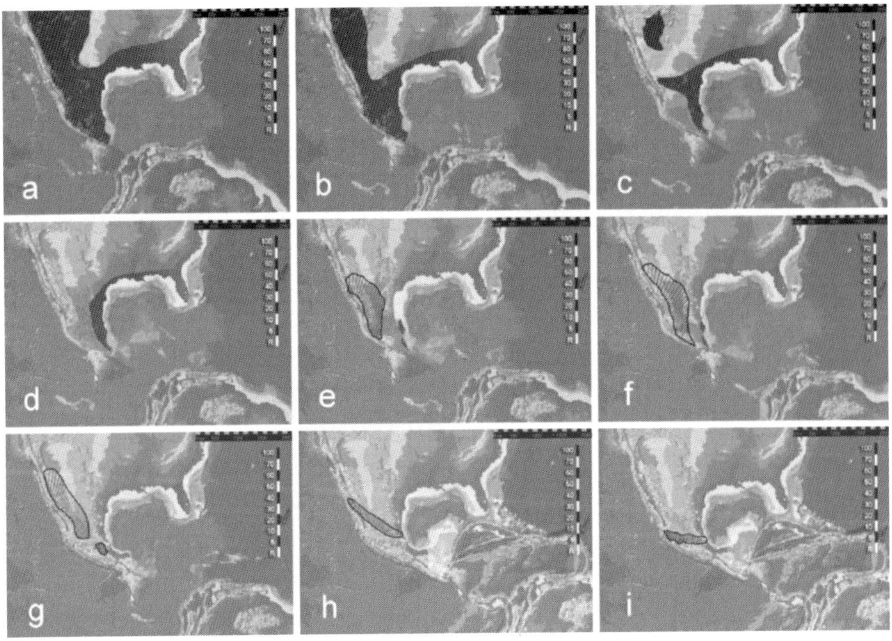

FIGURE 1. Schematic reconstruction of the geological history of Mexico during the last 100 million years, highlighting significant features. Black lines represent the approximate extension of the volcanic arch. a) Albian; generalization of marine transgressions, b) Maastrichtian; lifting of the northwest portion, c) Upper Paleocene; continuation of the marine regressions while the Laramide deformation is in progress, d) Middle Eocene; extensive volcanic activity e) Upper Eocene; displacement of the Chortis Block starts, widening of the Volcanic Arch, f) Lower Oligocene; ignimbritic flare up, g) Early Miocene; extensional pulses and change in orientation (now NW-SE) of the Volcanic Arch, h) Late Miocene; opening of the Gulf of California begins and formation of the Faja Volcánica Mexicana, i) Early Pliocene; emergence of the Yucatan Peninsula, the connection between the Pacific and the Caribbean is interrupted.

i) *Development of large magmatic provinces* (Fig. 1). Development of the Sierra Madre Occidental, which has a general N-S orientation along the Pacific coast, started in the north of the country spanning from Cretaceous to the early Miocene (Figs. 1a-g). During the Eocene volcanic rocks formed a wider Sierra Madre Occidental (Fig 1e, Aguirre-Díaz and McDowell, 1991). At the onset of the Oligocene (ca. 32-28 mybp) the volcanic activity of this province became especially intense, forming the largest ignimbrite deposit on the planet (Fig. 1f, Ortega *et al.* 1992; McDowell and Clabaugh, 1979). It has been calculated that most of these volcanic products, reaching average thickness of 1,000 m, were deposited in the span of 4 million years (Ortega *et al.* 1992; Ferrari *et al.* 2002).

The Faja Volcánica Mexicana is a volcanic arch with an extensive variety of stratovolcanos, monogenetic cones, etc. This volcanic province has been active since the Middle Miocene and it is composed by almost 8,000 volcanic centers in an area of approximately 1,000 km by 20-150 km (Velazco-Tapia and Verma, 2001). It is oriented in an E-W direction and divides Mexico into two regions: a drought and subtropical region towards the north; and a humid and tropical region to the south. The Sierra Madre del Sur is composed of plutonic and volcanic rocks of Paleocene to Middle Miocene age (Morán-Zenteno *et al.* 2000). However, the southern margin of the Mexican Pacific demonstrates that the region was exposed to intense lifting and erosion as there are extensive plutons exposed.

The aforementioned enormous volcanic activity contributed a colossal volume of materials that through time transformed the physiography of Mexico (e.g., morphology and hydrology). As igneous material piled up, new mountain chains with varying altitudes were formed, opening and/or closing basins with a consequent change in the hydrologic conditions. This common process most likely affected the magnitude of the natural barriers and selective forces that influenced the distribution of plants. Furthermore, gas emissions as well as the presence of suspended particles in the atmosphere, suggest the development of special atmospheric conditions in the region. For example, depending on the concentration of free gases in the stratosphere due to volcanic activity, SO and CO2 may have opposite effects and may perform differently at different time scales depending on how long these gases stay in the atmosphere. Thus, a regional warming, based on the greenhouse effect, would be expected due to the presence of SO during a relatively short period of time, but if the time span is bigger (months to ca. 10 years) the large eruptions may cause cooling instead. In contrast, the CO2 emissions may cause a regional warming in the scale from 10 to 10,000 years (Wignall, 2001).

ii) *Fragmentation and displacement of continental segments* (Fig.1). The fragmentation and displacement of continental blocks is a distinctive feature of the Cenozoic geology of Mexico. Two important processes can be recognized, the displacement of the Chortis Block, and the formation of the Baja California Peninsula.

The Chortis Block, a continental fragment that spans from Honduras to Nicaragua (Anderson and Schmidt, 1983), moved along the southwestern margin of the Mexican Pacific Coast in an interval of time during the Cenozoic. The oldest recognized position for this continental block during the Upper Cretaceous was at a maximum latitude similar to that of the region of Zihuatanejo (Fig. 1a-c, Schaaf *et al.* 1995). Its movement southwards began some 40 mybp (Fig. 1d). As the movement of this block proceeded (Fig 1e-i), three main effects that limited the biota distribution in the past could be predicted. First, some continental areas were gradually exposed or protected from the influence of the ocean. Second, in the southern margin of the Mexican Pacific an intense lifting of the continental crust was promoted, allowing the formation of large basins with continental deposits, altering the drainage, and promoting important processes of erosion. And finally, the establishment of a land connection between North and South America, approximately 3.5 mybp, as it contacted the portion that today is South Central America.

The formation of the Baja California Peninsula and the opening of the Gulf of California are also related to the fragmentation and displacement of a continental fragment. Approximately 5 million years ago (Fig. 1h-i), the Oriental Pacific Plate began spreading northwards, entering in the incipient Gulf of California and initiating a marine invasion (Martín-Barajas, 2000). From that moment on, the separation of continental blocks continued, generating extensional tectonic processes that began the formation of the basin and range system in the area. The opening of the Gulf of California is the consequence of a right lateral movement between the Pacific and North America plates.

iii) *Marine regressions* (Fig. 1). The marine regressions that extensively affected eastern Mexico during the Cenozoic contributed to the configuration of the region. They also were a determinant factor for the establishment of the flora and vegetation through the habilitation of positive exposed areas where organism became established.

iv) *Lifting of the oriental orogenic belt* (Fig. 1). The Laramide orogeny, a process linked to the marine regressions, started in the Late Cretaceous and ended in the Middle Eocene (Fig. 1a-d), prompting since then the retreat of the seas and influencing the current configuration of the area. The territory that emerged due to this process increased the surface of the country in ca. 100% (Ortega et al., 1992), offering the continental biota a considerable new space for its establishment and development. The retreat of the seas was a gradual process, with alternation of regressions and transgressions, each one of which had individual characteristics in different regions. As previously pointed out, the formation of diverse sedimentary basins is an important characteristic of the geology of Mexico, and in this respect, the Coastal Plain of the Gulf of Mexico is a prominent example. As seas withdrew and the orogenic system of the Sierra Madre Oriental conformed and lifted, one of the more important mountain systems of the current geography of Mexico became established.

5. Outstanding Aspects of the Fossil Record

The hypotheses that had been previously proposed to explain the origin of the vegetation of Mexico were mainly based on interpretations of the extant flora and vegetation. In a few exceptions, fossil plants collected in Mexico, -but little studied-, were considered (e.g., Rzedowski, 1978). Instead, fossils of higher latitude paleocommunities were frequently extrapolated in order to fill this gap (e.g., United States of North America, Europe; e.g., Graham, 1973, 1976). The limited integration of the flora and vegetation of Mexico and the misuse of other biological principles (e.g., the present is the key to the past) encouraged the extrapolations, to the point that the biological and geological processes that were taking place in the area were, at best, marginally considered.

It is a fact that the fossil record of the flowering plants of Mexico has not been extensively studied and that the majority of the reports are known from the palynological record (whose focus has been mainly biostratigraphic) limiting in that sense the knowledge of the taxonomic affinities of the plants (Cevallos-Ferriz and Ramírez, 1996; Martínez-Hernández and Ramírez-Arriaga, 1996). Nevertheless, the combined work on macro and microfossils of the flowering plants of Mexico, made in recent years, renders more than 200 identified taxa. Although this number of identified plants is still low to firmly support a conclusive explanation for the integration of the vegetation of Mexico, it is sufficient to promote reconsideration of some ideas. For example, *Ilex, Quercus* and *Pinus* grow natively in the tropical regions of Mexico today, but traditionally, their origin is considered to have occurred north of Mexico and they have been used as an example of plants that reached the southern part of North America when the ice sheets spread on the continent during the Plio-Pleistocene. However, *Ilex* and *Pinus* are now known from Eocene sediments of the NE of Mexico and *Pinus* and *Quercus* are important elements of the Mexican taphofloras since the Oligocene (Martínez-Hernández and Ramírez-Arriaga, 1996). Thus, their participation in the integration of the vegetation of Mexico has to be reconsidered. In a similar way, the presence of plants in the tropical forests of Mexico like *Cyrilla, Ticodendron, Alfaroa* and *Oreomuna* has been explained as northern elements that became established in the south when seeking shelter during the glacial maximum of the Plio-Pleistocene. Another example concerning these reinterpretations involves the recognition of fossils of certain groups such as Anacardiaceae and Leguminosae in Mexico (Figs. 8-18). Traditionally, Mexican presence of these plant families has been interpreted as a consequence of the establishment of the Isthmus of Panama, through which they supposedly arrived. However, the known fossil record in the Mexican territory of the former may go back to the Cretaceous and members of the latter have been collected in Eocene sediments in northern Mexico, suggesting once again the need for a reevaluation of our ideas on the integration of vegetation in low latitude North America (e.g., Magallón-Puebla, 1994a; González-Medrano, 1996; Calvillo-Canadell and Cevallos-Ferriz, 2001, 2005; Martínez-Millán, 2000; Martínez-Cabrera, 2002; Ramírez and Cevallos-Ferriz, 2000a, 2002).

Among the taxa that grew in Mexico during the Tertiary there are some that suggest relationships with plants of different regions of the world. For example, *Eucommia* (Fig. 19; Eucommiaceae) was widespread in the Northern Hemisphere during the Eocene-Miocene but today lives only in the humid forests of China (Magallón-Puebla and Cevallos-Ferriz, 1994b). *Cedrelospermum* (Fig. 21; Ulmaceae) is an extinct taxon that coexisted with *Eucommia* in the Miocene (Magallón-Puebla and Cevallos-Ferriz, 1994c). The presence of *Statzia* (Fig. 22; *incertae sedis*) in Mexico is extremely interesting, since aside from this report, it is only known from

Oligocene sediments of Roth, Germany (Magallón-Puebla and Cevallos-Ferriz, 1994d) suggesting again, relationships with the other regions of the Northern Hemisphere. Some flowers recovered from the amber of Simojovel, Chiapas, are comparable to the *Statzia* impression/compression fossils from Tepexi de Rodríguez, Mexico and Roth, Germany. Other Tertiary plants of Mexico suggest relationships with other regions of the world, *Berberis* and *Mahonia* (Figs. 20, 23, 27; Berberidaceae) may be related to extant plants of Asia and South America (Ramírez and Cevallos-Ferriz, 2000b) while *Copaiferoxylon* (Leguminosae) supports old relationships with Africa or South America (Cevallos-Ferriz and Barajas-Morales, 1994).

Some other fossil plants of the Tertiary of Mexico point to important links with the Caribbean and reinforce those with South America. Certainly, *Hymenaea* (Leguminosae), the plant that produced the amber of the Oligocene-Miocene of Simojovel, Chiapas, Dominican Republic and Puerto Rico leads the list (Calvillo-Canadell, 2000). Not less surprising is the presence of *Haplorhus and Loxopterygium* (Anacardiaceae) among the fossils collected in Mexico. The former is an endemic plant that grows at present in the dry zones of Peru and Chile (Ramírez *et al.*, 2000; Ramírez and Cevallos-Ferriz, 2002), while the latter grows naturally from Venezuela to Argentina (Colina-Hernández, pers. com. 2000); however, both of them are found as fossils in the Oligocene of Mexico and Miocene of Ecuador (Burnham and Carranco, 2004). Other plants that suggest old and complex historic floristic links between Mexico and South America are *Inga* and *Pithecellobium* (Figs. 16-17; Leguminosae). Although their current distribution includes the two areas, their diversity is greater in South America, but the oldest known fossil record comes from Eocene and Oligocene sediments in Mexico (Calvillo-Canadell and Cevallos-Ferriz, 2001, 2005).

The fossil record of *Tapirira* (Anacardiaceae) suggests, in a subtle way, a southwards movement of the lineage during the Tertiary; this pattern may also be reinforced by the fossil record of other taxa (Martínez-Cabrera, 2002). The oldest record of this lineage corresponds to wood from the Eocene of Wyoming (Manchester, 1977). More recently, woods of this taxon appear in Oligocene-Miocene sediments of Baja California and Baja California Sur, and its flowers are known from even younger rocks containing the amber of Simojovel (Miranda, 1963; McKeown *et al.* 1991; Martínez-Cabrera, 2002). Finally, today *Tapirira* is found growing naturally in Southern Mexico and Northern South America (Terrazas and Wendt, 1995).

A recent report of *Berberis* (Figs. 20, 23; Berberidaceae) from Oligocene sediments of Central Mexico (Ramírez and Cevallos-Ferriz, 2000a) introduces to the discussion some important aspects as information from both, fossils and extant plants are used in conjunction. Today, this genus grows naturally in North and South America, but not in Mexico. Its fossil record, nevertheless, demonstrates that during the Tertiary it was able to grow in Mexico and that the characters of the fossil leaves collected in this region link them to extant species of South America and Asia (Ramírez and Cevallos-Ferriz, 2000a).

Another important piece of information that can be extracted from the fossil

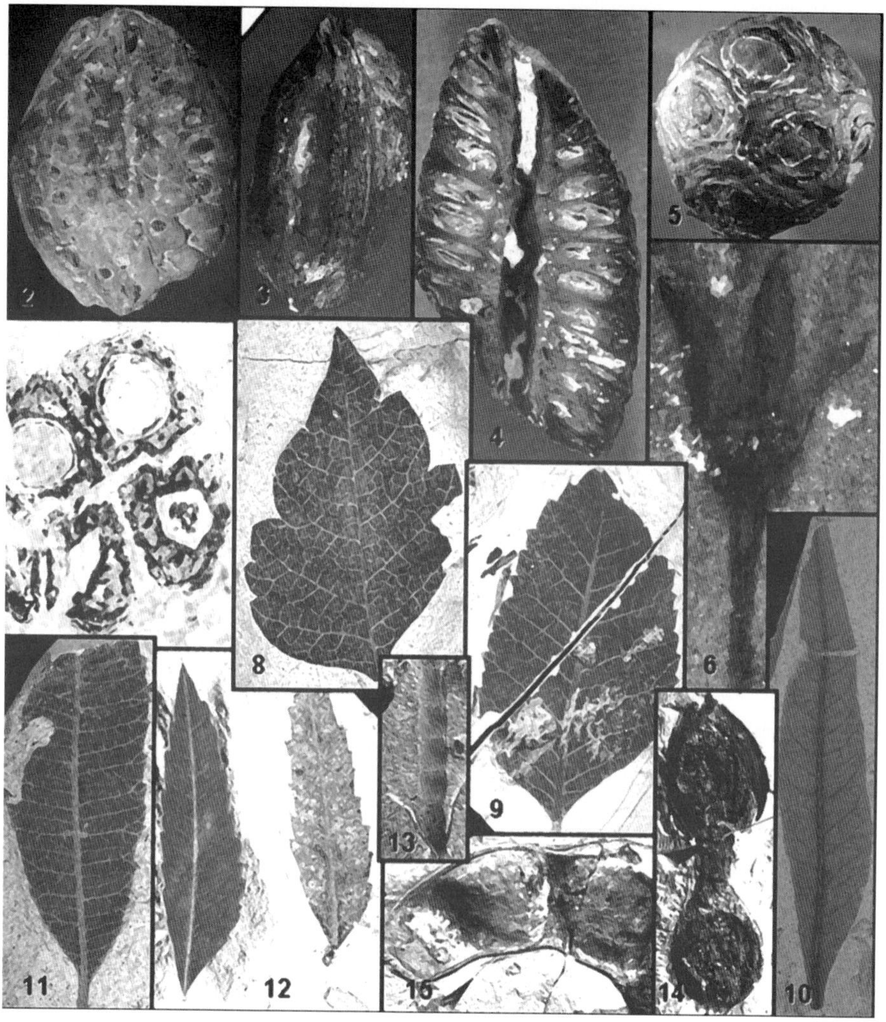

FIGURES 2-15. 2-6. Cretaceous fossils. 7-15. Tertiary fossils. 2.- *Striatornata sanantonensis* (Musaceae). 3.- *Tricostatocarpon silvapinedae* (Strelitziaceae). 4.- Hamamelidaceous fruit. 5.- Fruit of a probable Araceae. 6.- Flower of Rhamnaceae. 7.- *Tarahumara sophieae* (Haloragaceae). 8-12. Anacardiaceae. 8.- *Rhus toxicodendroides*. 9.- *Pseudosmodingium terrazasiae*. 10.- *Pistacia marquezii*. 11.- *Comocladia intermedia*. 12.- *Pseudomodingium mirandae* (right) and extant specimen (left), for comparison. 13-15. Leguminosae. 13.- *Reinweberia ornithopoides*. 14.- *Sophora sousae*. 15.- *Lysiloma mixtecana* .

record of Mexico is the recognition of the minimum age in which the precursors of endemic genera or species, such characteristic of the current vegetation of the country, first appeared. For example, in Oligocene sediments of the surroundings of Tepexi of Rodríguez, Puebla, which are assumed an age of 32 my, fossil fruits and leaves suggest that lineages that currently give the vegetation of the region a unique character were already present by that time. Among them are *Pseudosmodingium* (Figs. 9, 12, Anacardiaceae), *Comocladia* (Fig. 11, Anacardiaceae), *Mimosa* (Leguminosae), *Sophora* (Fig. 14, Leguminosae), *Lysiloma* (Fig. 15, Leguminosae), *Reinweberia* (Fig. 13; Leguminosae), and *Populus* (Fig. 24; Salicaceae) (Magallón-Puebla and Cevallos-Ferriz 1994a; Martínez-Millán, 2000; Ramírez *et al.* 2000; Ramírez and Cevallos-Ferriz, 2000b, 2002).

Detailed comparison between fossil and extant plants based on leaf architecture of Oligocene leaves collected in the Los Ahuehuetes locality within the surroundings of Tepexi of Rodríguez, Puebla, makes it evident that although the identified genera can be similar to those of the current vegetation of the area, at the specific level, there are important differences (Ramírez and Cevallos-Ferriz, 2000a, 2000b, 2002; Velasco de León *et al.* 1998; Velasco de León and Cevallos-Ferriz, 2000). Furthermore, similar studies with some fossil plants of the Miocene-Pliocene of Hidalgo and of the Plio-Pleistocene of Morelos suggest that the development of the extant species took place in relatively recent times. It is also extremely interesting that the assemblage of fossil plants in the outcrops so far studied show a combination of elements that are characteristic of different types of extant vegetation. This suggests that the types of vegetation now recognized have also changed, and therefore, it is necessary to understand how the patterns of plant association varied through time (Velasco de León, 1999; Ramírez-Garduño, 1999).

6. Integration of the Biodiversity

During the assemblage of Mexico as part of the North American Plate, the accretion of blocks has always been an important process, especially during the Paleozoic and Mesozoic; but it is during the Cenozoic that this process along with the displacement of lands, culminated in the current configuration of the region. A discussion of the vegetation of Mexico and the integration of the plants from an historical point of view is presented next. This discussion starts in the Upper Cretaceous, where the oldest plant macrofossils have been reported, since the majority of the continental land was formerly covered by sea. However, it is necessary to mention some enigmatic Lower Cretaceous pollen grain records; *Afropollis* (Winteraceae) and *Retimonocolpites* (Laurales) are found in Aptian/Albian sediments of Michoacán and Colima and, *Tricolpites* (Hamamelidaceae) and cf. *Spathiphyllum* (Araceae) in Durango (Martínez-Hernández and Ramírez-Arriaga, 1996). These records suggest the presence of exposed areas in the central part of Mexico, representing perhaps islands in which the angiosperms were able to become established (Martínez-Hernández and Ramírez-Arriaga, 1996). Stems or wood of *Palmoxylon and Mimoxylon* from Cretaceous sediments of Oaxaca had been reported (Müller-Stoll and Mädel, 1967; Cevallos-Ferriz and Ricalde-Moreno, 1994; Cevallos-Ferriz and Barajas-Morales, 1994), but unfortunately the locality where the material was collected has not been found again,

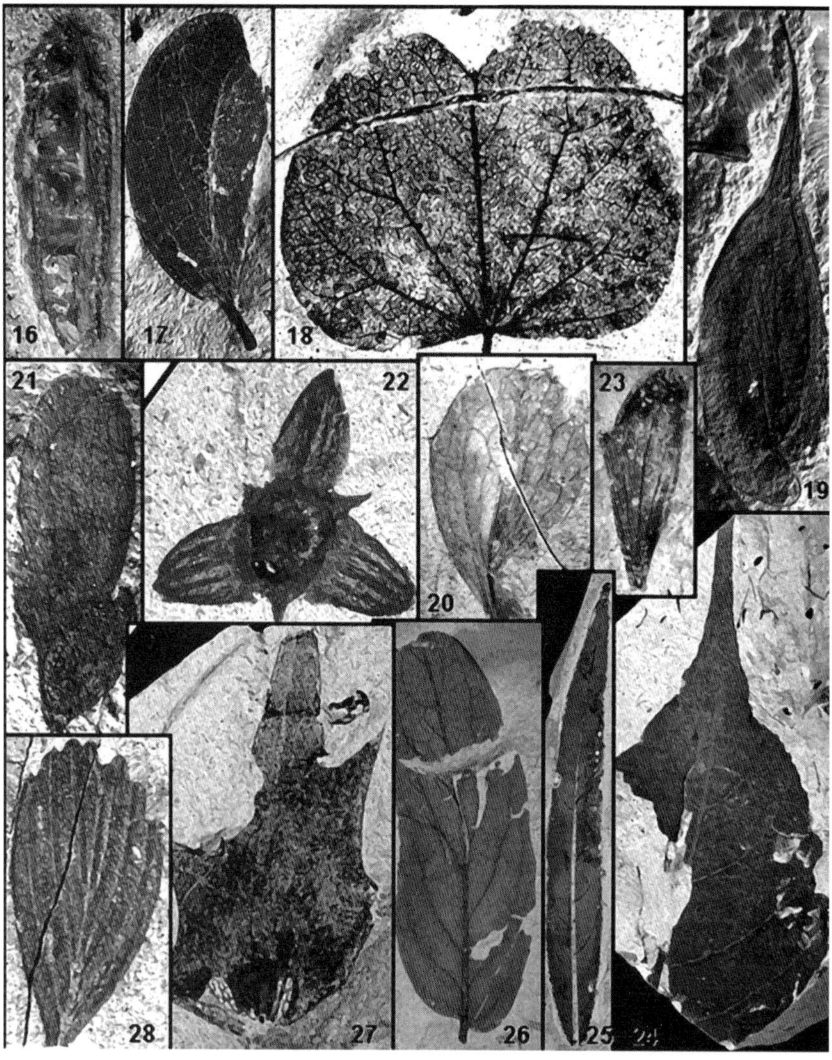

FIGURES 16-28. Tertiary fossils. 16.- Fruit similar to *Inga*. 17.- Leaflet similar to *Pithecellobium*. 18.- *Bauhcis moranii*. 19.- *Eucommia constans* (Eucommiaceae). 20.- *Berberis tepexiana* (Berberidaceae). 21.- *Cedrelospermum manchesteri* (Ulmaceae). 22.- *Statzia (Insertae sedis)*. 23.- *Berberis lozanofolia* (Berberidaceae). 24.- *Populus dentiacuminata* (Salicaceae). 25.- *Salix alencasterae* (Salicaceae). 26.- *Salix balsana* (Salicaceae). 27.- *Mahonia martinezii* (Berberidaceae). 28.- *Cercocarpus mixteca* (Rosaceae).

so the presence and age of these plants cannot be confirmed.

The most extensive and best known positive continental area of the southern portion of the North American Plate where life could have established in the Upper

Cretaceous corresponds to portions of the northern states of Mexico, in the so called Paleopeninsula of Coahuila (Fig. 1b). This paleopeninsula extended from the north-center portion of Tamaulipas to the southern region of Coahuila and then to the NE tip of Baja California. There is indication that some islands existed toward the south, but little is known about the life established on them. However, some fragments of wood collected in the surroundings of Huetamo, Michoacán, have been identified as *Araucarioxylon*. Some other conifers had been reported from the paleopeninsula of Coahuila, especially from its coal deposits, and in Sonora and Baja California (Rueda-Gaxiola, 1967; Martínez-Hernández *et al.* 1980; Weber, 1972).

Nevertheless, the angiosperms are more common in all these sediments. The macrofossils collected so far around Rincón Colorado in Coahuila, represent plants of the families Arecaceae, Musaceae (Fig. 2), Strelitziaceae (Fig. 3), Ranunculaceae, Menispermaceae, Hamamelideae (Fig 4), Phytolaccaceae, Rhamnaceae (Fig. 6) and probably Araceae (Fig. 5) (Rodríguez de la Rosa and Cevallos-Ferriz, 1994; Rodríguez de la Rosa *et al.* 1998; Pérez-Hernández *et al.* 1997). From Sonora, members of Arecaceae and Haloragaceae (Fig. 7) have been reported (Cevallos-Ferriz, 1983; Hernández-Castillo and Cevallos-Ferriz, 1999); and in both sites *Paraphyllanthoxylon*, a wood type that may represent Lauraceae, Anacardiaceae or Burseraceae, among other families, has been described.

These plants show that in the Upper Cretaceous, the biota of the positive continental parts of Mexico was represented by groups that today are more common either in the Northern (e.g., Ranunculaceae, Menispermaceae) or in the Southern Hemisphere (e.g., Musaceae, Strelitziaceae, Anacardiaceae) (Raven and Axelrod, 1974; Taylor, 1990). Furthermore, the fossil record of some of the families that are currently more diverse in the Southern Hemisphere, suggests that during the Cretaceous, and possibly the Tertiary, they were more diverse in the Northern Hemisphere. This observation highlights the need of a cautious interpretation of the biogeographic patterns of the past, especially if they are used as a reference to explain the current distribution of plants.

Unfortunately, macrofossils from the Paleocene have not been collected in Mexico. Nevertheless, the pollen record of the Burgos Basin, Coahuila, suggests that a community comparable to those reported from the Mississippi Embayment in United States was capable of establishing itself within Northwestern Mexico during this time (Martínez-Hernández and Ramírez-Arriaga, 1996). By the end of the Cretaceous, plants related to those from the Mississippi Basin followed the retreat of the Epicontinental Sea establishing themselves in the south; in a similar way during the Tertiary, they followed the transgression-regression cycles of Eastern Mexico. At the same time, the Sierra Madre Occidental was developing southwards promoting the exposition of central Mexico. In general terms, the region at this point, was in the shape of a great peninsula (Figs. 1a-1c).

By the Eocene, part of the elements reported in the Burgos Basin flora, (e.g., *Ilex,* Sapotaceae, Juglandaceae, Betulaceae) had expanded southwards, reaching some areas of Chiapas (Martínez-Hernández and Ramírez-Arriaga, 1996). Also, by the Eocene, Leguminosae (Fig. 16) is recognized through well preserved fruits and seeds in La Popa Basin, Nuevo León (Calvillo-Canadell and Cevallos-Ferriz, 2001). The Leguminosae, a major component of the extant vegetation of Mexico, seems to have expanded its distribution during the Eocene and Oligocene, since it is well represented in Oligocene and Miocene sediments of Baja California Sur, Puebla, Tlaxcala, Oaxaca and Chiapas (Figs. 13-17; e.g., Miranda, 1963; Magallón-Puebla and Cevallos-Ferriz, 1994a; Cevallos-Ferriz and Barajas-Morales, 1994; Müller-Stoll and Mädel, 1967; Calvillo-Canadell and Cevallos-Ferriz, 2001, 2005; Castañeda-Posadas pers. com. 2002). These examples further support the idea that by this time, the vegetation of Mexico represented an extension of the Northern Hemisphere Tertiary flora (e.g. Rzedowski, 1991; Wendt, 1998). However, it also suggests that by this time, representatives of Mexican endemic lineages were already present in the region, giving to the southern Tertiary vegetation of North America a distinctive character. It is during the Oligocene too that the rose family (Rosaceae) also appears in the fossil record of the region (Fig. 28).

The Oligocene was an important moment for plant life in Mexico (Fig. 1f). For ca. 4 million years (32–28 mybp) ignimbrites flare up in the Sierra Madre Occidental building the largest structure of this kind in the world. This volcanic arch subsequently (from the Middle Miocene on) evolved and changed its orientation to conform to the Faja Volcánica Mexicana (Fig. 1g; Ferrari *et al.* 1999). During this time also, the Yucatan Peninsula was still covered by sea and what today is the Baja California Peninsula was part of the continent. Most likely, their subsequent geological evolution generated important and variable physiographic and climatic conditions that influenced biological selection, radiation and evolution (Fig. 1h-i). The southwards movement of the Chortis Block is particularly important to generate an idea of the geological setting where the plants were establishing themselves (Fig. 1h-i). Mexico had the configuration of a peninsula, but along its southern western coast, the displacement of the Chortis Block generated some degree of continentality, modifying the environmental conditions under which plants were able to establish, grow, radiate, etc.

The diversity of some groups of plants attained during the Oligocene contrasts with today's plant flora. For example, the presence of six genera of Anacardiaceae and eleven species of Salicaceae in the Los Ahuehuetes locality in Puebla does not have counterparts amongst the current vegetation (Figs. 8-12, 24-26; Ramírez Garduño, 1999; Ramírez and Cevallos-Ferriz, 2002). Although the diversity found in the fossil record is contradictory with respect to the current communities, most likely, these differences reflect the existence in the past of different paths that the plant communities followed, paths determined by the variable dynamics of the geological setting and the biological response of the plants to the changing environment. Definitely, these dynamic biological-geological interactions significantly contributed to the relatively sudden appearance of endemic plants in Mexico. That through the evolution of the stage (geology) and the actors (plants) during the Neogene, the megadiversity of plants and environments that are currently observed in the landscape of the region generated and established.

The fossil plant record, in increasingly younger sediments, continues documenting types of vegetation composed of elements that currently would be associated with different communities. Some researchers have suggested that this mixture of plants in the fossiliferous outcrops is due to the occurrence of alochthonous and indigenous elements (e.g., Martínez-Hernández and Ramírez-Arriaga, 1996; Graham, 1976). They predict the presence of mountains close to the fossil localities and assume an altitudinal gradient in which it would be possible to locate different communities (Martínez-Hernández and Ramírez-Arriaga, 1996; Graham, 1976). While this scenario cannot be discarded, one would expect that these strong physiographic alterations would be reflected in the geological record, especially in post-Miocene times.

To fully understand the complexity of the integration of the biota and communities of Mexico, it is important to discuss the geological process along with the plants inhabiting the different settings, and the way these were associated. An integral view would be incomplete if some climatic parameters that had a predominant role in this history were not mentioned. For example, the climatic belts that are recognized today did not always have the same extension; they expanded or contracted according to the position of the continents, the presence/absence of ice on the poles and the circulation of air and marine currents as well as their temperature, among other factors. With regard to the integration of the vegetation of Mexico during the last 70 million years, the bridge connecting North and South America, the Isthmus of Panama had only existed for the last 3.5 million years. Before that time, Mexico had the shape of a peninsula and most likely all the factors discussed above performed differently if compared with their extant behavior. Before the Isthmus of Panama formed, the double climatic cell that today characterizes both hemispheres was not well established. Shortly before Mexico became fully connected to South America, the shape of the country changed from a prominent peninsula to its current configuration by the addition of the Yucatan Peninsula to its southeastern region and the Baja California Peninsula to its northwestern region. Most likely, the addition of these peninsulas altered the physical factors of the environment to some extent, generating in the main continental mass new conditions for the establishment of plants. Furthermore, to this new conformation of Mexico we must add an important factor for the integration of its vegetation; the ice sheets that progressed upon the continents during the Plio-Pleistocene. Certainly, in the latest 3.5 million years the diversity of the flora and vegetation of Mexico that had established and developed in the region through the Upper Cretaceous and Tertiary became "modernized" and started resembling more and more the present day vegetation.

Finally, it is important to highlight that this long historical process explains how some southern and northern lineages were able to arrive into Mexico during the Cretaceous and Tertiary while it still explains how other lineages could have arrived into Central America through the Chortis Block and, how this step helped some plants in their journey to South America. To continue supporting hypothesis on the integration of the flora and vegetation of Mexico, like the one outlined here, the historical point of view should always be central in the discussion, seeking mechanisms that explain how the floras of various continents interacted through processes similar to those outlined here.

Acknowledgments

We would like to thank the commentaries to improve this work and its discussion by L. Calvillo-Canadell, Instituto de Geología, UNAM, and M. Martínez-Millán, Cornell University, their comments certainly improved contents, readability and conciseness.

References

Aguirre-Díaz, G., and McDowell, F. W., 1991, The volcanic section at Nazas, Durango, Mexico and the possibility of widespread Eocene volcanism within the Sierra Madre Occidental, *J. Geoph. Res.* 96:173-188.

Anderson, T., and Schmidt, R., 1983, The evolution of Middle America and the Gulf of Mexico-Caribbean Sea region during Mesozoic time, *Geol. Soc. Am. Bull.* 94:941- 966.

Axelrod, D. I., 1950, Evolution of desert vegetation in western North America, *Car. Inst. Wash. Publ.* 590:215-306.

Bray, W. L., 1898, On the relation of the flora of the Lower Sonoran zone in North America to the flora of arid zones in Chile and Argentina, *Bot. Gaz.* 26:121-147.

Bray, W. L., 1900, The relations of the North American flora to that of South America, *Science n.s.* 12:712-716.

Burnham, R. J., and Graham, A., 1999, The history of Neotropical vegetation: new developments and status, *Ann. Missouri Bot. Gard.* 86:540-589.

Burnham, R. J., and Carranco, N. I., 2004. Miocene winged fruits of *Loxopterigium* (Anacardiaceae) from the Ecuatorian Andes, *Am. J. Bot.* 91:1767-1773.

Calvillo-Canadell, L., 2000, Determinación taxonómica de estructuras reproductivas y vegetativas de plantas fósiles de Tepexi de Rodríguez, Puebla. Unpublished M.S. thesis, Universidad Nacional Autónoma de México.

Calvillo-Canadell, L., and Cevallos-Ferriz, S. R. S., 2001, Tertiary legumes from central and northern Mexico, *IV International Legume Conference,* Canberra, Abst. p. 13.

Calvillo-Canadell, L., and Cevallos-Ferriz, S. R. S., 2005, Diverse assemblage of Eocene and Oligocene Leguminosae from Mexico, *Int. J. Plant. Sci.* 166 (4): 671-692.

Cevallos-Ferriz, S. R. S., 1983, Descripción de una madera de angiosperma cretácica de Cananea, Son., Méx. Los xilitos en el estudio del origen de las angiospermopsida, *An. Inst. Biol. UNAM* 54:97-112.

Cevallos-Ferriz, S. R. S., and Barajas-Morales, J., 1994, Fossil woods from the El Cien Formation in Baja California Sur: Leguminosae, *IAWA Bull.* 15:229-245.

Cevallos-Ferriz, S. R. S. and Ramírez, E., 1996, Aspectos de la historia de la vegetación de México con base en macrofósiles, *Bol. Soc. Bot. Méx.* 58:99-111.

Cevallos-Ferriz, S. R. S. and Ricalde-Moreno, O. S., 1994, Palmeras fósiles del Norte de México. *An. Inst. Biol. UNAM* 66:1-106.

Engler, A., 1882, *Versuch eine Entwicklungsgeschichte der Pflanzenwelt, insbesondere der Florengebiete seit der Tertiaerperiode. Part. II. Die extratropischen Florengebiete der suedlichen Hemisphaere and die tropischen Gebiete.* W. Engelmann. Liepzig. 386 pp.

Engler, A., 1905, Ueber floristische Verwandschaft zwischen dem tropischen Afrika und Amerika. *Stizungsb. Koenigl. Preuss. Akad. Wiss.* 6: 1-52.

Engler, A., 1914, Ueber Herkunft Alter und Verbeitung extremer xerothermer Pflanzen. *Sitzungbs. Koenigl. Preuss. Akad. Wiss.* 1914:564-621.

Ferrari, L., López-Martínez, M., Aguirre-Díaz, G., and Carrasco-Nuñez, G., 1999, Space-time patterns of Cenozoic arc volcanism in central Mexico – from the Sierra Madre Occidental to the Mexican Volcanic Belt, *Geology* 27:303-307.

Ferrari, L., López-Martínez, M., and Rosas-Elguera, G., 2002, Ignimbrite Flare-Up and deformation in the southern Sierra Madre Occidental, western Mexico: Implications for the late subduction history of Farallon Plate, *Tectonics* 21: 17-24.

González-Medrano, F., 1996, Algunos aspectos de la evolución de la vegetación de México, *Bol. Soc. Bot. Méx.* 58:129-136.

Graham, A., 1973, History of the arborecent temperate element in the northern Latin American biota, in: *Vegetation and vegetational history of northern Latin America.* (A. Graham, ed.), Elsevier, Amsterdam, pp. 301-314.

Graham, A., 1976, Studies in neotropical Paleobotany, part 2, The Miocene communities of Veracruz, Mexico, *Ann. Missouri Bot. Gard.* 63:787-842.

Hemsley, W. B., 1879-1888, Botany, in: Godwin, F.D. and O. Salvin. *Biologia Centrali-Americana.* R. H. Porter. London. 5 vols.

Hernández-Castillo, G. R., and Cevallos-Ferriz S. R. S., 1999, Reproductive and vegetative organs with Haloragaceae affinity from the Huepac chert, Upper Cretaceous of Sonora, Mexico. *Am. J. Bot.* 86:1717-1734.

Johnston, I. M., 1924, Expedition of the California Academy of Sciences to the Gulf of California in 1921, *Proc. Cal. Acad. Sci. IV.* 12:951-1218.

Johnston, I. M., 1940, The floristic significance of shrubs common to North and South American deserts, *J. Arnold Arb.* 21:356-363.

Krueger, O., 1934, *Die Gramineenflora im nordamerikanischen und sudamerikanischen Trockengebiet.* Tesis. Philos. Fakultaet Friedrich Wilhelms-Universitaet. Berlin. 63 p.

Magallón-Puebla, S., and Cevallos-Ferriz, S. R. S., 1994a, Fossil legume fruits from Tertiary strata of Puebla, Mexico, *Can. J. Bot.* 72:1027-1038.

Magallón-Puebla, S., and Cevallos-Ferriz, S. R. S., 1994b, *Eucommia constans* n. sp. fruits from upper Cenozoic strata of Puebla, Mexico: Morphological and anatomical comparison with *Eucommia ulmoides* Oliver, *Int. J. Plant Sci.* 155:80-95.

Magallón-Puebla, S., and Cevallos-Ferriz, S. R S., 1994c, Latest occurrence of the extint genus *Cedrelospermum* (Ulmaceae) in North America: *Cedrelospermum manchesteari* from Mexico, *Rev. Paleobot. Palyn.* 81:115-128.

Magallón-Puebla, S., and Cevallos-Ferriz, S. R. S., 1994d, An enigmatic fruit from Oligocene strata of Puebla, Mexico, *BSA, Am. J. Bot.* 81:96.

Manchester, S. R., 1977, Wood of *Tapirira* (Anacardiaceae) from the Paleogene Clarno Formation of Oregon, *Rev. Paleobot. Palyn.* 23:119-127.

Martín-Barajas, A., 2000, Volcanismo y extensión en la provincia extensional del Golfo de California, *Bol. Soc. Geol. Mex.* LIII, No. 2000, p.72-83.

Martínez-Cabrera, H. I., 2002, Las maderas del Oligoceno-Mioceno de la Formación El Cien, B.C.S., México. Unpublished Professional thesis, Universidad Nacional Autónoma de México.

Martínez-Hernández, E. and Ramírez-Arriaga, E. 1996. Paleocorología de angiospermas de la flora Mexicana durante el Mesozoico y Terciario: algunas evidencias palinológicas, *Bol. Soc. Bot. Mexico* 58:87-97.

Martínez-Hernández, E., Hernández-Campos, H., and Sánchez-López, M., 1980, Palinología del Eoceno en el Noreste de México, *Rev. Inst. Geol., UNAM* 4:155-166.

Martínez-Millán, M., 2000, Biogeografía histórica (Terciario y Cuaternario) de Anacardiaceae con base en caracteres anatómicos de la madera. Unpublished Professional thesis, Universidad Nacional Autónoma de México.

Matuda, E., 1953, Plantas asiáticas en México, *Mem. Cong. Cient. Mex.* 6:230-248.

McDowell, F., and Clabaugh, S, 1979, Ignimbrites of the Sierra Madre Occidental and their relation to tectonic history of western Mexico, *Geol. Soc. Am. Spec. Pap.* 180:113-124.

McKeown, J., Leon de la Luz, J. L., and Jones, J. H., 1991, Fossil wood from the Miocene Comondú Formation of Baja California Sur, *PaleoBios* 13(50 suppl. 50): 7.

Miranda, F., 1952, *La vegetación de Chiapas*. Ediciones del Gobierno del Estado. Tuxtla Gutiérrez. 2 vols.

Miranda, F., 1958, Estudios acerca de la vegetación, in: *Los recursos naturales del sureste y su aprovechamiento,* Inst. Mex. Rec. Nat. Renov. México D.F., vol. 2, pp. 215-271.

Miranda, F., 1960, Posible significación del porcentaje de géneros bicontinentales en América tropical, *An. Inst. Biol. UNAM* 30:117-150.

Miranda, F., 1963, Two fossil plants from the amber of Simojovel, Chiapas, México, *J. Pal.* 37:611-614.

Morán-Zenteno, D., Martiny, B., Tolson, G., Solís-Pichardo, G., Alba-Aldave, L., Hernández-Bernal, M., Macías-Romo, C., Martínez-Serrano, R., Schaaf, P., and Silva-Romo, G. 2000, Geocronología y características geoquímicas de las rocas magmáticas terciarias de la Sierra Madre del Sur, *Bol. Soc. Geol. Mex.*, LIII, No. I (2000):27-58.

Müller-Stoll, W. R., and Mädel, E., 1967, Die Fossilen Leguminose-Hölzer. Einen Revision der mit Leguminosen Verglichenen Fossilen Hölzer und Beschreibung Älterer und Arten, *Paleontographica* 119B:43-97.

Ortega, F., Mitre, L. M., Roldan, J., Aranda, J., Moran, D., Alanis, S., and Nieto, A., 1992, *Texto explicativo de la quinta edición de la Carta Geológica de la República Mexicana Escala 1:2,000,000.* UNAM, Inst. Geol., y Sec. Ener. Min. Ind. Paraestatal, Cons. Rec. Min., 74p.

Pérez-Hernández, B. R., Rodríguez de la Rosa R. A., and Cevallos-Ferriz, S. R. S., 1997, Permineralized infructescence from the Cerro del Pueblo Formation (Campanian) near Saltillo, Coahuila, Mexico: Phytolaccaceae, *Am. J. Bot.* 84:139.

Ramírez-Garduño, J. L., 1999, Analisis foliar de Anacardiaceae, Berberidaceae y Salicaceae en Los Ahuehuetes (Oligoceno), Tepexi de Rodríguez, Puebla. Unpublished M.S. thesis, Universidad nacional Autónoma de México.

Ramírez, J. L., and Cevallos-Ferriz S. R. S., 2000a, Leaves of Berberidaceae (*Berberis* and *Mahonia*) from Oligocene sediments, near Tepexi de Rodríguez, Puebla, *Rev. Palaeobot. Palyn.* 110:247-257.

Ramírez, J. L., and Cevallos-Ferriz S. R. S., 2000b, Leaves of Salicaceae (*Salix* and *Populus*) from Oligocene sediments, near Tepexi de Rodríguez, Puebla, *Int. J. Plant Sci.* 161:521-534.

Ramírez, J. L., and Cevallos-Ferriz, S. R. S., 2002, A diverse assemblage of Anacardiaceae from Oligocene sediments, Tepexi de Rodríguez, Puebla, Mexico, *Am. J. Botany* 89:535-545.

Ramírez, J. L., Cevallos-Ferriz, S. R S., and Silva-Pineda, A., 2000, Reconstruction of the leaves of two new species of *Pseudosmodingium* from Oligocene strata of Puebla, Mexico, *Int. J. Plant Sci.* 161: 509-519.

Raven, P. H., 1963, Amphitropical relationships in the floras of North and South America, *Quart. Rev. Biol.* 38:151-177.

Raven, P. H., and Axelrod, I. 1974, Angiosperm biogeography and past continental movements, *Ann. Missouri Bot. Gard.* 61:539:673..

Rodríguez de la Rosa, R. A., and Cevallos-Ferriz S. R. S., 1994, Upper Cretaceous Zingiberalean fruits with in situ seeds from southeastern Coahuila, Mexico, *Int. J. Plant Sci.* 155:786-805.

Rodríguez de la Rosa, R. A., Cevallos-Ferriz, S. R. S., and Silva-Pineda, A., 1998, Paleobiological implications of Campanian coprolites, *Palaeog. Palaeoclimat. Palaeoecol.* 142:231-254.

Rueda-Gaxiola, J., 1967, *Contribution a l'étude palynologique et pétrographique du charbon cretacé du Bassin de Sabinas, Coahuila, Mexique.* Unpublished thesis, Univertité de Lille.

Rzedowski, J., 1965, Relaciones geográficas y posibles orígenes de la flora de México, *Bol. Soc. Bot. Mex.* 29:121-177.

Rzedowski, J., 1966, Vegetación del Estado de San Luis Postosí, *Acta Cien. Potosina* 5:5-291.

Rzedowski, J., 1972, Contribución a la fitogeografía florística e histórica de México. II. Afinidades geográficas de la flora fanerogámica de diferentes regiones de la República Mexicana, *An. Esc. Nac. Cien. Biol. Méx.* 19:45-48.

Rzedowski, J., 1978, *Vegetación de México,* Limusa. México. 432 p.

Rzedowski, J., 1991, Diversidad y orígenes de la flora fanerogámica de México, *Acta Bot. Mex.* 14:3-21.

Sarukhán, K. J., 1968, Los tipos de vegetación arbórea de la zona cálido-húmeda de México in: *Manual para la identificación de los principales árboles tropicales de México* (Pennington, T.D. and J. Sarukhán K. eds.), Instituto Nacional de Investigaciones Forestales y FAO. México D.F. Pp. 3-46.

Schaaf, P., Morán-Zenteno, D., Hernández-Bernal, M. S., Solís-Pichardo, G., Tolson, G., and Kohler, H. 1995, Paleogeno continental margin truncation in southwestern Mexico: Geochronological evidence, *Tectonics* 14:339–350.

Sharp, A. J., 1951, The relation of the Eocene Wilcox flora to some modern floras, *Evolution* 5:1-5.

Sharp, A. J., 1953, Notes on the flora of Mexico; world distribution of the woody dicotyledonous families and the origin of the modern vegetation, *J. Ecol.* 41:374-380.

Standley, P. C., 1936, Las relaciones geográficas de la flora mexicana, *An. Inst. Biol. UNAM* 7:9-16.

Taylor, D. W., 1990, Paleobiogeographic relationships of angiosperms from the Cretaceous and early Tertiary of the North American area, *Bot. Rev.* 56:279-417.

Terrazas, T., and Wendt, T., 1995, Systematic wood anatomy of the genus *Tapirira* Aublet (Anacardiaceae) – a numerical approach, *Brittonia* 47:109-129.

Toledo, V. M., 1982, Pleistocene changes of vegetation in tropical México, in: *Biological diversification in the tropics*, Columbia University Press, New York pp. 93-111.

Trejo-Vázquez, I. and González-Medrano, F., 2001, Nomenclatura de los tipos de vegetación de México, *XV Congreso Mexicano de Botánica.* Querétaro, Méx.

Velasco de León, P., 1999, Estudio paleoecológico de una comunidad del Terciario en el estado de Puebla, México. Unpublished Ph.D. thesis, Universidad Nacional Autónoma de México.

Velasco de León, P., and Cevallos-Ferriz, S. R. S., 2000, Leaves of *Cercocarpus mixteca* n. sp. (Rosaceae) from Oligocene sediments, near Tepexi de Rodríguez, Puebla, *Rev. Paleobot. Palyn.* 111:285-294.

Velasco de León, P., Cevallos-Ferriz, S. R. S., and Silva-Pineda, A., 1998. Leaves of *Karwinskia axamilpense* n. sp. (Rhamnaceae) from Oligocene sediments, near Tepexi de Rodríguez, Puebla, *Can. J. Bot.* 76:410-419.

Velazco–Tapia, F., and Verma, S., 2001, Estado actual de la investigación geoquímica en el campo monogenético de la Sierra Chichinautzin: análisis de información y perspectivas, *Rev. Mex. Cien. Geol.* 18:1-36.

Weber, R., 1972, La vegetación maestrichtiana de la Formación Olmos de Coahuila, México, *Bol. Soc. Geol. Méx.* 33:5-19.

Wendt, T., 1998, Composición, afinidades florísticas y orígenes de la flora arbórea del dosel de los bosques tropicales húmedos de la vertiente mexicana del Atlántico, In: Lot, A., Ramamoorthy, P.Fa. J. and Bye, R. (eds.) *Diversidad Biológica de México: orígenes y distribución.* Instituto de Biología, UNAM, pp. 581-664.

Wiggins, I. L., 1960, The origins and relationships of the land flora, *In*: The biogeography of Baja California and adjacent seas. Part. III. Terrestrial and fresh-water biotas. *Syst. Zool.* 9:148-165.

Wignall, P. B., 2001, Large igneous provinces and mass extinctions. *Earth Sci. Rev.* 53:1-33.

Chapter 2

Tertiary Palynofloristic correlations between Mexican Formations with emphasis in dating the Balsas Group

ENRIQUE MARTÍNEZ-HERNÁNDEZ and ELIA RAMÍREZ-ARRIAGA

Instituto de Geología, UNAM. Departamento de Paleontología. Laboratorio de Palinología Ciudad Universitaria 04510, Coyoacán, México, D.F. emar@servidor.unam.mx

F. Vega et al. (eds.), Studies on Mexican Paleontology, 19–45.

1. Abstract

The present work includes a palynofloristic synthesis of Tertiary research carried out at marine and continental basins with particular emphasis on stratigraphy of the Balsas Group *sensu* Fries (1960). Also it includes palynostratigraphic information related with San Gregorio Formation (Baja California Sur State) and Cuayuca Formation (Puebla State) assemblages. Taxa of chronostratigraphic and ecologic importance are emphasized at every paleobasin. The numerical and cladistic correlation showed that palynological assemblages of units belonging to the Balsas Group are closely related to San Gregorio flora. The results indicate that since the Paleocene, under a tropical regime, an evergreen forest was present together with warm temperate elements. Later, in the Eocene epoch, the tropical floras remained as the principal components of the vegetation. In the Late Eocene - Early Oligocene, the assemblages are a mixture of temperate and desert taxa prevailing over tropical elements at the Sierra Madre Occidental Province. But the subtropical and tropical floras continued as the main components in the Gulf Coast Province where the tropical vegetation stayed up to the Holocene. In contrast, in the western basin localities (Sierra Madre Occidental Province) it was detected a desertification since the Oligocene where savannas and grasses became established. Afterwards, in the Miocene, at the same region, the xerophitic flora dominates the assemblages, proving the existence of a desert at that time.

It is evident that the coniferous forest and oak communities inhabit the mountain ranges since the Eocene becoming more abundant in the Oligocene, but they became the dominant taxa in the pollen assemblages from the Neogene to Quaternary.

According with the Tertiary flora analyzed in this survey the actual vegetation present in Mexico is a response to climatic changes and active volcanism throughout the Tertiary. Finally, different types of vegetation began to be recognizable in the fossil assemblages since the Oligocene. The historical development of Mexico (tectonic, rifting, volcanism) explains the richness and endemism of Mexican flora.

2. Introduction

Most palynostratigraphic work has been developed from North American Tertiary basins (Elsik and Dilcher, 1974; Frederiksen, 1984, 1988; Srivastava, 1984; Tschudy, 1973). In this sense, important palynofloristic information was generated at different formations. Besides the relevance that palynomorphs have in stratigraphic geology, palynology likewise gives information related to paleocommunities, paleoclimates and paleoenvironments of Tertiary basins.

The Balsas Group was described by Fries in 1960, and it represents the oldest Tertiary continental rocks deposited over Cretaceous formations. Basins that belong to Balsas Group have a tectonic origin probably related to the Chortis block displacements.

Since Fries (1960) published his outstanding geological work related to the meridional central region of Mexico, there has not been and new geological framework published on the area, nevertheless, paleopalynological studies have generated transcendental stratigraphic finds dating some formations in southern Mexico as part of the Balsas Group (Martínez-Hernández and Ramírez-Arriaga, 1999, 2000; Ramírez-Arriaga and Martínez –Hernández, 2003).

The present paper is a synthesis of the Tertiary palynostratigraphy carried out in Mexican basins with emphasis in their time-correlation with continental basins of the Balsas Group.

3. Methodology

The present contribution includes a revision of previously published palynologic studies on the Tertiary basins of Mexico and also includes new information related to research carried out within the San Gregorio Formation at La Purísima, Baja California Sur and Cuayuca Formation, near Cuayuca village, Puebla (Figure 1 and Table 1).

The palynologic assemblages were analyzed by NTSYS program which provides a qualitative correlation based in similarity between the different formations by the UPGMA method (Unweighted pairgroup method using arithmetic averages). The numerical analysis assumes the formations as OTUs and the characters selected

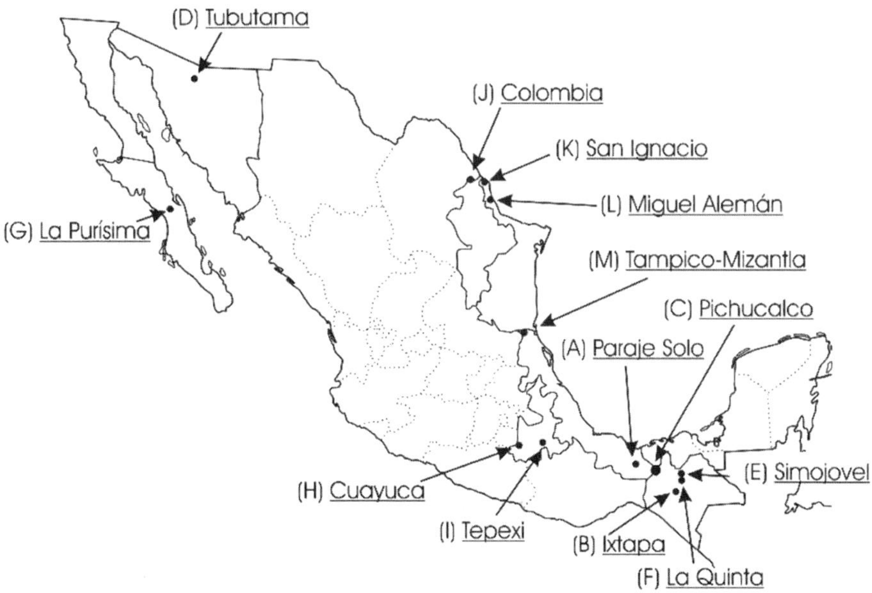

FIGURE 1. Tertiary localities included in this survey.

TABLE 1. Published paleopalynology research within Mexican Tertiary Formations. Some localities (*) were considered for the analysis of similarity and parsimony analysis of endemicity (PAE). Research related to the Balsas Group are marked (û).

	Authors	***Formations / Status***	***Environment***
Pliocene	*Graham, 1976 (A)	Paraje Solo Fm.: Veracruz	Lagoon - nearshore
Miocene	*Martínez-Hernández, 1992 (B)	Ixtapa Fm.: Ixtapa, Chiapas	Mangrove – nearshore
	*Palacios and Rzedowski, 1993 (C)	Pichucalco: Chiapas	Mangrove – nearshore
	Martínez-Hernández and Tomasini, 1986 (D)	Tubutama Fm.: Sonora	Continental
Oligocene – Miocene	Langenheim *et al.* 1967 (E)	Simojovel Group: Chiapas	Lagoon - nearshore
	*Biaggi, 1978 (F)	La Quinta Fm.: Chiapas	Lagoon - nearshore
Oligocene	*Martínez-Hernández and Ramírez-Arriaga (in process) (G)	San Gregorio Fm.: La Purísima, Baja California Sur	Marine
Eocene – Oligocene	* Ramírez-Arriaga and Martínez- Hernández, 2003 (H)	Cuayuca Fm.: Puebla.	Continental
	* Martínez-Hernández and Ramírez-Arriaga, 1999 (I)	Pie de Vaca Fm. (PVF): Tepexi, Puebla	Continental
	*Tomasini-Ortíz and Martínez-Hernández, 1984 (E)	Esperanza Group: Simojovel, Chiapas	Lagoon - nearshore
Eocene	*Martínez-Hernández *et al.* 1980 (J, K and L)	Claiborne Fm.: Tamaulipas	Marine to continental
Paleocene – Eocene	Rosales-Lomelí *et al.* 1992 (M)	Tampico Mizantla basin: Veracruz	Marine

were palynomorphs. A data matrix is then produced, characterized by OTUs in the columns and characters in the rows, which include the presences (1) or absences (0) of palynomorphs. The phenogram has a numeric scale from 0 to 1 that corresponds to the coefficient of similarity with the number one representing the highest similarity (Méndez and Villaseñor, 1997).

Another method applied was parsimony analysis of endemicity "PAE" (Brooks and van Veller, 2003; Luna and Alcántara, 2003; Morrone, 1994) that produce a cladogram of relationships among Tertiary formations based on shared palynomorphs. In order to achieve this analysis a data matrix was made where columns correspond to taxa and rows correspond to formations (areas). Furthermore, there were coding present taxa as "1" and absence taxa as "0". Taxa that are exclusive of only one area were excluded from the matrix. A hypothetical formation coded with "0" for all columns was added to root the tree. The cladistic programs used were WINCLADA (Nixon, 2002) and NONA (Goloboff, 1993) that perform a heuristic searching of the data.

4. Geological Settings of Balsas Group

According to Fries (1960), the Balsas Group includes rocks of Tertiary age that lie unconformably over Cretaceous or older formations located throughout the hydrologic basin of the Mexcala-Balsas river located within central meridional and southwestern Mexico. Geographically, this Group is found in the states of Michoacán, Guerrero, Puebla, Morelos, Hidalgo, Oaxaca and México. Fries (1960) did not make a detailed differentiation of lithostratigraphic units that constitute the Balsas Group. Nevertheless, he assured that several formations could be differentiated based upon detailed cartographic and geological framework.

The Balsas Group is characterized by a great lithologic heterogeneity among the clastic rocks, the most abundant are the red conglomerate, mudstones and marls. Also chert, gypsum with different grades of purity, limestones, and volcanic rocks are present.

In general, the estimated age of the Balsas Group is from Maastrichtian to Late Oligocene. According to Fries (*1960*), the majority of the Balsas Group was deposited from the Late Eocene to Middle Oligocene (Fig. 2); additional radiometric dates support this hypothesis (Cserna and Fries, 1981; Ortega-Gutiérrez, 1980; Morán-Zenteno *et al.* 2004).

Recent biostratigraphic studies have allowed the inclusion of two formations as part of the Balsas Group, "Pie de Vaca" Formation and Cuayuca Formation, age determination from palynostratigraphy and previous chronostratigraphic assignments of these units (Martínez-Hernández and Ramírez-Arriaga, 1999, 2000; Ramírez-Arriaga and Martínez-Hernández, 2003).

5. Paleopalynology, Environments and Chronostratigraphic Inference throughout the Tertiary of Mexico

5.1. Paleocene-Eocene

Within the Paleocene-Eocene time interval Rosales-Lomelí *et al.* (1992) analyzed a borehole from Petróleos Mexicanos from the Tampico-Mizantla basin. In order to determine the age and the environment of deposition the authors studied six kinds of palynomorphs (spores, pollen, dinoflagelates, algae, scolecodontes and thecamoebians). The palynomorphs recovered range from Paleocene to the

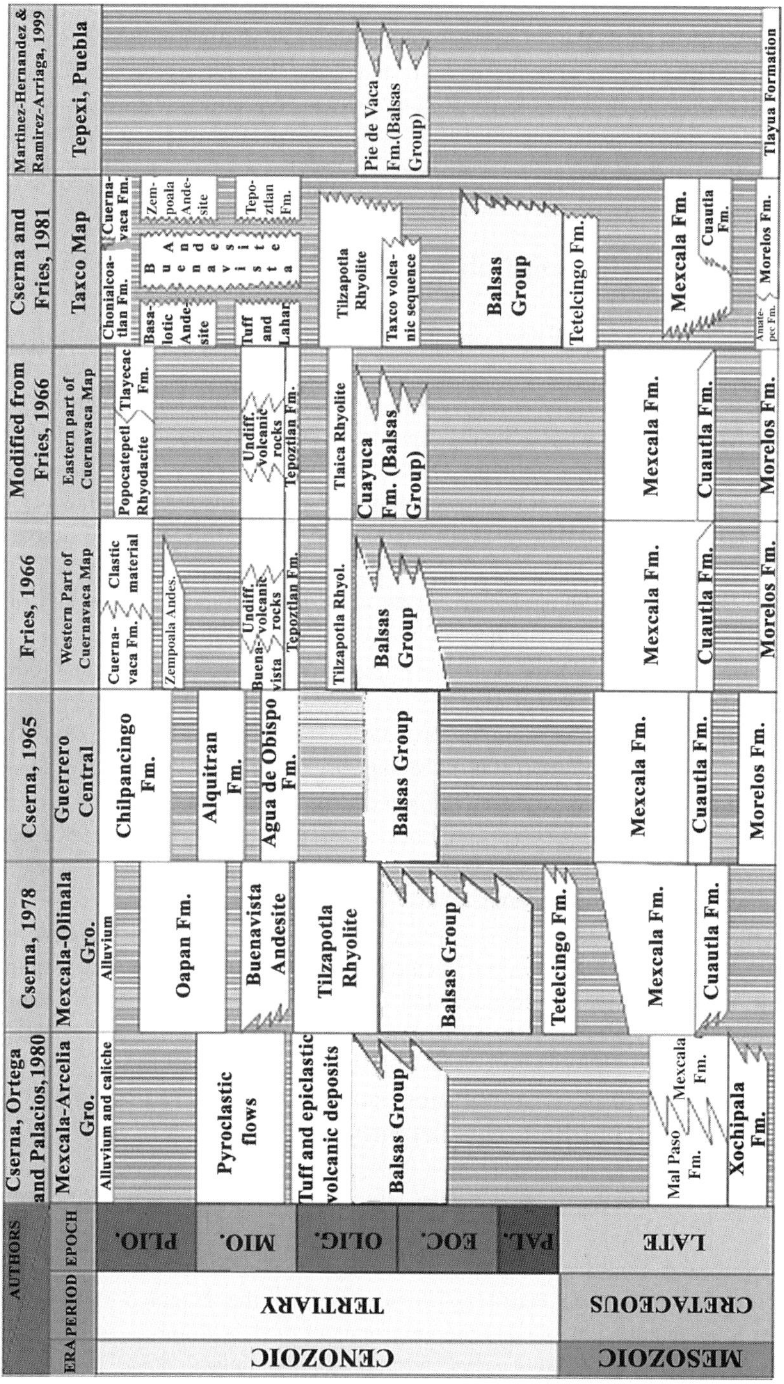

FIGURE 2. Composite sections of some localities where Balsas Group has been reported.

Lower Eocene – Middle Eocene boundary. The dominant palynomorphs were the dinoflagelates as evidence of a marine basin and also there were registered pollen grains of chronostratigraphic importance such as *Momipites tenuipolus, Tricolpopollenites hians, Pandaniidites, Nudopollis terminalis, Retitrescolpites anguloluminosus, Pistillipollenites* cf. *mcgregorii, Anacolosidites claibornensis, Cicatricosisporites dorogensis, Ephedripites claricristata,* among others.

5.2. Eocene

Within the northeastern Mexican Burgos basin two kinds of paleoenvironments, estuarine and marine, have been identified (Martínez-Hernández *et al.*, 1980). A borehole survey for lignites included exploration at three localities: Colombia (Nuevo León State), San Ignacio and Miguel Alemán (both from Tamaulipas State).

Based on palynostratigraphic results, Colombia wells belong to the Claiborne Group with an assemblage dominated by *Bombacacidites claibornensis, Choapollenites eximius, Sernapollenites duratus, Proxapertites* spp. and *Castanea* spp.; these taxa are index fossils for the Early Eocene and Middle Eocene. Lignites at the San Ignacio area, also belong to the Claiborne Group, with an assemblage dominated by *Platycarya swasticoidus* and *Nudopollis terminalis*; these taxa are index fossils for the middle Eocene (Martínez-Hernández *et al.* 1980). Finally, Miguel Alemán wells belong chronostratigraphically to the Middle to Upper Eocene, due to the acme zone of *Cicatricosisporites* spp., *Momipites* spp., Sapotaceae and *Polypodium heteroverrucatus.* The age assignment is corroborated by the presence of the dinoflagelate cysts *Charlesdowniea* (*Wetzelliella*) *coleothrypta* and *Rhombodinium draco*, which indicate a near shore marine environment. From the floristic point of view, the palynological assemblages indicate the existence of several types of vegetation: a cloud forest (*Momipites*) and a tropical forest (*Bombacacidites, Cicatricosisporites* cf. *dorogensis* and *Sapotaceae*) as well as warm temperate flora from the abundance of *Castanea* and *Ilex* (Elsik, 1974).

5.3. Eocene – Oligocene

Paleopalynological research includes the following formations: El Bosque, La Trinidad and Rancho Berlín, at Simojovel, Chiapas. Palynologic assemblages place these formations within the Eocene – Oligocene boundary and environmentally represent neritic facies (Tomasini-Ortíz and Martínez-Henrández, 1984). In the nearshore facies, mangrove communities were well developed with abundant pollen of *Zonocostites* and *Pellicieroipollis* together with a diversity of dinocysts.

Within these formations three acme zones were defined: *Inapertisporites vulgaris*, a fungospore abundant in the Trinidad Formation (Upper Eocene); *Inapertisporites* cf. *circularis*, a fungospore characteristic of Rancho Berlín (Lower Oligocene); and *Pellicieroipollis* sp. an abundant zone established at La Quinta Formation (Upper

Oligocene). In addition, five range-zones were established: *Psilatricolporites crassus*, with a taxon range-zone from Late Eocene to Early Oligocene; *Cicatricosisporites* sp. with a taxon range-zone from Early Oligocene to Early late Oligocene; and finally three range zones with the taxon *Polypodiisporites* cf. *alienus; Momipites* sp. and *Diporisporites* sp. belonging to the Late Oligocene. In general, this part of the Isthmus saline basin was overrepresented by mangrove elements. In addition, the palynological assemblages indicate the presence of other types of vegetation such as the existence of a cloud forest with *Momipites* (*Engelhardtia*) and conifer forest with *Abietineaepollenites* and *Cedripites*.

5.3.1. Balsas Group

It is important to review previous palynologic research related to the Balsas Group, which is the main aim of this survey. The first case study for the differentiation of the Balsas Group was developed near Tepexi de Rodríguez village, Puebla. Here the Pie de Vaca Formation (PVF) was assigned to the Paleogene (Martínez-Hernández and Ramírez-Arriaga, 1996b; 1999). Fries (1960) dated the PVF as Late Eocene to Early Oligocene placing it as a part of the Balsas Group sensu. PVF was deposited in a continental lacustrine environment where several plant paleocommunities were recognized following altitudinal gradient with different mosaics of vegetation; in the upper part of the mountain range a coniferous forest with *Pinus* and *Picea* was common. In contrast, at lowlands, grasses and Ephedraceae were present. Also, in some places, the development of thorn woodland with *Bursera*, Compositae, *Acacia* and others Leguminosae was possible (Fig. 5.8). Taxa belonging to the cloud forest indicate higher humidity at ravine where several species of *Engelhardtia* grew (*Momipites coryloides, M. microcoryphaeus* (Fig. 5.7) and *M. triorbicularis*). In this case study the key elements that give chronostratigraphic information were the *Momipites* group, *Aglaoreidia cyclops* (Fig. 5.5) and *Eucommia* (Fig. 5.10). In this part of southern North America, chronostratigraphic assignment was difficult to establish due to the early abundance of grasses (Fig. 5.6) (Martínez-Hernández and Ramírez-Arriaga, 2003).

Paleobotanic studies have also generated important floristic contributions at the PVF. Magallón-Puebla and Cevallos-Ferriz (1994) described *Eucommia constans* and, although this species is similar to *E. montana* from the Oligocene of North America, the authors considered *E. constans* as a Neogene fossil. Other paleobotany results do not make any contribution to the chronostratigraphy of PVF.

Another formation that has been reassigned to the Balsas Group is the Cuayuca Formation that initially Fries (1966) described as belonging to the Miocene-Pliocene age. Based on palynological studies, Ramírez-Arriaga and Martínez-Hernández (2003) assigned the age of the Cuayuca Formation as Late Eocene to Early Oligocene, suggesting that it was part of the Balsas Group. A total of 100 taxa were assigned with a paleopalynogical assemblage having 3% algae, 15% spores of pteridophytes and bryophytes, 8% gymnosperms and 74% angiosperms (5% monocotyledons and 69% of dicotyledones). Abundant pollen grains at the Cuayuca locality were *Bursera*, Chenopodiaceae – Amaranthaceae, *Cyatheaceae, Ephedripites* (Figs. 3.2 and 3.3), grasses, *Momipites coryloides, Momipites microcoryphaeus, Picea* (Fig. 3.4) and *Pinus*. At the Izúcar de Matamoros locality *Cicatricosisporites dorogensis* (Fig. 1.1), Cyatheaceae, Compositae, *Leiosphaeridia, Leptolepidites, Liliacidites, Momipites coryloides, Pinus,* Tiliaceae and *Stereisporites* (*Sphagnum* spp.) were common (Ramírez-Arriaga and Martínez-Hernández, 2003).

The presence of pollen and spores indexes such as *Aglaoreidia pristina, Cicatricosisporites dorogensis, Eucommia* and *Momipites* group allow to propose a Late Eocene to Early Oligocene age for the Cuayuca Formation. With the diverse assemblage of this formation it was possible to recognize different types of vegetation probably growing throughout a paleoaltitudinal gradient, such as pine forest, cloud forest, tropical deciduous forest, tropical scrub, grasses and chaparral (Ramírez-Arriaga and Martínez-Hernández, 2003).

5.4. Oligocene

Martínez–Hernández and Ramírez-Arriaga (1996a) studied boreholes of the San Gregorio Formation, a marine paleobasin dated as Late Oligocene by means of K-Ar dating (Hausback, 1984) and diatom flora (Kim and Barron, 1986). The age of the San Gregorio Formation has been dated to the Oligocene from the presence of dinocysts, of which, *Chiropteridium lobospinosum* (Fig. 6.8) has chronostratigraphic importance (Martínez-Hernández and Ramírez-Arriaga, 1996a; 1998).

The pollen assemblages of the San Gregorio Formation are cited in Table 2 and some taxa are shown in Figs. 4.1-4.7. The assemblages were dominated by herbs such as Chenopodiaceae-Amaranthaceae (10-80%), Ephedraceae (5-40%) and *Graminidites* (2-33%), which suggests an early evolution of low biomass vegetation, i.e. grassland, savanna and chaparral flora, with also tropical elements present: Sterculiaceae, Bombacaceae and Melastomataceae. Finally, there are taxa indexes such as *Eucommia* and *Corsinipollenites*, also present at PVF and Cuayuca Formation (Balsas Group).

5.5. Oligocene-Miocene

Langenheim *et al.* (1967) made a palynologic study of amber-bearing strata that occurred at Simojovel, belong to the Esperanza Group, Chiapas. The aim of that study was to give information related to the environment of the amber deposition and the vegetation during the Middle Tertiary in southern Mexico. The analyzed samples correspond to estuarine environments, where noticeable abundance and diversity of *Rhizophora* species is present such as: *R. mangle, R. racemosa, R. samoensis, R. sarrisonii* and *Pelliciera* sp. These *Rhizophora* species represent a complex tropical

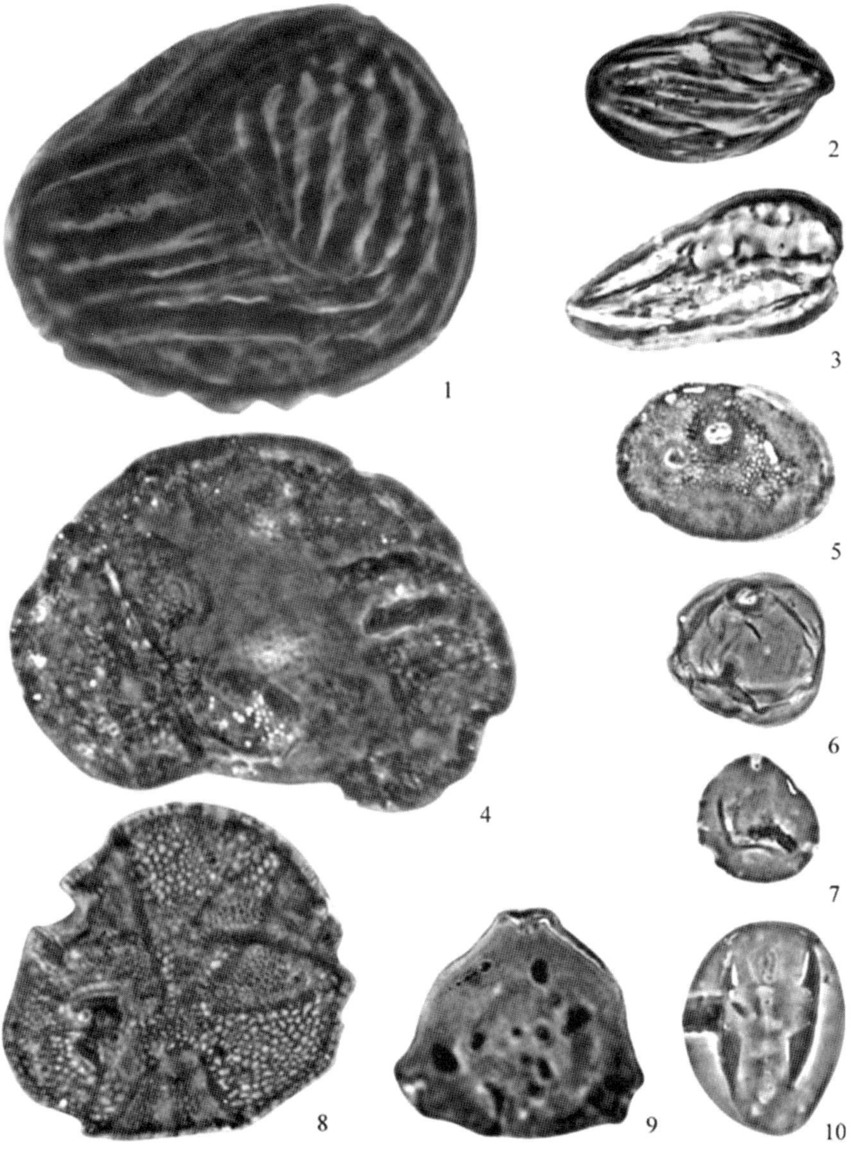

FIGURE 3.1. *Cicatricosisporites dorogensis* [Cuayuca Fm. Pb-8869(8): 86.8/14.6]; 2. *Ephedripites* [Cuayuca Fm. Pb-9340(4): 90.3/16.9]; 3. *Ephedra claricristata* [Cuayuca Fm. Pb-9349(4): 97.2/15.3]; 4. *Picea* [Pie de Vaca Fm. Pb-3713(2): 98.7/7]; 5. *Aglaoreidia* [Pie de Vaca Fm. Pb-3713(4): 96.8/12.4]; 6. Gramineae [Pie de Vaca Fm. Pb-3713(2): 116.6/17]; 7. *Momipites microcoryphaeus* [Pie de Vaca Fm. Pb-3713(2): 105.2/13.2]; 8. *Caesalpinia* [Pie de Vaca Fm. Pb-3713(2): 101.6/6.7]; 9. *Corsinipollenites* [Pie de Vaca Fm. Pb-3713(6): 110.2/14.8]; 10. *Eucommia* [Pie de Vaca Fm. Pb-3713(2):115.6/18.8].

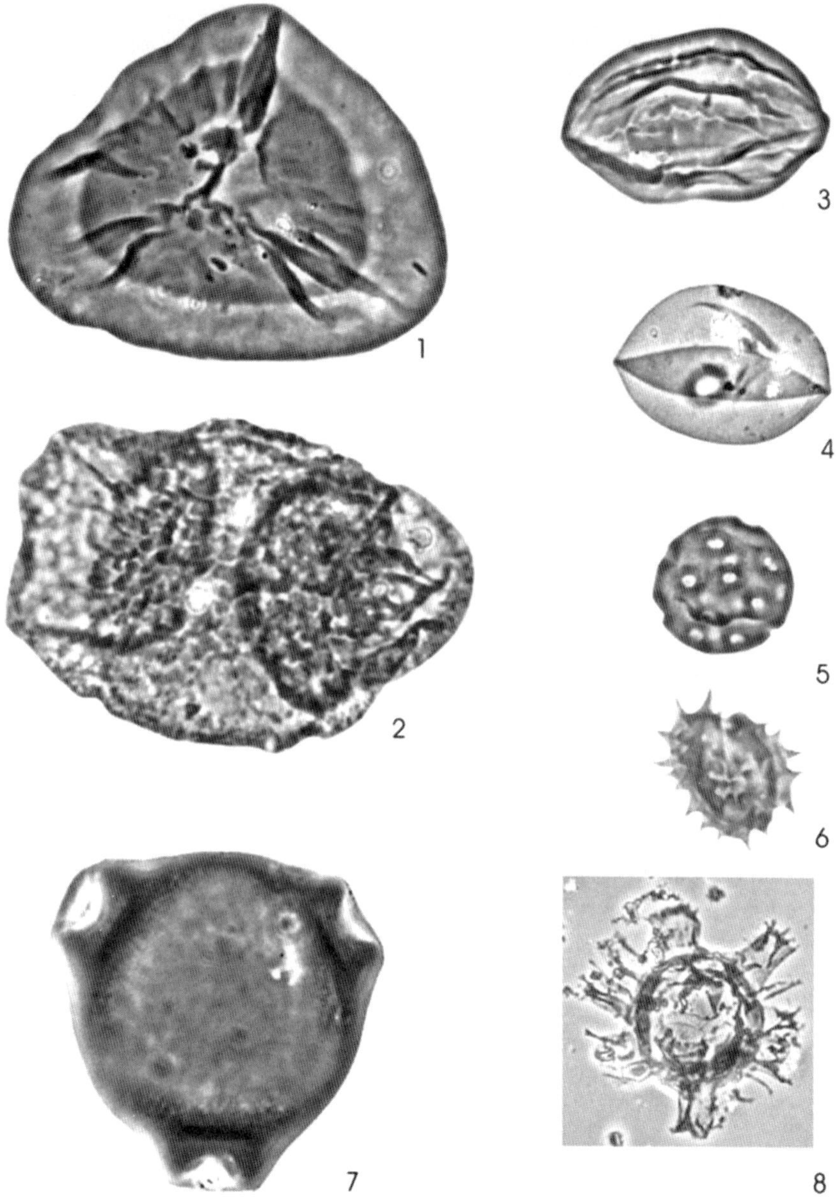

FIGURE 4. All taxa from San Gregorio Formation. 1. *Selaginella* [Pb-6129: 97.3/6]; 2. *Pinus* [Pb-6125 (3): 106.1/4.2]; 3. *Ephedra* [Pb-7288(3): 98.5/20.2]; 4. *Graminidites* [Pb-6125(A): 96.6/16.7]; 5. *Chenopodipollis* [Pb-6125(A): 95.4/17]; 6. Compositae [Pb-7288(1): 103.2/16.9]; 7. *Corsinipollenites* [Pb-6137(A): 100/9.4]; 8. *Chiropteridium lobospinosum* [Pb-7300(9): 93/20.1].

mangrove community. Elements with low frequencies were *Engelhardtia*, *Podocarpus* and *Pinus*. According to the biostratigraphy of the region, the authors assigned a Late Oligocene to Early Miocene age for the palynologic assemblages.

Another important contribution was made by Biaggi (1978), where palynostratigraphic research at "La Quinta" Formation, south Simojovel, Chiapas corresponds to the Upper Oligocene to Lower Miocene. The sediments studied correspond to shallow marine environments with presence of dinoflagelates and pollen grains, associated with microforaminifera, coralline algae and invertebrates. Biaggi's (1978) study revealed four plant communities, two of them developed at lowland: a) a mangrove vegetation with *Rhizophora, Pelliciera, Nypa, Acrostichum* and *Thespesia*; b) a tropical-subtropical forest including *Bombax, Terminalia* and *Ficus*; and two communities growing at uplands: c) subtropical to warm temperate community and 4) temperate to cool temperate community with *Pinus, Quercus, Liquidambar, Nyssa, Fagus* and *Picea*.

5.6. Miocene

Sediments from a borehole at Pichucalco, north of Chiapas, were dated by means of foraminifera as Lower Miocene to middle Miocene. The pollen assemblages were analyzed and compared with modern pollen rain of cloud forests and tropical evergreen forests (Palacios and Rzedowski 1993). The sediments analyzed yielded palynomorphs of mangrove vegetation of *Rhizophora* and *Pelliciera* indicative of nearshore marine environment where pollen representatives of cloud forest were abundant (*Engelhardtia-Oreomunnea*). The cloud forest certainly did not grow near the coast over particular edaphic conditions as stated by Palacios and Rzedowski (1993) but rather in the highlands where ravine are present with high humidity as occurs today at the modern cloud forest at the Sierra Madre. Other taxa reported by Palacios and Rzedowski (1993) are *Acer, Alchornea, Alfaroa, Billia, Brunellia, Calatola, Calyptranthes, Carya, Chiranthodendron, Clethra, Cornus, Deppea, Fagus, Gunnera, Heberdenia, Hedyosmum, Hoffmania, Ilex, Liquidambar, Liriodendron, Magnolia, Meliosma, Myrica, Nothofagus, Nyssa, Oreopanax, Phyllonoma, Podocarpus, Smilacina, Styrax, Symplocos, Ternstroemia, Tilia, Viburnum, Weinmannia*, spores of Cyatheaceae, pteridophytes and Melastomataceae. In addition, pollen grains from highlands coniferous forest were abundant (*Pinus* and *Picea*). Another type of vegetation was the riparian forest with the presence of *Taxodium, Alnus, Platanus, Populus, Salix, Carya, Fraxinus, Hedyosmum* and *Nyssa*.

In contrast to the climatic conditions of high temperatures and humidity that prevailed during the Miocene at Chiapas there is a palynologic study of similar age within Tubutama Formation, northeast of Sonora, where dry conditions prevailed with the presence of xeric elements like Compositae, Gramineae, Chenopodiaceae-Amaranthaceae, Leguminosae and *Boerhavia*, which are characteristic taxa of desert vegetation and thorn scrub forest (Martínez-Hernández and Tomasini, 1986).

5.7. Pliocene

Graham (1976) launched a program on the Tertiary Palynology of the Caribbean region. The first study was done on lignites at Paraje Solo Formation, near Coatzacoalcos, Veracruz. The assemblage of middle Pliocene age contained 110 different palynomorphs, a high number of taxa considering that those assemblages came from lignites that represent mostly a local flora. Although Graham (1976) did not consider this a definitive proof that tropical evergreen forest was present in Veracruz at least since 5.5 millions years ago. He mentioned that tropical rain forest was poorly developed at Paraje Solo Fm. (Graham, 1999b). The presence of mangrove vegetation, tropical deciduous forest and pine-oak forest was mentioned as part of the communities zonation from the shoreline to the mountain range. Graham (1976) inferred that due to the presence of *Picea* the existence of more cold and humid conditions must have existed.

6. Palynofloristic Affinities of the Balsas Group with other Tertiary Formations

The only formal study of the Balsas Group was accomplished by Fries (1960). Since that time nobody has accomplished a detailed stratigraphic work. Consequently the formations that constitute the Balsas Group remain unknown. This lack of geological interest probably was caused by the absence of chronostratigraphic control. It wasn't until the palynostratigraphic studies were completed when the differentiation of the Balsas Group into several formation began as was initially postulated by Fries (1960). Among the first palynostratigraphic studies are those of Martínez-Hernández and Ramírez-Arriaga (1999) and Ramírez-Arriaga and Martínez-Hernández (2003), where according to age determination the two formations *i.e.* Pie de Vaca Formation "PVF" (Eocene – Oligocene) and Cuayuca Formation (Oligocene) could be incorporated into the Balsas Group, according with the upper limit of this unit established by Fries (1960) as Late Oligocene (26 Ma) for the Tilzapotla rhyolite that overlie the Group.

It is important to mention that due to the absence of traditional fossils (vertebrates, invertebrates, etc.) and geochronologic data, PVF was considered as Pliocene-Pleistocene (Pantoja *et al.*, 1988) and the Cuayuca Formation was erroneously assigned to Miocene-Pliocene (Fries, 1966). With the palynostratigraphyc studies, PVF and Cuayuca Formation are now considered Upper Eocene to Lower Oligocene (Martínez-Hernández and Ramírez-Arriaga, 1999; Ramírez-Arriaga and Martínez-Hernández, 2003).

The aim of this section is to carry out a qualitative comparison between the different palynologic assemblages that occur throughout the Tertiary of Mexico. This approach asserts the existence of different types of floras and their succession throughout the Tertiary time as a response to geologic and climatic changes.

Comparison of the palynoflora of the Balsas Group (PVF and Cuayuca Formation) with other Tertiary palynological assemblages reported previously, makes possible the determination of common taxa, which reflects the paleochorology and evolution of Tertiary ecosystem in southern North America.

In general, when comparing PVF and the Cuayuca Formation (both belonging to Balsas Group) a high number of elements (n = 47) had in common: *Acacia, Aglaoreidia, Anacardiaceae, Anthoceros, Armeria,* Bombacaceae, *Bursera, Caesalpinia, Carya, Compositae, Corsinipollenites, Cupressus-Juniperus-Taxodium, Cyatheaceae, Cyperaceae, Chenopodiaceae-*Amaranthaceae, *Ephedra claricristata, Ephedripites, Eucommia,* Gramineae, *Juglans, Labiatae, Laevigatosporites,* Leguminosae, *Liliacidites, Liquidambar, Lycopodium, Malvacipollis, Meliaceae, Momipites coryloides, Momipites microcoryphaeus, Momipites triradiatus,* Moraceae, Myrtaceae, *Picea, Pinus, Platanus, Polygalaceae, Polygonaceae, Pterocarya, Quercus, Rosaceae,* Sapindaceae, *Selaginella,* Tiliaceae, *Typha, Ulmus* and Urticaceae. The variation between the PVF and Cuayuca Formation could be caused by difference in age and lithology, whereas Cuayuca Formation presents dominance of evaporitic facies that were deposited over xeric conditions, PVF has predominantly a clastic sedimentation.

PVF and phosforites of Baja California Sur (San Gregorio Formation) share 37 taxa: *Acacia*, Agavaceae, Anacardiaceae, *Betula*, Bignoniaceae, Bombacaceae, *Bursera, Caesalpinia, Carya*, Clethraceae, Compositae, *Corsinipollenites, Cupressus-Juniperus-Taxodium,* Cyatheaceae, Chenopodiaceae-Amaranthaceae, *Ephedripites, Eucommia,* Gramineae, *Juglans, Labiatae, Liliacidites, Lycopodium,* Meliaceae, *Momipites coryloides*, Myrtaceae, Palmae, *Pinus*, Polygalaceae, Polygonaceae, *Quercus*, Rosaceae, *Selaginella*, Solanaceae, Tiliaceae, *Typha, Ulmus* and Urticaceae.

Comparing PVF with Oligo-Miocene sediments (Biaggi, 1978) 19 taxa are in common: *Acacia*, Caryophyllaceae, *Corylus*, Chenopodiaceae-Amaranthaceae, Gramineae, *Ilex*, *Juglans*, Labiatae, *Liliacidites, Liquidambar, Lycopodium,* Myrtaceae, *Pinus*, Polypodiaceae, *Pterocarya, Quercus, Selaginella, Ulmus* and Urticaceae.

Despite the difference in age, the PVF and Pichucalco, Chiapas (Palacios and Rzedowski, 1993), basins share 29 taxa: *Abies, Acacia, Agavaceae, Bursera, Carya,* Compositae, *Corylus, Cupressus-Juniperus-Taxodium*, Cyathaceae, Chenopodiaceae-Amaranthaceae, *Ephedripites, Gramineae, Ilex, Juglans, Laevigatosporites, Liliacidites, Liquidambar, Lycopodium,* Malpighiaceae, Meliaceae, *Picea, Pinus, Platanus, Podocarpus, Quercus, Selaginella, Sphagnum, Ulmus* and Urticaceae.

It is worth noting the resemblance between the Cuayuca Formation (Balsas Group) and San Gregorio Formation assemblages which have 38 taxa in common: *Acacia, Alchornea, Alnus*, Anacardiaceae, Bombacaceae, *Bursera, Caesalpinia, Carya, Casuarinidites,* Compositae, *Corsinipollenites, Cupressus-Juniperus-Taxodium*, Cyatheaceae, Chenopodiaceae-Amaranthaceae, *Ephedripites, Eucommia,*

Euphorbiaceae, Fabaceae, Gramineae, *Juglans*, Labiatae, *Liliacidites*, Lycopodium, Meliaceae, *Momipites coryloides*, Myrtaceae, *Pinus*, Polygalaceae, Polygonaceae, Pteridophyta, *Quercus*, Rosaceae, *Selaginella*, Sterculiaceae, Tiliaceae, *Typha*, *Ulmus* and Urticaceae. This close association indicates similar flora growing at the same paleolatitude, because the Peninsula of Baja California during the Oligocene was part of Mexican mainland and was also located farther south at the time (Lyle and Ness, 1991).

Between La Quinta Formation (Oligo-Miocene), Chiapas (Biaggi, 1978) and Cuayuca Formation, Puebla (Ramírez-Arriaga and Martínez-Hernández, 2003) there are 25 taxa in common: *Acacia, Alchornea, Alnus, Arecipites, Casuarinidites, Celtis,* Chenopodiaceae-*Amaranthaceae, Ficus, Gramineae, Juglans, Labiatae, Liliacidites, Liquidambar, Lycopodium,* Myrtaceae, *Corisinipollenites, Pinus, Pterocarya, Quercus, Salix, Selaginella, Sphaeropteris, Sterculiaceae, Ulmus* and Urticaceae.

Despite the dissimilar age between Pichucalco (Palacios and Rzedowski, 1993) and Cuayuca Formation, they share 32 taxa: *Acacia, Alchornea, Alnus, Annonaceae, Bursera, Carya, Celtis,* Compositae, *Cupressus-Juniperus-Taxodium,* Cyatheaceae, Chenopodiaceae-Amaranthaceae, *Ephedripites*, Gramineae, *Juglans, Leiotriletes, Laevigatosporites, Liliacidites, Liquidambar, Lycopodium, Meliaceae, Mimosa, Picea, Pinus, Platanus, Platycarya, Quercus, Rhamnus, Salix, Selaginella, Thomsonipollis, Ulmus* and Urticaceae.

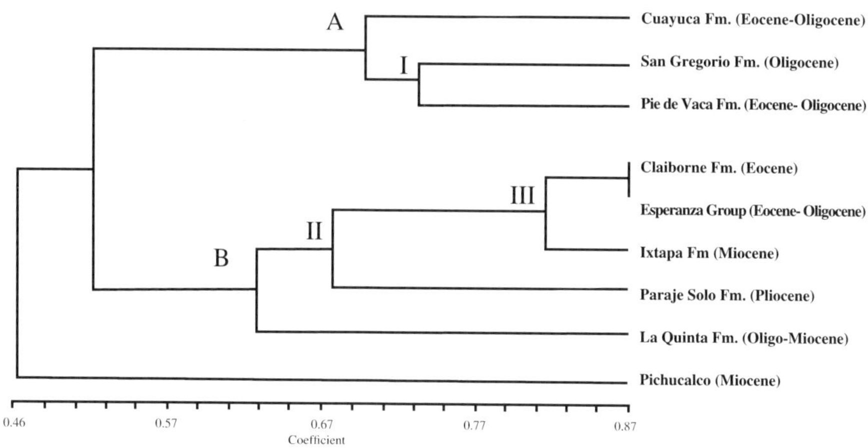

FIGURE 5. Qualitative similarity tree obtained with NTSYS program based on data of Claiborne Formation (Martínez-Hernández *et al.* 1980), Esperanza Group: El Bosque, La Trinidad and Rancho Berlín Formations (Tomasini-Ortíz and Martínez-Hernández, 1984), Pie de Vaca Formation (Martínez-Hernández and Ramírez-Arriaga, 1999), Cuayuca Formation (Ramírez-Arriaga and Martínez-Hernández, 2003), San Gregorio Formation (Martínez-Hernández and Ramírez-Arriaga in process), La Quinta Formation (Biaggi, 1978), Pichucalco (Palacios y Rzedowski, 1993), Ixtapa Formation (Martínez-Henrández, 1992) and Paraje Solo Formation (Graham,1976).

On the other hand, seventeen taxa were shared between Paraje Solo, Veracruz (Middle Pliocene), Graham (1976) and Cuayuca Formation, Oligocene, Puebla: *Alchornea, Alnus, Bursera, Celtis*, Chenopodiaceae- Amaranthaceae, Compositae, Cyperaceae, Gramineae, *Juglans, Liquidambar, Lycopodium, Mimosa, Picea, Pinus, Quercus, Selaginella* and *Ulmus*. This indicates different types of vegetation. As stated earlier, the lignites reflect that the tropical evergreen forest grew near the site of deposition and at Cuayuca Formation, a continental basin with regional pollen rain indicates different plant communities in an altitudinal gradient.

In order to understand numerically the similarity between the assemblages it was necessary to apply a qualitative analysis. By means of a numerical analysis, a tree of qualitative similarity was obtained (Fig. 5). This is a phenogram with hypothetic relationships between Tertiary basins.

The phenogram shows two interesting clusters A and B (Fig. 5). The first association (Fig. 5: A), with 0.69 of similarity includes the Balsas Group (Cuayuca Formation and PVF). This phenogram shows that between the PVF and San Gregorio Formation there is a 0.76 coefficient of similarity (Fig. 5: I). This suggests that PVF and SGF have palynological affinities and share taxa of chronostratigraphic importance such as *Eucommia, Corsinipollenites* and *Momipites* group. In addition, San Gregorio Formation also has dinocyst indexes that confirm an Oligocene age. The previous association (Fig. 5:I) is clustered with Cuayuca Formation at level "A", revealing generally the Balsas Group (PVF and Cuayuca Fm.) is very palynologically close to the San Gregorio Formation. This cluster is a confirmation of similar floras that were growing in southern Mexico in the Lower Tertiary. With the presence of a low biomass vegetation i.e. savanna or dry thorn woodland where grasses were abundant, corroborates a previous report (Martínez-Hernández and Ramírez-Arriaga, 1999) of early abundance of grasses in North America (Martínez-Hernández and Ramírez-Arriaga, 2003).

The second association (Fig. 5: B) has 0.63 as a coefficient of similarity; it includes the southeastern Tertiary formations with exception of the Claiborne Formation. This clustering reveals that the Tertiary northeastern palynoflora of Mexico (Claiborne Fm.) Mexico (Claiborne Fm.) has affinities with southeastern Mexican formations (Fig. 5: II). The Esperanza Group, Chiapas (Eocene-Oligocene) and the Claiborne Formation (Tamaulipas) are joined at level III with 0.87 of similarity (Fig. 5). It shows that despite the slight discrepancy in age, the assemblage reflects similar coastal vegetation probably growing under a tropical climate growing near the shore of Gulf of Mexico, although the mangrove vegetation is not present yet during Claiborne time (Middle Eocene).

Considering the total assemblages it was difficult to delimit the relationships between all formations by means of a phenetic analysis. Nevertheless, the Balsas Group relationships could be confirmed using a cladistic method.

7. Historical Biogeographic Analysis between Mexican Tertiary Palynofloras

In order to make inferences of historic connection between areas and floras it is necessary to employ a cladistic method based on the number of species shared amongst different areas. In this sense, "parsimony analysis of endemicity" (PAE) could identify ancient historical connection (Rosen, 1988; Brooks *et al.* 2003).

This method proposes that the clustered areas were caused by a common historic development that could be interpreted as regions, domains or biogeographic provinces (Espinosa *et al.* 2003). In this survey the PAE analysis only considered the common taxa shared at the different basins (Annex 1, Table 2), in contrast with the numerical analysis of similarity that included all the taxa found among the different formations. Therefore, the PAE cladistic analysis cluster the formations which are historically related throughout the Tertiary.

The analysis produced only one cladogram (198 steps, consistency index = 0.56, retention index = 0.54) (Fig. 6).

The cladogram shows hypothetic clusters at different degrees of historical relationship. The group results with the closest congruence (Fig. 6: I) include the Cuayuca and Pie de Vaca formations, belonging to the Balsas Group, which are unifying by the presence of *Anthoceros, Ephedra claricristata, Aglaoreidia, Armeria, Malvacipollis, Momipites microcoryphaeus, M. triradiatus* and *Sapindaceidites*. Among them, the taxa of chronostratigraphic importance are *E. claricristata, Aglaoreidia, Armeria, M. microcoryphaeus* and *M. triradiatus*.

The next level of historical association exists between the Balsas Group and the San Gregorio Formation (Fig. 6: II) defined by *Typha, Bombacacidites, Corsinipollenites, Eucommia, Heliocarpus, M. coryloides, Polygalacidites* and *Reevesia*. The important fossil indexes are *Bombacacidites, Corsinipollenties, Eucommia* and *M. coryloides*. The floristic association is important because since the Tertiary they have shared a close floral historical relationship when the Baja California peninsula was southerly placed and attached to the mainland of Mexico, despite actually being in disjunct areas.

Another congruence association (Fig. 6: III) shows similar environments between the Paraje Solo Formation and Pichucalco which are Neogene in age and very close geographically. This cluster is defined by *Cedrela, cf. Malpighia, Coccoloba, Cuphea, Faramea, Guarea, Hedyosmum, Meliosma Myrica, Passiflora, Populus, Serjania, Smilax* and *Struthanthus*.

When group III is associated with the La Quinta Formation (Fig. 6: IV), a smaller number of taxa, *Combretum, Eugenia* and *Tournefortia*, are common between the two of them.

Most of the Paleogene formations grouping at level II and the Neogene formations associated at level IV share *Acacia, Alchornea, Juglans, Quercus* and *Rhamnus* (Fig. 6: V).

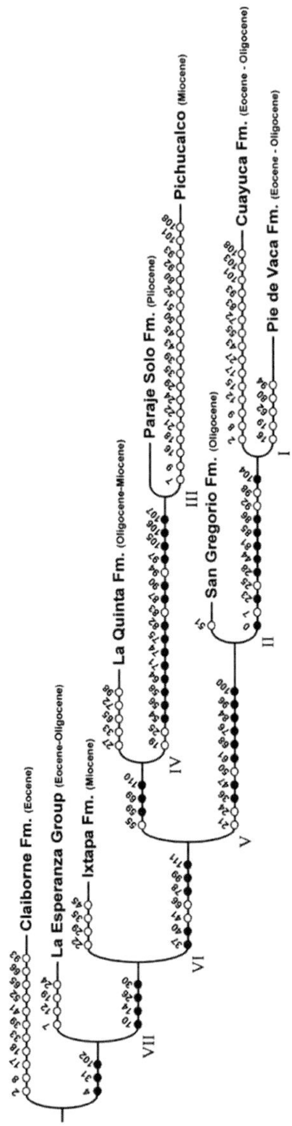

FIGURE 6. Cladogram where branches are defined by presence of taxa. Numbers correspond to taxa cited on table 2. Formations were studied by Martínez-Hernández *et al.* (1980: Claiborne Fm.), Tomasini-Ortíz and Martínez-Hernández (1984: El Bosque, La Trinidad and Rancho Berlín Fms.), Martínez-Hernández and Ramírez-Arriaga (1999: Pie de Vaca Fm), Ramírez-Arriaga and Martínez-Hernández, (2003: Cuayuca Fm.), Martínez-Hernández and Ramírez-Arriaga (in process: San Gregorio Fm.), Graham (1976: Paraje Solo Fm.) Biaggi (1978: La Quinta Fm.), Palacios and Rzedowski (1993: Pichucalco), Martínez-Hernández (1992: Ixtapa Fm.) and Graham (1976: Paraje Solo Fm.).

Ixtapa Formation is related with all the previous formations cited (Fig. 6: VI) because it shares *Lycopodium, Selaginella, Pinus* and *Graminidites*, with them and all these taxa have been reported from Paleogene to the present.

The Esperanza Group (Fig. 6: VII) shares with the preceding formations *Cyathea*, and *Liliacidites*. On the other hand, the Claiborne Formation has only palynomorphs that could be common in different Tertiary paleobasins: *Camarozonosporites, Leiosphaeridae, Stereisporites, Verrucatosporites, Monocolpopollenites, Aesculidites, Alnus, Castanea, Cupuliferoipollenites, Chenopodipollis* and *Platycarya*.

Claiborne assemblages indicate a unique flora growing under asteady environment with a subtropical climate, far away from the volcanism of western and southern Mexico. Consequently, the historical relationship of the Claiborne assemblages with the Balsas Group of similar age is obscure because the adaptation of the Balsas flora to frequent explosive volcanism that produced natural selection of certain taxa. In this case, the effects of volcanism produced a different flora during the Eocene-Oligocene between northeastern and southern assemblages of Mexico (Martínez-Hernández and Ramírez-Arriaga, 1996b).

Subsequent palynological studies of Ixtapa, La Esperanza and Claiborne Formations will improve the pollen fossil record, allowing for a better historical relationships.

In summary, sequential changes in Mexican fossil floras can be produced by global climatic changes and regional volcanism through time.

8. General Climatic Trends during Tertiary Period

The climatic trends in the Tertiary of North America were analyzed recently by Graham (1999a) based mainly on in the palynologic studies done in the United States. In this Mexican survey, climatic changes detected by means of several palynofloras do correspond, in general, with climatic mode established by Graham (1999a). Emphasis is placed in the Paleogene because during this period the Balsas Group was deposited in southern Mexico and the enhancement of climatic zonation had began to establish in North America subcontinent.

During the Paleocene, the North American paleophytogeography indicate a close relationship with Asia and Europe. This is due to similar climate regimes and flora migration by a land connection between eastern Asia and western North America and the close proximity of Western Europe with eastern North America (Briggs, 1987).

At the same time, the Mexican territory began its differentiation of floral provinces between the Pacific Coast where the "Sierra Madre Occidental Province" was subject to constant and explosive volcanisms throughout the Tertiary and the Province of the Gulf Coast plains that suffered the Laramide orogeny (Martínez-Hernández and Ramírez-Arriaga, 1996b).

It is well known that in the Paleocene there was a world wide global warming, with a high pluvial regime which allowed the establishment of a tropical rain forest with angiosperm predominance (Frakes *et al.* 1992; Graham, 1999a). In Mexico, there are few reports of Paleocene age. In northeastern Mexico Rosales-Lomelí *et al.* (1992), included genera that belong to the Wilcox Flora with tropical affinities. At the same time in northwestern Mexico the flora indicated a warm temperate forest with predominance of wind pollinated plants (Cross and Martínez, 1980).

During the Early Eocene, the tropical mode persisted with high temperatures and high precipitation. The rainforest continued to inhabit the low plain environments. However from about the Middle Eocene there is a drastic climatic change with a decrease in precipitation and lowering temperature (Graham, 1999a). In consequence, throughout the Middle Eocene in northeastern Mexico, except for the boreotropical taxa of the Wilcox flora: Bombacaceae, Sapotaceae (Martínez *et al.* 1980), certain temperate taxa such as *Pinus* and *Quercus* began to have presence within this Gulf Coast Province. In contrast, at the western localities, the Eocene floras were an admixture of temperate and warm temperate floras with the presence of certain plant families Compositae and Ephedraceae (Cross and Martínez, 1980), which indicated periodical dryness and a few tropical taxa.

In the Oligocene dramatic paleogeographic changes occurred when the Turgai strait was closed and the super continent Euro-Asiatic was formed. As a consequence, patterns in oceanic currents worldwide triggered climatic changes toward dryness and increasing seasonableness.

This climatic change caused the development of a low biomass-vegetation (Leopold *et al.*, 1992). This is supported by palynological evidence from the Cuayuca Formation and PVF where there is a diverse and abundant *Graminidites* (grasses) assemblage for the first time in North America (Martínez-Hernández and Ramírez-Arriaga, 1996b; 2003). Additionally the climate change is supported by the presence of *Ephedripites* spp. and other taxa found in the thorn forest and chaparral, although within the highlands grew the cloud forest and the coniferous forest. At San Gregorio Formation (Late Oligocene) high frequencies of grasses and Chenopodiaceae - Amaranthaceae denote the existence of a savanna with a high density of grasses (Gramineae) unique to North America. Data that contrasts with the *statusquo* which considers the grass ecosystems dominated in North America until the mid-Miocene (Leopold *et al.*, 1992), despite the opinions of other authors that stated an early occurrence of grasses ecosystem in North America (Retalack, 1997).

The palynologic data from the Eocene to the Early Miocene are concentrate in the southeastern region of the Gulf Coast Province. The survey of the Esperenza Group done by Biaggi (1978) and Tomasini-Ortíz and Martínez-Hernández (1984) clearly show that since the Late Eocene, there were several palynofloras that represented a mixture of neotropical elements, which were probably relicts of the North America boreotropical flora *sensu* Wolfe (1985) such as *Rhizophora, Pelliceri*a, Bombacaceae and *Ficus* which grew in low plains with other communities toward higher altitudes as cloud forest and *Quercus* and *Pinus* forest. At the same time, the communities that grew toward the west (PVF, Cuayuca and San Gregorio) indicate a climatic regime colder and dryer with flora similar to the Rocky Mountains with thorn forest, chaparral at low elevations and *Quercus*, *Pinus* and *Picea* in the ranges.

Also in the Gulf Coast Province within the Pichucalco region (northern Chiapas) the palynological spectra of middle Miocene age demonstrated the influence of a cool mode that prevail in the Miocene with temperate taxa such as *Picea, Pinus, Alnus, Quercus* together with a great diversity of taxa belonging to the tropical rain forest that inhabited the plains (Palacios and Rzedowski, 1993).

Contemporaneously to the development of the evergreen forest in southern Mexico in the state of Sonora (northeastern Mexico), the Middle Miocene Tubutama Formation of Sonora, which is enclosed in the Sierra Madre Occidental Province, the palynologic data show quite different climatic regime with pollen assemblages dominated by Compositae, Gramineae, Chenopodiaceae – Amaranthaceae. This suggests an arid climate characterized by strong seasonal dryness and the beginning of desertification of northwestern México (Martínez-Hernández and Tomasini, 1986) and supporting previous hypothesis of Axelrod (1950, 1979).

Acknowledgments

The authors wish to acknowledge the valuable numeric and cladistic methods assistance carried out by Dra. Hilda Flores Olvera, Instituto de Biología, UNAM. Cuayuca Fm. survey was supported by DGAPA with number: IN110798 and the results are part of a Doctorate thesis developed by Elia Ramírez-Arriaga.

References

Axelrod, D. I., 1950, Evolution of desert vegetation in western North America. *Carnegie Inst. Wash. Pub.* 590: 216-306.

Axelrod, D. I., 1979, Age and origin of Sonoran desert vegetation. *Occ. Pap. California Acad. Sci.* 132: 73p.

Biaggi, R. E., 1978, *Palynology and Paleoecology of some Oligo-Miocene sediments from Chiapas, Mexico.* M. S. Thesis, Walla Walla College, California. U.S.A. 92p.

Briggs, J. S., 1987, Biogeography and plate tectonics developments. *Pal. and Strat.* 10: 195.

Brooks, D. R., and Van Veller, M. G. P., 2003, Critique of parsimony analysis of endemicity as a method of historical biogeography. *J. Biogeog*. 30:819-825.

Cross, A. T., and Martínez-Hernández, E., 1980, Compositae pollen in early Tertiary rocks, Baja California, Mexico, *Abstracts 5th. International Palynological Conference, Cambridge* p. 97.

Cserna, Z. de, 1965, Reconocimiento geológico en la Sierra Madre del Sur de México, entre Chilpaningo y Acapulco, Estado de Guerrero, *UNAM, Inst. Geol. Bol.* 62, 76p.

Cserna, Z. de, 1978, Relaciones de Facies de las Rocas cretácicas en el noreste de Guerrero y en áreas colindantes de México y Michoacán. *UNAM, Inst. Geol. Rev.* 2:8-18.

Cserna, Z. and Fries, C. Jr., 1981, Hoja Taxco 14 Q-h(7). Resumen de la Geología de la hoja Taxco, Estados de Guerrero, México y Morelos. *UNAM. Inst. Geol., Carta Geológica de México, Serie 1:1000, 000,* 47 p.

Cserna, Z. de, Ortega-Gutiérrez, F., and Palacios-Nieto, M., 1980, Reconocimiento Geológico de la parte central de la cuenca del alto río Balsas, Estados de Guerrero y Puebla: *In* Libro-Guía de la excursión geológica a la Parte Central de la Cuenca. *Soc. Geol. Mex.* p. 1-31.

Elsik, W. C., 1974, Characteristic Eocene palynomorphs in the Gulf Coast, U.S.A. *Palaeont. Abt.* B 149. 1-4: 90-111

Elsik, W. S., and Dilcher, D. L., 1974, Palynology and age of clays exposed Lawrence clay pit, Henry County, Tennessee. *Palaeont. Abt.* B 146:65-87.

Espinosa, O. D., Aguilar, Z. C., and Escalante, E. T., 2003, Endemismo, áreas y reorganización biogeográfica, in: *Introducción a la Biogeografía en Latinoamérica: Teorías, conceptos, métodos y aplicaciones* (Llorente B.J. and Morrone J.J., eds.), *Las Prensas de Ciencias, Fac. Ciencias, UNAM,* 273- 277.

Frakes, L. A., Francis, J. E., and Syktus, J. I., 1992, *Climate modes of Phanerozoic*. Cambridge University Press, 274 p.

Frederiksen, N. O., 1984, Stratigraphic, paleoclimatic and paleobiogeographic significance of Tertiary sporomorphs from Massachusetts. *U. S. Geol. Surv. Prof. Pap.* 1308:1-25.

Frederiksen, N. O., 1988, Sporomorph Biostratigraphy, floral changes and paleoclimatology, Eocene and earliest Oligocene of the eastern Gulf Coast. *U. S. Geol. Sur. Prof. Pap.* 1448:1-68.

Fries, C. Jr., 1960, Geología del Estado de Morelos y de partes adyacentes de México y Guerrero, región central meridional de México. *UNAM, Inst. Geol. Bol.* 60:1-236.

Fries, C. Jr., 1966, Hoja Cuernavaca 14Q-h(8). Resúmen de la geología de la Hoja Cuernavaca, Estado de Morelos. *Univ. Nac. Autón. México. Inst. Geología, Carta Geológica de México. Serie 1: 100, 000, mapa con texto.*

Goloboff, P.A. 1993. NONA version 2.0 Program and documentation. Computer program distributed by J.M. Carpenter, Dept. of Entomology. Ameriacan Museum of Natural History, New York.

Graham, A., 1976, Late Cenozoic evolution of tropical lowland vegetation in Veracruz, México. *Evolution* 29:723-735.

Graham, A., 1999a, *Late Cretaceous and Cenozoic History of North American Vegetation*. Oxford University Press. 350p.

Graham, A., 1999b, Terrestrial Pliocene climates in northern Latin America, in: *The Pliocene: Time of change* (Wrenn, J.H., Suc, J.-P. and Leroy, S.A.G., eds.), *Am. Assoc. Strat. Palyn. Foundation* p. 209-216.

Hausback, B. P., 1984, Cenozoic volcanic and tectonic evolution of Baja California Sur. Mexico, in: *Geology of the Baja California Peninsula. San Diego, Pacific Section* (Frizell, V.A. Jr., ed.), S.E. P.M. 39:219-236.

Kim, W. H., and Barron, J. A., 1986, Diatom biostratigraphy of the Upper Oligocene to lowermost Miocene San Gregorio Formation, Baja California Sur, Mexico. *Diatom Res.* 1:169-187.

Langenheim, J. H., Hackner, B. L., and Bartlett, A., 1967, Mangrove pollen at the depositional site of Oligo-Miocene amber from Chiapas, México. *Bot. Mus. Leaflets, Harvard Univ.* 21:289-324.

Leopold, E. B., Gengwu, L., and Scott, C. P., 1992, Low-biomass vegetation in the Oligocene? in: *Eocene-Oligocene climatic and biotic evolution* (D.R. Prothero and W.A. Berggren, eds.): 399-420. Princeton University Press.

Luna, V. I. and Alcántara, A. O., 2003, Análisis de simplicidad de endemismos (PAE) para establecer un modelo de vicarianza preliminar del Bosque Mesófilo de Montaña mexicano in: *Introducción a la Biogeografía en Latinoamérica: Teorías, conceptos, métodos y aplicaciones. Llorente B. J. and Morrone J. J. (eds.). Las Prensas de Ciencias, Facultad de Ciencias, UNAM:* 273- 277.

Lyle, M. and Ness, G.E., 1991, The opening of the southern Gulf of California. *In:* Dauphin J.P. and Simoneit B.R.T. (eds.) The Gulf and Peninsular Province of the Californias. *Am. Assoc. Pet. Geol. Mem.* 47:403-427.

Magallón-Puebla, S., and Cevallos-Ferriz, S. R. S., 1994, *Eucommia constans* n. sp. fruits from Upper Cenozoic strata of Puebla, México: Morphological and anatomical comparason with *Eucommia ulmoides* Oliver. *Int. J. Plant Sci.* 155:80-95.

Martínez-Hernández, E., 1992, Caracterización ambiental del Terciario de la región de Ixtapa, estado de Chiapas – un enfoque palinoestratigráfico. *UNAM, Inst. Geol. Rev.* 10:54-64.

Martínez-Hernánez, E., and Ramírez-Arriaga, E., 1996a, Oligocene phosphorite sequence and presence of low biomass vegetation from San Gregorio Fm., Baja California, Mexico. IX IPC Meeting. Houston,

Texas, U.S.A. June 23-28. Abstracts: p. 98.

Martínez-Hernánez, E., and Ramírez-Arriaga, E., 1996b, Paleocorología de angiospermas de la flora mexicana durante el Mesozoico y Terciario. Algunas evidencias palinológicas. *Bol. Soc. Bot. Méx.* 58: 87-97.

Martínez-Hernández, E., and Ramírez-Arriaga, E., 1998, Paleogeographic implications of Oligocene dinocysts from the phosphorites deposits of Baja California Sur: opening of the Gulf of California. Sixth Internacional Conference on Modern and Fossil Dinoflagellates. DINO 6. Trondheim, Norway, June 7-12.

Martínez-Hernández, E., and Ramírez-Arriaga, E., 1999, Palinoestratigrafía de la región de Tepexi de Rodríguez, Puebla, implicaciones cronoestratigráficas. *Rev. Mex. Cienc. Geol.* 16:187-207

Martínez-Hernández, E., and Ramírez-Arriaga, E., 2000, Palinoestratigrafía y ambientes de la Fm. Cuayuca, Izúcar de Matamoros. *GEOS Boletín informativo. Época II. Unión Geof. Mex.* 20:159-160.

Martínez-Hernández, E., and Ramírez-Arriaga, E., 2003, Early abundance of grasses (Poaceae) during the Oligocene in southern North America. *Am. Assoc. Strat. Palyn.. St. Catharines Ontario. Abstracts.* p.15.

Martínez-Hernández, E., and Tomasini, O. A. C., 1986, Palinoestratigrafía preliminar del Terciario continental de Sonora. 1er Simposio Geología Regional de México. Instituto de Geología, Ciudad Universitaria. Programas y Resúmenes: 19-21.

Martínez-Hernández, E., Hernández-Campos, H., and Sánchez-López, M., 1980, Palinología del Eoceno en el Noreste de México. *UNAM, Inst. Geol. Rev.* 4:155-166 p.

Méndez, L.I. and Villaseñor R.J.L. 1997. Breve guía para el uso de NTSYS-PC en estudios de Taxonomía numérica. Instituto de Biología UNAM. Departamento de Botánica. 22p.

Morán-Zenteno D. J., Alba-Aldave, L. A., Solé, J., and Iriondo, A. 2004. A major resurgent caldera in southern Mexico: the source of the late Eocene Tilzapotla ignimbrite. *J. Vol. Geoth. Res.* 136: 97-119.

Morrone, J. J., 1994, On the identification of areas of endemism. *Sist. Biol.* 43:438-441.

Nixon, K. C. 2002. WinClada ver. 1.00.08 Published by the autor, Ithaca, N.Y.

Ortega-Gutiérrez, F., 1980, Rocas volcánicas del Maestrichtiano en el área de San Juan Tetelcingo, Estado de Guerrero: *in* Libro-guía de la excursión geológica a la parte central de la cuenca del alto Río Balsas, México, D. F., *Soc. Geol. Mexicana,* p.34-38.

Palacios, Ch. R., and Rzedowski, J., 1993, Estudio palinológico de las floras fósiles del Mioceno inferior y principios del Mioceno medio de la región de Pichucalco, Chiapas, México. *Acta Bot. Mex.* 24:1-96.

Pantoja-Alor, J., Malpica-Cruz, V. M., and Galguera, G. A., 1988, Geología de los alrededores de la cantera de Tlayua, Tepexi de Rodríguez, Puebla. UNAM, Tercer Simp. Geol. Reg. México. Mem. p. 61-69.

Ramírez-Arriaga, E., and Martínez-Hernández, E., 2003, Synthesis of Cuayuca Formation and correlation with palynoflora of Balsas Group: Paleogene Lacustrine environments from central of Mexico. *Am. Assoc. Strat. Palyn. Palynology* 27:264.

Retallack, G. J., 1997, Neogene expansion of the North American Praire. *Palaios* 12:380-390.

Rosales-Lomelí, J., Ayala-Nieto, M., and Martínez-Hernández, E., 1992, Investigación palinológica de fósiles terciarios en una columna de la Cuenca Tampico-Misantla. Proyecto CAO-1405, 27 p. (Unpub. report).

Rosen D. E. 1988. From fossils to earth history: applied historical biogeography. *In*: Myers A.A. and P.S. Giller eds. Analytical biogeography an integrated approach to the study of animal and plant distributions. Chapman, London & New York. 437-481.

Srivastava, S. K., 1984, Palynology of the Monterey Formation (Miocene) Phosphatic facies at Lions Head, Santa Maria area, California. *Palynology* 8:33-49.

Tomasini-Ortíz, A. C., and Martínez-Hernández, E., 1984, Palinología del Eoceno-Oligoceno de Simojovel, Chiapas. *UNAM, Inst. Geol. Pal. Mex.* 50:61 p.

Tschudy, R. H., 1973, Stratigraphic distribution of significant Eocene palynomorphs of the Mississippi embayment. Contributions to Paleontology. *U. S. Geol. Surv. Prof. Pap.* 743-B:B1 – B24.

Wolfe, J. A., 1985, Distribution of major vegetational types during the Tertiary. *Am. Geoph. Union Mon.* 32:357-375.

Appendix: List of Palynomorphs

TABLE 2. List of Palynomorphs shared between Tertiary basins. Identification number (Id) used in the parsimony analysis of endemicity (PAE). Taxa with (FAMILY) or [BOTANICAL AFFINITY] Martínez-Hernández *et al.* (1980: J-L), Tomasini and Martínez (1984: E), Martínez-Hernández and Ramírez-Arriaga (1999: I), Ramírez-Arriaga and Martínez-Hernández (2003: H), Martínez-Hernández and Ramírez-Arriaga in process (G) Biaggi (1978: F), Palacios and Rzedowski,(1993: C), Martínez-Hernández (1992: B) and Graham (1976: A).

		Eoc	Eoc– Oli			Oli	Oli-Mio	Mio		Plio
Id	**TAXA**	**J-L**	**E**	**I**	**H**	**G**	**F**	**C**	**B**	**A**
PTERIDOPHYTA and BRYOPHYTA spores										
0	*Anthoceros* (Anthocerotaceae)			″	″					
1	*Asplenium* (Aspleniaceae)						″	″		
2	*Camarozonosporties* [Lycopodiaceae]	″			″					
3	*Cicatricosisporites* [Shizaeaceae]	″	″				″			
4	*Cyathea* (Cyatheaceae)		″		″		″	″	″	″
5	*Equisetum* (Equisetaceae)						″	″		
6	*Jamesonia* (Adiantaceae)						″	″		
7	*Laevigatosporites* [*Insertae sedis*]		″	″	″			″		
8	*Leiosphaeridae* [*Insertae sedis*]	″			″					
9	*Leiotriletes* [Schizaeaceae]				″			″		
10	*Lycopodium* (Lycopodiaceae)			″	″	″	″	″	″	″
11	*Lygodium* (Schizaeaceae)						″	″		
12	*Polypodiisporites* [Polypodiaceae]		″		″					
13	*Pteris* (Adiantaceae)						″	″	″	″
14	*Selaginella* (Selaginellaceae)			″	″	″	″	″	″	″
15	*Sphaeropteris* (Cyatheaceae)				″		″	″		
16	*Sphagnum* (Sphagnaceae)			″				″		
17	*Stereisporites* [Bryophyta]	″			″					
18	*Verrucatosporites* [Polypodiaceae]	″						″		
GYMNOSPERMS										
19	*Abies* (Pinaceae)		″	″				″		″
20	*Cedrus* (Pinaceae)						″	″		
21	*Cupressus/Juniperus/Taxodium*			″	″	″		″		
22	*Cycadopodites* [Cycadaceae]							″	″	
23	*Ephedra claricristata* (Ephedraceae)			″	″					
24	*Ephedripites* [Ephedraceae]		″	″	″	″		″		
25	*Picea* (Pinaceae)			″	″			″		″
26	*Pinus* (Pinaceae)			″	″	″	″	″	″	″

TABLE 2. Continued.

		EOC	EOC– OLI			OLI	OLI-MIO	MIO		PLIO
Id	**TAXA**	**J-L**	**E**	**I**	**H**	**G**	**F**	**C**	**B**	**A**
ANGIOSPERMAE – MONOCOTILEDÓNEA										
27	*Arecipites* [PALMAE]				″		″			
28	*Aglaoreidia* (RESTIONACEAE)			″	″					
29	*Catopsis* (BROMELIACEAE)							″	″	
30	*Graminidites* (GRAMINEAE)			″	″	″	″	″	″	″
31	*Liliacidites* [LILIACEAE, PALMAE, BROMELIACEAE]		″	″	″	″	″	″	″	
32	*Liliacidites variegatus* [LILIACEAE, PALMAE, BROMENIACEAE]	″	″							
33	*Monocolpopollenites* [*Insertae sedis*]	″					″			
34	*Nypa* (PALMAE)						″	″		
35	*Tillandsia* (BROMELIACEAE)							″	″	
36	*Typha* (THYPHACEAE)			″	″	″				
DICOTILEDÓNEA										
37	*Acacia* (LEGUMINOSAE)			″	″	″	″	″		
38	*Acer* (ACERACEAE)						″	″		
39	*Aesculidites* [HIPPOCASTANACEAE]	″						″		
40	*Alchornea* (EUPHORBIACEAE)			″	″	″	″	″		″
41	*Alnus* (BETULACEAE)	″			″	″	″	″		″
42	*Striatricolporites* (ANACARDIACEAE)			″	″	″				
43	*Annona* (ANNONACEAE)				″			″		
44	*Armeria* (PLUMBAGINACEAE)			″	″					
45	*Ayenia* (STERCULIACEAE)							″	″	
46	*Betula* (BETULACEAE)			″		″				
47	*Bombacacidites* [BOMBACACEAE]			″	″	″				
48	*Bombax* (BOMBACACEAE)						″	″		
49	*Bursera* (BURCERACEAE)			″	″	″		″		″
50	*Carya* (JUGLANDACEAE)			″	″	″		″		
51	*Cassia* (LEGUMINOSAE)					″		″		
52	*Castanea* (FAGACEAE)	″						″		
53	*Casuarinidites* [CASUARINACEAE]				″	″	″			
54	*Cedrela* (MELIACEAE)							″		″
55	*Celtis* (ULMACEAE)				″		″	″		″
56	cf. *Malpighia* (MALPIGHIACEAE)							″		″
57	*Clethra* (CLETHRACEAE)						″	″		
58	*Coccoloba* (POLYGONACEAE)							″		″
59	*Combretum* (COMBRETACEAE)						″	″		″

TABLE 2. Continued.

		EOC	EOC– OLI			OLI	OLI-MIO	MIO		PLIO
Id	**TAXA**	**J-L**	**E**	**I**	**H**	**G**	**F**	**C**	**B**	**A**
60	*Tubulifloridites* (COMPOSITAE)			"	"	"		"		"
61	*Corsinipollenites* [ONAGRACEAE]			"	"	"				
62	*Corylus* (BETULACEAE)			"			"	"		
63	*Cupania* (SAPINDACEAE)						"			"
64	*Cuphea* (LYTHRACEAE)							"		"
65	*Cupuliferoipollenites* [FAGACEAE]	"					"			
66	*Chenopodipollis* (CHENOPODIACEAE-AMARANTHACEAE)	"		"	"	"	"	"		"
67	*Engelhardia* (JUGLANDACEAE)	"	"				"	"	"	"
68	*Eucommia* (EUCOMMIACEAE)			"	"	"				
69	*Eugenia* (MYRTACEAE)						"	"		"
70	*Fagus* (FAGACEAE)						"	"		
71	*Faramea* (RUBIACEAE)							"		"
72	*Ficus* (MORACEAE)			"	"		"			
73	*Fraxinus* (OLEACEAE)						"	"		
74	*Guarea* (MELIACEAE)							"		"
75	*Hedyosmum* (CHLORANTHACEAE)							"		"
76	*Heliocarpus* (TILIACEAE)			"	"	"				
77	*Ilex* (AQUIFOLIACEAE)	"		"			"	"	"	"
78	*Juglans* (JUGLANDACEAE)			"	"	"	"	"		"
79	*Liquidambar* (HAMAMELIDACEAE)			"	"		"	"		"
80	*Perisyncolporites* (MALPIGHIACEAE)				"	"				
81	*Malvacipollis* [MALVACEAE]			"	"					
82	*Meliosma* (SABIACEAE)							"		"
83	*Mimosa* (LEGUMINOSAE)				"			"		"
84	*Momipites coryloides* [JUGLANDACEAE]			"	"	"				
85	*Momipites microcoryphaceous* [JUGLANDACEAE]			"	"					
86	*Momipites triradiatus* [JUGLANDACEAE]			"	"					
87	*Myrica* (MYRICACEAE)							"		"
88	*Myrtaceidites* (MYRTACEAE)			"	"	"	"			
89	*Nyssa* (NYSSACEAE)						"	"		
90	*Passiflora* (PASSIFLORACEAE)							"		"

TABLE 2. Continued.

		Eoc	Eoc– Oli			Oli	Oli-Mio	Mio		Plio
Id	**TAXA**	**J-L**	**E**	**I**	**H**	**G**	**F**	**C**	**B**	**A**
91	*Pelliciera* (Pellicieraceae)						″	″	″	
92	*Platanus* (Platanaceae)			″	″			″		
93	*Platycarya* (Juglandacea)	″			″			″		
94	*Podocarpus* (Podocarpaceae)			″				″		″
95	*Polygala* (Polygalaceae)						″	″		
96	*Polygalacidites* (Polygalaceae)			″	″	″				
97	*Populus* (Salicaceae)							″		″
98	*Pterocarya* (Juglandaceae)			″	″		″			
99	*Quercus* (Fagaceae)			″	″	″	″	″		″
100	*Reevesia* (Sterculiaceae)			″	″					
101	*Rhamnus* (Rhamnaceae)				″			″		
102	*Rhizophora* (Rhizophoraceae)		″				″	″	″	″
103	*Salix* (Salicaceae)				″		″	″		
104	*Sapindaceidites* (Sapindaceae)			″	″					
105	*Serjania* (Sapindaceae)							″		″
106	*Smilax* (Smilacaceae)							″		″
107	*Struthanthus* (Loranthaceae)							″		″
108	*Thomsonipollis* [Ulmaceae]				″			″		
109	*Tilia* (Tiliaceae)						″	″		
110	*Tournefortia* (Boraginaceae)						″	″		″
111	*Ulmus* (Ulmaceae)			″	″	″	″	″		″

Chapter 3

Mesozoic Corals of Mexico

HARRY F. FILKORN

Department of Invertebrate Paleontology, Natural History Museum of Los Angeles County, 900 Exposition Boulevard, Los Angeles, California 90007, USA. hfilkorn@nhm.org

1. Abstract

The majority of the Mesozoic scleractinian corals reported from Mexico, 117 species or about 80 percent, are from Cretaceous strata. Comparatively little is known of the earlier Mesozoic corals. The Jurassic corals, a total of 17 species, have been described from five localities in Mexico, all in Upper Jurassic rocks. The 10 Triassic coral species described from Mexico are from the Upper Triassic Antimonio Formation of northwestern Sonora. The majority of the Jurassic and Triassic coral species reported from Mexico have been described previously from occurrences in Europe.

2. Introduction

The Mesozoic rocks of Mexico contain an important part of the fossil record of scleractinian corals. The vast majority of the Mesozoic coral species and occurrences known from Mexico have been reported from the Cretaceous (Filkorn, 2003). However, corals also have been described from the older Mesozoic periods, including some from the Jurassic and a diverse Triassic fauna.

F. Vega et al. (eds.), Studies on Mexican Paleontology, 47–59.

The objective of this report is to provide a current synopsis of the Mesozoic corals of Mexico. The occurrences and history of research of the Cretaceous corals of Mexico recently have been reviewed (Filkorn, 2003), so only supplementary Cretaceous coral data are provided herein. An earlier compilation of the published studies of Mesozoic corals (Álvarez, *et al.* 1994) was published more than ten years ago and it included only 16 articles pertaining to the corals from the entire Mesozoic of Mexico. However, no compilations of the coral species reported from the older Mesozoic periods in Mexico have ever been published. The compilation of coral species reported from the Mesozoic of Mexico presented in this report should help to advance further investigations of their taxonomy and evolution as well as their geographic and stratigraphic distributions.

The reported occurrences of corals from the Mesozoic of Mexico are reviewed in the following discussion. This discussion was organized in three parts, one part for each of the three periods, and arranged from youngest to oldest: Cretaceous, Jurassic, and Triassic. The published studies discussed in each part were reviewed in chronologic order, based on year of publication, and any coral species described or illustrated in that work were listed in Table 1. Therefore, the resultant list of studies and coral species in Table 1 are in a generalized stratigraphic order with those from the Cretaceous at the top of the table and those from the Triassic at the bottom. The majority of the studies that described Mexican Cretaceous corals and the coral species described in each of those works are listed in the table of an earlier report (Filkorn, 2003, Table 1). The table of studies and coral species presented herein contains the same kinds of information, in the same order, as presented in the earlier report: 1) author(s) of the study and year of publication; 2) name of the geologic formation and age of the occurrence; 3) the name of the Mexican state; and 4) a list of the coral species described in each study. In the studies that utilized newly collected material or presented new information on existing collections, the institution that served as the repository for the specimens was stated along with the discussion of that study.

The institutional abbreviations that were used in the following discussion are: IGM = Museo de Paleontología, Instituto de Geología, Universidad Nacional Autónoma de Mexico (UNAM), Mexico City. UNAM = Universidad Nacional Autónoma de Mexico.

3. Mesozoic Corals of Mexico: Recent Additions to the Cretaceous Faunas

Subsequent to a recent review of the Cretaceous corals of Mexico (Filkorn, 2003), there have been few additional published studies of these faunas. Late in 2003, a total of 16 coral species collected from the Maastrichtian upper member of the Cárdenas Formation, San Luis Potosí, were listed in a study of the depositional environment of the unit's coral - rudist associations by A. Schafhauser and colleagues (Schafhauser, *et al.*, 2003, p. 190), but none of the species was described or illustrated therein.

TABLE 1. List of the coral species described from the Mesozoic of Mexico, exclusive of the Cretaceous coral species described in studies prior to 2003, which are presented in an earlier report (see Filkorn, 2003). The author(s) of each study, the formation name and age of each occurrence, if known, the Mexican state, and the coral species described therein are listed in chronologic order based on year of publication. Refer to the text for discussion.

Period, Author	Formation and Age	State	Described coral species
Cretaceous			
Filkorn and Pantoja Alor (2004)	Mal Paso Formation, upper Albian	Guerrero	*Blastozopsammia guerreroterion* Filkorn and Pantoja Alor, 2004
Jurassic			
Wells (1946)	La Casita Formation, Upper Jurassic, Kimmeridgian	Durango	*Actinastrea kellumi* Wells, 1946
Alencáster de Cserna and Buitrón (1965)	Mapache Formation, Upper Jurassic, Oxfordian(?)	Puebla	*Isastrea* sp.
Reyeros de Castillo (1974)	Upper Jurassic	Chihuahua	*Actinastrea crassoramosa* (Michelin, 1843) *Actinastrea tenuisepta* (Koby, 1885) *Thamnasteria Mcoyi* (Milne Edwards and Haime, 1851) *Thamnasteria* sp. cf. *T. lamourouxii* Lesauvage, 1823 *Placophyllia chihuahuensis* Reyeros de Castillo, 1974 *Cladophyllia dichotoma* (Goldfuss, 1826) *Montastrea cyclops* (Felix, 1891)
Beauvais and Stump (1976)	Upper Jurassic, probably upper Oxfordian - lower Kimmeridgian	Sonora	*Allocoeniopsis bernensis* (Koby, 1885) *Lepidophylliopsis* cf. *L. longimana* (Quenstedt, 1882) *Macgeopsis sonorensis* Beauvais and Stump, 1976 *Stylina lobata* (Goldfuss, 1826) *Montlivaltia* sp. indet. *Brachyseris chellalensis* Beauvais, 1969 *Latiastraea variabilis* (Etallon, 1858)

TABLE 1 Continued.

Beauvais and Stump (1976)	Upper Jurassic, probably upper Oxfordian - lower Kimmeridgian	Sonora	*Allocoeniopsis bernensis* (Koby, 1885) *Lepidophylliopsis* cf. *L. longimana* (Quenstedt, 1882) *Macgeopsis sonorensis* Beauvais and Stump, 1976 *Stylina lobata* (Goldfuss, 1826) *Montlivaltia* sp. indet. *Brachyseris chellalensis* Beauvais, 1969 *Latiastraea variabilis* (Etallon, 1858)
Reyeros de Castillo (1978)	Zuloaga Formation, Upper Jurassic, upper Oxfordian	San Luis Potosí	*Stylosmilia michelini* Milne Edwards and Haime, 1848
Triassic			
Stanley, *et al.* (1994)	Antimonio Formation, Upper Triassic, Norian	Sonora	*Distichomeandra austriaca* (Frech, 1890) *Retiophyllia norica* (Frech, 1890) *Retiophyllia* sp. *Alpinophyllia flexuosa* Roniewicz, 1989 *Astraeomorpha sonorensis* Stanley, *et al.*, 1994 *Chondrocoenia waltheri* (Frech, 1890) *Pamiroseris rectilamellosa* (Winkler, 1861)
Stanley and González León (1998c)	Antimonio Formation, Upper Triassic, Norian – Rhaetian	Sonora	*Anthostylis acanthophora* (Frech, 1890) Anthostylis sp. *Meandrostylis antimoniensis* Stanley and González León, 1998c *Cuifastraea granulata* Melnikova, 1983 (= *Pamiroseris rectilamellosa* (Winkler, 1861) of Stanley, *et al.*, 1994)

A detailed taxonomic study of the coral species from the Cárdenas Formation is needed in order to help elucidate the paleobiogeographic distributions and Cretaceous - Tertiary extinctions of Maastrichtian corals within the Caribbean region.

A colonial coral from the upper Albian - lower Cenomanian upper member of the Mal Paso Formation in the State of Guerrero, southwestern Mexico, has recently been described as a new genus and species: *Blastozopsammia guerreroterion* Filkorn and Pantoja Alor, 2004. This coral is the earliest known colonial member of the Dendrophylliidae, a family with worldwide distribution in modern oceans. This new coral is most unusual because of the configuration of its dimorphic corallites, specifically the continuous axial corallite at the core of each branch and their evenly spaced, radially arrayed daughter corallites, which is a corallite arrangement similar in appearance to many modern species of the reef-building coral *Acropora*. *Blastozopsammia guerreroterion* is the first Cretaceous coral species to be described from southwestern Mexico in more than a century. The last person known to have described a species of Cretaceous coral from this region was M. Urquiza in 1883 (Filkorn, 2003). The type specimens of *Blastozopsammia guerreroterion* are in the collections of the Museo de Paleontología of the Instituto de Geología (IGM), Universidad Nacional Autónoma de Mexico (UNAM), Mexico City.

3.1. Jurassic Corals of Mexico

The Jurassic corals are perhaps the least studied and most poorly known group of Mesozoic invertebrates in Mexico. Although Jurassic corals have been mentioned in numerous geologic investigations, relatively few of these works have presented detailed taxonomic treatments of the coral species. Jurassic marine deposits were believed to exist in Mexico since the mid to late 1800s (for examples, see: Nyst and Galeotti, 1840; Nikitin, 1890; Felix, 1891; Heilprin, 1891; Aguilera, 1897; Felix, 1899), but the earliest mention of Mexican Jurassic corals located thus far in the literature did not occur until the early 1900s. In a guidebook for the Tenth International Geologic Congress, Carlos Burckhardt reported the first Jurassic corals from Mexico, specifically the occurrences of Upper Jurassic coral banks (below the Kimmeridgian, probably Sequanian, i.e., upper Oxfordian) at Picacho de la Abra, in the Sierra de Concepción del Oro (Burckhardt, 1906a, p. 2, 18), and in the Sierra de Santa Rosa, Sierra de Mazapil, Zacatecas (Burckhardt, 1906b, p. 5, 20, table 1). The occurrence of corals at the latter locality was derived from, and also discussed in, his detailed report on the Jurassic fauna of Mazapil (Burckhardt, 1906c, p. 161, 172, table). Astonishingly, subsequent to these initial reports, several decades pass before the first coral species was described from the Mexican Jurassic by John W. Wells in 1946 (see discussion below).

The Jurassic coral occurrences first reported by Burckhardt were later noted by José G. Aguilera (1907, p. 233) in his general summary of the geology of Mexico. A few years later, Burckhardt (1910, p. 310, and table; 1912, p. 203, 210, and table following p. 228) reported Upper Jurassic (Oxfordian) corals from Cerro del Volcán in the vicinity of San Pedro del Gallo, Durango, and correlated the coral-bearing stratum with the one at Mazapil, Zacatecas. Nearly a decade later, Wilhelm Freudenberg (1921, p. 79, 83) included a review of Burckhardt's studies of the Jurassic stratigraphy and paleontology of Mazapil and San Pedro del Gallo in his book, "Geologie von Mexiko". Subsequently, in an analysis of the stratigraphy and structure of northeastern Mexico, Emil Böse (1923, p. 196-198) also mentioned the Jurassic coral occurrences reported earlier by Burckhardt and in addition noted the same coral-bearing bed at a newly discovered locality in the valley of San José de Boquillas, Sierra de las Vigas, Coahuila and Nuevo León (Böse, 1923, p. 203, 320).

Later, Burckhardt, in his extensive "Etude synthétique sur le Mésozoïque mexicain", discussed the regional extent of the Jurassic coral bed and noted that it likely was lower Oxfordian (Burckhardt, 1930, for examples see p. 46, 51, 58, 61, 62, 65, 71, table 6, 86, table 8).

Late in the 1930's, investigation of the geology in the Melchor Ocampo area of northern Zacatecas - southern Coahuila by Ralph W. Imlay resulted in the discovery of poorly preserved corals in some beds of the Upper Jurassic (Oxfordian) Zuloaga Limestone in the Sierra Sombreretillo, northern Zacatecas. Imlay noted that this formation was the same as the one that Burckhardt (1930, p. 46, and his previous articles) had referred to in his earlier works in this region as the "calcaires à Nérinées", or the *Nerinea*-bearing limestone (Imlay, 1938, p. 1659). A few years later in the early 1940's, Charles L. Baker (1941, p. 2) noted the Upper Jurassic (Oxfordian) corals in the anticline of San José de Boquillas that he and Emil Böse had discovered decades earlier during reconnaissance trips between Monterrey and Saltillo (see above, Böse, 1923). Friedrich K. G. Müllerried (1942, p. 131) later noted the regional distribution of the Upper Jurassic coral-bearing rocks in northeastern Mexico in a report that summarized the Mesozoic geology of Mexico and northwestern Central America. Imlay (1943, p. 1482-1485), in a study of the Jurassic formations of the Gulf of Mexico region, subsequently discussed the stratigraphy, correlation, and depositional environment of the Zuloaga Limestone in northeastern Mexico and noted the same occurrences of corals that he had mentioned in his earlier report in 1938.

In 1946, John W. Wells described one Jurassic coral species from the Kimmeridgian strata of the La Casita Formation at a locality southwest of Las Cuevas Ranch, State of Durango: *Astrocoenia kellumi* Wells, 1946 (p. 2, pl. 1, fig. 1). At that time, Wells realized that this coral species "...seems to be the first scleractinian coral to be described from the Mexican Jurassic, ..." (Wells, 1946, p. 1). Based on the present literature review, his statement appears to be correct. The type specimens of A. *kellumi* reportedly are in the collections of the University of Michigan, Museum of Paleontology, Ann Arbor. This Jurassic species has since been reassigned to another genus: *Actinastrea kellumi* (Wells, 1946). Notably, six species of Cretaceous corals from various other horizons and localities in Mexico also were described by him in the same study.

In a field excursion guide for the Twentieth International Geological Congress held in Mexico in 1956, A. Calderón García (1956a, p. 29; 1956b, p. 81, fig. 18) reported the occurrence of Upper Jurassic (Oxfordian?) corals in the "Caliza con *Cidaris*" at an outcrop along the southwest side of the highway at a point 4.85 km northwest of Petlalcingo, southern Puebla. Heinrich K. Erben considered the "limestone with *Cidaris*" at the latter locality to be comparable and laterally equivalent to the Oxfordian Teposcolula Limestone (Erben, 1956, p. 57; 1957, p. 51 and pl. 2). The fossil corals from this horizon have never been studied, but one Jurassic species was described from this area several years later (see below). That same year, C. L. Rogers, with Mexican colleagues Z. de Cserna, E. Tavera, and S. Ulloa, reported coral-bearing horizons in the Zuloaga Limestone of the Concepción del Oro district, Zacatecas (Rogers, *et al.*, 1956, p. 14-15). They listed one coral species by name, *Thamnasteria imlayi* Wells, 1942, and assigned an upper Oxfordian age to the formation based on its stratigraphic position. The following year, this information was reiterated in a Spanish version of the article by the same authors (Rogers, *et al.*, 1957, p. 17-18). In an expanded study of the phosphate deposits in the same region of Mexico published a few years later, Rogers and his collaborators repeated much of the same information about the Zuloaga Limestone that was given in the earlier reports and identified two coral species from the formation, *T. imlayi* and *Stylina* sp. (Rogers, et al., 1961, p. 44-45). The occurrence of corals in the Zuloaga Limestone in the vicinity of Concepción del Oro, Zacatecas, that was reported by Rogers and colleagues (Rogers, *et al.*, 1957, p. 17-18) was later acknowledged and verified by E. Mapes Vázquez and co-workers in a study of the geology and mineral deposits of that area (Mapes Vázquez, *et al.*, 1964, p. 20-21). Farther to the south, subsequent investigations of the stratigraphy and paleontology of the area west and northwest of Petlalcingo, southern Puebla (Pérez Ibargüengoitia, *et al.*, 1965; Alencáster de Cserna and Buitrón, 1965, p. 7, 12), resulted in the discovery of corals in two Upper Jurassic formations: the Caliza Chimeco (Oxfordian), which was considered to be equivalent to the Caliza con *Cidaris* beds discussed earlier, and the Formación Mapache. One coral species was described, but not illustrated, in the latter report on the Upper Jurassic fauna from that area: *Isastrea* sp. from the Mapache Formation at Arroyo Agua Escondida (Alencáster de Cserna and Buitrón, 1965, p. 12). The specimen is in the collections of the Museo de Paleontología of the Instituto de Geología (IGM), Universidad Nacional Autónoma de Mexico (UNAM), Mexico City (Perrilliat, 1981). The occurrence of this coral was noted later in a study of the stratigraphy of the Tehuacán area of Puebla by J. Barceló Duarte (1978, p. 37, 38). There have been no additional studies of the Jurassic corals from this locality.

Several occurrences of corals were reported in investigations of the Mexican Jurassic during the 1970's and some of these studies provided descriptions and illustrations of the coral species. Charles L. Baker (1971, p. 46) noted the occurrence of corals and other fossils of Upper Jurassic (Portlandian, i.e., Tithonian) age in a limestone near Miquihuana, southwestern Tamaulipas. The corals from this locality have never been studied. A few years later, María M. Reyeros de Castillo (1974) described seven coral species from the Upper Jurassic of the Sierra de Águila, northern Chihuahua. Six of the seven coral species had been described previously from Europe and the remaining one was described as new: *Placophyllia chihuahuensis* Reyeros de Castillo, 1974. The coral species described by Reyeros de Castillo (1974) are listed in Table 1. The specimens are in the collections of the Museo de Paleontología of the Instituto de Geología (IGM), Universidad Nacional Autónoma de Mexico (UNAM), Mexico City (Perrilliat, 1981). Two years later, Louise Beauvais and Tom E. Stump (1976) reported on the coral species of a diverse Upper Jurassic invertebrate fauna from Cerro Pozo Serna, northwestern Sonora. They described seven coral species (see Table 1), most of which were known previously from the European Jurassic, a paleobiogeographic relationship that had been noted earlier by Beauvais (1974, p. 466). The seven coral taxa also included one newly described species, *Macgeopsis sonorensis* Beauvais and Stump, 1976, and the newly erected the genus *Lepidophylliopsis* Beauvais and Stump, 1976. The coral specimens that they described are deposited in the Stump Collection, University of California, Davis (Beauvais and Stump, 1976).

One species each of coral and chaetetid from the Upper Jurassic (upper Oxfordian) Zuloaga Limestone near Charcas, northwestern San Luis Potosí, were described by M. M. Reyeros de Castillo in 1978. The single scleractinian coral species, *Stylosmilia michelini* Milne Edwards and Haime, 1848, was known previously from the Upper Jurassic of Europe, but the sponge was described as new: *Ptychochaetetes potosiensis* Reyeros de Castillo, 1978. The specimens are in the collections of the Museo de Paleontología of the Instituto de Geología (IGM), Universidad Nacional Autónoma de Mexico (UNAM), Mexico City (Perrilliat, 1992). This was the first coral species to be described from the Zuloaga Limestone.

In the 1980's and later, there were relatively few studies of Mexican Jurassic corals. Geologic investigations by D. A. Barnes and J. J. Hickey in the vicinity of the Vizcaino Peninsula, northern Baja California Sur, noted corals in the carbonate fossil debris and clasts of the conglomerates of the Eugenia Formation (Barnes, 1984, p. 122; Hickey, 1984, p. 157). The overall age range of this formation was constrained to within the latest Jurassic - earliest Cretaceous based on microfossils (radiolaria and foraminifera), therefore the corals from the stratigraphically lower beds are pre-Cretaceous and most likely Upper Jurassic. No further studies of the corals from the Eugenia Formation have been made. To the east in the Saltillo - Monterrey area, A. Frame and W. C. Ward recognized corals in the section of the Upper Jurassic La Casita Formation exposed in San Juan Canyon, Nuevo León (Frame and Ward, 1987, p. 35). In this region, the coral-bearing beds of the La Casita Formation are conformably overlain by the coral-rich San Juan lentil (Berriasian) of the Lower Cretaceous Taraises Formation. The corals from these two formations have not been studied. Lastly, in a study of the Jurassic corals from the circum-Pacific region, L. Beauvais (1989) listed the corals that had been described previously from the Upper Jurassic of Sonora (see Beauvais and Stump, 1976) and additionally noted the occurrence of one other species, *Columnocoenia* sp.

3.2. Triassic Corals of Mexico

Triassic rocks of marine origin had been known to exist in western North America since the 1800's (see Smith, 1904, 1912, 1927), but comparative beds of this age in Mexico were not widely recognized by geologists until the 1900's (see Burckhardt and Scalia, 1905; Keller, 1928, 1973). Although Carlos Burckhardt and Salvador Scalia (1905, p. 35) noted the occurrence of poorly preserved corals in the

Upper Triassic strata of Zacatecas, the corals of these beds have not been studied. The Triassic corals of Mexico are thus far known only from investigations in the northwestern state of Sonora.

George D. Stanley, Jr. and colleagues (Stanley, *et al.*, 1994) presented a comprehensive study of the invertebrate fauna from the Upper Triassic (Norian) Antimonio Formation in the Sierra del Álamo, near El Antimonio, northwestern Sonora, which included descriptions of seven scleractinian species (see Table 1). Their coral specimens are in the Paleontology Museum of the University of Montana, Missoula. They noted that all seven of the coral species also had been identified in the Triassic of Nevada, but that most of them were different from the species known from the displaced terranes at other locations in the North American Cordillera. The following year, six of the same coral species were listed in an analysis and discussion of the paleobiogeographic and tectonic implications of the biota from the Antimonio Formation (Stanley and González León, 1995). Subsequent research on the geology of northwestern Sonora (Stanley and González León, 1998a, and articles therein) produced new reports of Triassic corals, notably in the Sierra Santa Teresa (Stewart, *et al.*, 1998, p. 133) and the Barra los Tanques (Goodwin and Stanley, 1998), and added three more species to the list of coral taxa known from that region (Stanley and González León, 1998b, 1998c). The specimens of the coral species described in the latter study (see Table 1) are deposited in the collections of the Instituto de Geología, UNAM, Hermosillo, Sonora. The taxonomy of the seven scleractinian coral species reported earlier also was updated: the specimens that were described as *Pamiroseris rectilamellosa* (Winkler, 1861) by Stanley, *et al.* (1994) were determined to actually belong to *Cuifastraea granulata* Melnikova, 1983.

4. Conclusions

To summarize, a total of 144 scleractinian coral species have been described from the Mesozoic rocks of Mexico, the majority of them (117, or about 80 percent) from the Cretaceous (see Filkorn, 2003). The older Mesozoic rocks of Mexico have yielded only 27 scleractinian coral species: 17 from the Upper Jurassic strata in Sonora, Chihuahua, Durango, San Luis Potosí, and Puebla, and 10 from the Upper Triassic Antimonio Formation of northwestern Sonora.

Acknowledgments

The author thanks Francisco J. Vega for the invitation to contribute this synopsis to the present volume on Mexican paleontology. Deb J. Lyons, Kent State University, helped the author to acquire copies or loans of much of the literature cited in this report and her efforts are appreciated.

References

Aguilera, J.G., 1897, Sinopsis de geología mexicana: *Bol. Inst. Geol. México*, 4-6, 187-250.

Aguilera, J.G., 1907, Aperçu sur la géologie du Mexique pour servir d'explication à la carte géologique de l'Amérique du Nord: Congrès Géologique International, Compte Rendu de la Dixième Session, Mexico, 1906, v. 1, p. 227-248.

Alencáster de Cserna, G., and B. E. Buitrón. 1965. Estratigrafía y paleontología del Jurásico Superior de la parte centromeridional del Estado de Puebla. Parte II. Fauna del Jurásico Superior de la región de Petlalcingo, Estado de Puebla. Paleont. Mex., 21, Parte II, 53 p., 14 pls., 1 fig.

Álvarez, G., F. Bosellini, H. Eliášová, G. Gill, B. Lathuilière, H. Löser, D. Marchal, D. Turnšek, J. Stolarski, and A. Zhuravlev. 1994. The Mesozoic corals. Bibliography 1758-1993. In H. Löser (ed.), Coral Res. Bull., 1:1-97.

Baker, C. L. 1941. Upper Jurassic deposits and structures of the Monterrey - Saltillo area, 4 p. In South Texas Geological Society, Thirteenth Annual Meeting, October 31st and November 1st and 2nd, 1941, Monterrey, Mexico. South Texas Geological Society, San Antonio.

Baker, C. L. 1971. Geologic Reconnaissance in the Eastern Cordillera of Mexico. *Geol. Soc. Amer., Spec. Pap.* 131, 85 p., 20 pls.

Barceló Duarte, J. 1978. Estratigrafía y Petrografía Detallada del Área de Tehuacán - San Juan Raya, Edo. de Puebla. Tesis Profesional, Universidad Nacional Autónoma de México, Facultad de Ingeniería, 143 p., 11 tables, 8 figs., 11 pls.

Barnes, D. A. 1984. Volcanic arc derived, Mesozoic sedimentary rocks, Vizcaino Peninsula, Baja California Sur, Mexico, p. 119-130. In V. A. Frizzell, Jr. (ed.), Geology of the Baja California Peninsula. *The Pacific Section of the Society of Economic Paleontologists and Mineralogists*, 39, Los Angeles, California.

Beauvais, L. 1974. Quelques exemples pris dans le Malm montrant le rôle des Madréporaires dans la théorie de la mobilité continentale. *Bull. Soc. Géol. France, Sér.* 7, 16(4):465-469.

Beauvais, L. 1989. Jurassic corals from the circum Pacific area, p. 291-302. In P. A. Jell and J. W. Pickett (eds.), Fossil Cnidaria 5. Proceedings of the Fifth International Symposium on Fossil Cnidaria including Archaeocyatha and Spongiomorphs held in Brisbane, Queensland, Australia, 25-29 July 1988.

Beauvais, L., and T. E. Stump. 1976. Corals, molluscs, and paleogeography of Late Jurassic strata of the Cerro Pozo Serna, Sonora, Mexico. *Palaeogeog. Palaeoclimatol. Palaeoecol.*, 19(4):275-301.

Böse, E. 1923. Vestiges of an ancient continent in northeast Mexico. *Amer. J. Sci., (Fifth Series),* 206:127-136, 196-214, 310-337.

Burckhardt, C., 1906a, Géologie de la Sierra de Concepción del Oro: Guide des Excursions du Dixième Congrès Géologique International, Mexico, XXIV (Excursions du Nord), 24 p., 1 map.

Burckhardt, C. 1906b. Géologie de la Sierra de Mazapil et Santa Rosa. Guide des Excursions du Dixième Congrès Géologique International, Mexico, XXVI (Excursion du Nord), 40 p., 1 table, 2 maps, 1 profiles sheet, 14 pls.

Burckhardt, C. 1906c. La Faune Jurassique de Mazapil avec un appendice sur les Fossiles du Crétacique Inférieur. *Bol. Inst. Geol. México,* 23, 216 p., 43 pls.

Burckhardt, C. 1910. Estudio geológico de la región de San Pedro del Gallo (Durango). *Parergones del Inst. Geol. México*, 3(6):305-357, 1 table, pls. 49-51.

Burckhardt, C. 1912. Faunes Jurassiques et Crétaciques de San Pedro del Gallo. *Bol. Inst. Geol. México,* 29, 264 p., and Atlas, 46 pls.

Burckhardt, C. 1930. Etude synthétique sur le Mésozoïque mexicain. *Abh. Schweiz. Palaeont. Ges., Mém. Soc. Paléont. Suisse,* 49(4):1-123, and 50(1):125-280.

Burckhardt, C., and S. Scalia. 1905. La faune marine du Trias Supérieur de Zacatecas. *Bol. Inst. Geol. México,* 21, 44 p., 8 pls.

Calderón García, A. 1956a. Estratigrafía de la Carretera Panamericana entre Petlalcingo, Pue. y Puebla, Pue., p. 29-33. In M. Maldonado Koerdell (ed.), Estratigrafía del Mesozoico y Tectónica del sur del Estado de Puebla; Presa de Valsequillo, Sifón de Huexotitlanapa y Problemas Hidrológicos de Puebla. Congreso Geológico Internacional, Vigésima Sesión, México, Excursión A–11.

Calderón García, A. 1956b. Guía de Campo, p. 35-91, figs. 2-20. In M. Maldonado Koerdell (ed.),

Estratigrafía del Mesozoico y Tectónica del sur del Estado de Puebla; Presa de Valsequillo, Sifón de Huexotitlanapa y Problemas Hidrológicos de Puebla. Congreso Geológico Internacional, Vigésima Sesión, México, Excursión A–11.

Caratini, C., and L. Beauvais. 1969. Les Polypiers du Kimmridgien inférieur de Chellala – Reibell (Département de Médea Algérie). Publications du Service Géologique de Algérie, 39:19-38.

Erben, H. K. 1956. El Jurásico Medio y el Calloviano de México. Congreso Geológico Interna-cional, Vigésima Sesión, Ciudad de México, 1956, Contribución al Congreso, del Instituto de Geología de la Universidad Nacional Autónoma de México, vii + 140 p., 19 pls., 5 maps.

Erben, H. K. 1957. New biostratigraphic correlations in the Jurassic of eastern and south-central Mexico. Congreso Geológico Internacional, Vigésima Sesión, Ciudad de México, 1956, Sección 2, El Mesozoico del Hemisferio Occidental y Sus Correlaciónes Mundiales, p. 43-52, 2 pls.

Étallon, A. 1859. Études paléontologiques sur le Haut-Jura. Rayonnés du Corallien. *Mém. Soc. Émul. Départ. Doubs, Série 3*, 3(for 1858):401-553.

Felix, J. 1891. Versteinerungen aus der mexicanischen Jura- und Kreide-Formation. *Palaeontographica*, 37(5-6):140-199, pls. 22-30.

Felix, J. 1899. Uebersicht über die Entwicklung der geologischen Formationen in Mexico, nebst einem Anhang über die Höhlenbildungen dieses Landes, p. 155-186. In J. Felix and H. Lenk, Beiträge zur Geologie und Paläontologie der Republik Mexico, Theil 2. Arthur Felix, Leipzig.

Filkorn, H. F. 2003. The Cretaceous corals of Mexico: occurrences and history of research. *Rev. Mex. Cien. Geol.*, 20(1):52-78.

Filkorn, H. F., and J. Pantoja Alor. 2004. A new Early Cretaceous coral (Anthozoa; Scleractinia; Dendrophylliina) and its evolutionary significance. *J. Paleont.*, 78(3):501-512.

Frame, A., and W. C. Ward. 1987. Lowermost Cretaceous coral-rich limestone in Nuevo Leon and Coahuila, Mexico, p. 33-39. In J. M. Barbarín Castillo, H.-J. Gursky, and P. Meiburg (eds.), El Cretácico de México y América Central, Resúmenes, Simposio Internacional, 24-26 de Noviembre de 1987, Linares, México. Actas de la Facultad de Ciencias de la Tierra de la Universidad Autónoma de Nuevo León, Linares, Volume 2.

Frech, F. 1890. Die Korallenfauna der Trias. *Palaeontographica*, 37(1):1-32, pls. 1-6, and 37(2-4):33-116, pls. 7-21.

Freudenberg, W. 1921. Geologie von Mexiko. Gebrüder Borntraeger, Berlin, 232 p., 1 pl., 29 figs.

Goldfuss, G. A. 1826-1833. Petrefacta Germaniae tam ea, Quae in Museo Universitatis Regiae Borussicae Fridericiae Wilhelmiae Rhenanae Servantur Quam Alia Quaecunque in Museis Hoeninghusiano Muensteriano Aliisque Extant, Iconibus et Descriptionibus Illustrata. Abbildungen und Beschreibungen der Petrefacten Deutschlands und der angränzenden Länder, unter Mitwirkung des Herrn Grafen Georg zu Münster. Erster Theil. Arnz & Comp., Düsseldorf. 254 p., 71 pl.

Goodwin, D. H., and G. D. Stanley, Jr. 1998. Norian sponge and coral biostromes in the Antimonio Formation, northwestern Sonora, Mexico. *Rev. Mex. Cienc. Geol.,* 14(2):160-166.

Heilprin, A. 1891. The geology and paleontology of the Cretaceous deposits of Mexico. *Proc. Acad. Nat. Sci. Philadelphia,* 1890:445-469, pls. 12-14.

Hickey, J. J. 1984. Stratigraphy and composition of a Jura–Cretaceous volcanic arc apron, Punta Eugenia, Baja California Sur, Mexico, p. 149-160. In V. A. Frizzell, Jr. (ed.), Geology of the Baja California Peninsula. *The Pacific Section of the Society of Economic Paleontologists and Mineralogists,* 39, Los Angeles, California.

Imlay, R. W. 1938. Studies of the Mexican Geosyncline. *Bull. Geol. Soc. Am.*, 49(11):1651-1694, 7 pls.

Imlay, R. W. 1943. Jurassic formations of Gulf Region. *Bull. Am. Assoc. Pet. Geol.*, 27(11):1407-1533.

Keller, W. T. 1928. Stratigraphische Beobachtungen in Sonora (Nordwest-Mexico). *Ecl. Geol. Helvetiae,* 21(2):327-335.

Keller, W. T. 1973. Observaciones estratigraficas en Sonora, Noroeste de México. *Bol. Asoc. Mex. Geól. Petr.,* 25(1-3):1-22.

Koby, F. 1885. Monographie des polypiers jurassiques de la Suisse, cinquième partie. *Abh. Schweiz. Paläontol. Gesell. (Mém. Soc. Paléont. Suisse),* 12:213-304, pls. 63-88.

Lesauvage, M. 1823. Mémoire sur un nouveau genre de Polypier fossile. *Mém. Soc. Hist. Natur. Paris,* 1(2):241-244, pl. 14.

Mapes Vázquez, E., S. Zamora Montero, and J. Gerónimo Godoy. 1964. Geología y yacimientos minerales del Distrito de Concepción del Oro y Avalos, Zacatecas. Consejo de Recursos Naturales No Renovables, Publicación 10 E, 133 p., 23 pls.

Melnikova, G. K. 1983. New Upper Triassic Scleractinia from the Pamir region. *Paleont. J.,* 17(1):41-49.

Michelin, H. 1847. Iconographie Zoophytologique, Description par Localités et Terrains des Polypiers Fossiles de France et Pays Environnants. P. Bertrand, Paris, xii + 348 p., 79 pls.

Milne Edwards, H., and J. Haime. 1848. Observations sur les Polypiers de la famille des Astréides. *Comp. Rend. Hebd. Séanc. Acad. Sci. Paris,* 27(19):465-470.

Milne Edwards, H., and J. Haime. 1851. Corals from the oolitic formations, p. 72-145, pls. 12-30. Second Part In A Monograph of the British Fossil Corals. Palaeontographical Society, London.

Müllerried, F. K. G. 1942. The Mesozoic of México and northwestern Central America, p. 125-147. In P. H. Oehser (ed.), Proceedings of the Eighth American Scientific Congress, Volume 4, Geological Sciences. Department of State, Washington, D. C.

Nikitin, S. 1890. Einiges über den Jura in Mexico und Centralasien. *Neu. Jahr. Miner., Geol. Paläont.,* 2:273-274.

Nyst, H., and H. Galeotti. 1840. Description de quelques fossiles du calcaire jurassique de Tehuacan, au Mexique. *Bull. Acad. Roy. Sci. Belles-Lettres de Bruxelles,* 7(2):212-221, 2 pls.

Pérez-Ibargüengoitia, J. M., A. Hokuto Castillo, and Z. de Cserna. 1965. Estratigrafía y paleontología del Jurásico Superior de la parte centromeridional del Estado de Puebla. Parte I. Reconocimiento geológico del área de Petlalcingo-Santa Cruz, Municipio de Acatlán, Estado de Puebla. *Paleont. Mex.,* 21, Parte I, 22 p., 1 pl., 2 figs.

Perrilliat, M. C. 1981. Catálogo de ejemplares tipo invertebrados fósiles en el Museo de Paleontología del Instituto de Geología, UNAM, México. *Paleont. Mex.,* 45, 183 p.

Perrilliat, M. C. 1992. Catálogo de ejemplares tipo invertebrados fósiles en la colección paleontológica del Instituto de Geología, UNAM, México, de los años 1977 a 1988. *Paleont. Mex.,* 58, 55 p.

Quenstedt, F. A. 1881 (1878-1880). Korallen (Röhren- und Sternkorallen). Petrefactenkunde Deutschlands, Abtheilung 1, Band 6. 1093 p., with: Atlas zu den Röhren- und Sternkorallen, pls. 143-184.

Reyeros de Castillo, M. M. 1974. Corales del Jurásico Superior de Chihuahua. *Paleont. Mex.,* 40, p. 1-21, pls. 1-9.

Reyeros de Castillo, M. M. 1978. Invertebrados (Porifera y Coelenterata) del Jurásico Tardio del Estado de San Luis Potosí. *Univ. Nal. Autón. México, Inst. Geol., Rev.,* 2(1):69-74.

Rogers, C. L., Z. de Cserna, E. Tavera, and S. Ulloa. 1956. General geology and phosphate deposits of Concepción del Oro District, Zacatecas, Mexico. *U. S. Geol. Surv., Bull.* 1037-A, 102 p., 2 pls.

Rogers, C. L., Z. de Cserna, E. Tavera Amezcua, and S. Ulloa. 1957. Geología general y depósitos de fosfatos del Distrito de Concepción del Oro, Estado de Zacatecas. *Inst. Nal. Invest. Rec. Miner., Bol.* 38, 129 p., 2 pls.

Rogers, C. L., Z. de Cserna, R. van Vloten, E. Tavera Amezcua, and J. Ojeda Rivera. 1961. Reconocimiento geológico y depósitos de fosfatos del norte de Zacatecas y areas adyacentes en Coahuila, Nuevo León y San Luis Potosí. *Cons. Rec. Nat. No Renov., Bol.* 56, 322 p., 3 pls.

Roniewicz, E. 1989. Triassic scleractinian corals of the Zlambach Beds, Northern Calcareous Alps, Austria. *Denk. Österr. Akad. Wiss., Math.-Naturwiss. Klasse, Wien,* 126, 152 p., 43 pls.

Schafhauser, A., S. Götz, R. Baron-Szabo, and W. Stinnesbeck. 2003. Depositional environment of coral-rudist associations in the Upper Cretaceous Cardenas Formation (central Mexico). *Geol. Croatica,* 56(2):187-198.

Smith, J. P. 1904. The comparative stratigraphy of the marine Trias of western America. *Proc. California Acad. Sci., Third Series, Geology,* 1(10):323-430, pls. 40-49.

Smith, J. P. 1912. The occurrence of coral reefs in the Triassic of North America. *Amer. J. Sci., Fourth Series,* 33(194):92-96.

Smith, J. P. 1927. Upper Triassic Marine Invertebrate Faunas of North America. *U. S. Geol. Surv., Prof. Pap.* 141, 262 p., 121 pls.

Stanley, G. D., Jr., C. González León, M. R. Sandy, B. Senowbari Daryan, P. Doyle, M. Tamura, and D. H. Erwin. 1994. Upper Triassic Invertebrates from the Antimonio Formation, Sonora, Mexico.

Paleont. Soc. Mem. 36 (*J. Paleont.*, 68(4), Suppl., part 2 of 3), 33 p.

Stanley, G. D., Jr., and C. M. González León. 1995. Paleogeographic and tectonic implications of Triassic fossils and strata from the Antimonio Formation, northwestern Sonora, p. 1-16. In C. Jacques Ayala, C. M. González León, and J. Roldán Quintana (Eds.), Studies on the Mesozoic of Sonora and Adjacent Areas. *Geol. Soc. Amer., Spec. Pap.* 301.

Stanley, G. D., Jr., and C. M. González León (eds). 1998a. Special issue dedicated to the International Workshop on the Geology of Northwestern Sonora, Mexico. *Rev. Mex. Cien. Geol.*, 14(2)(for 1997), iii-viii, 115-260.

Stanley, G. D., Jr., and C. M. González León. 1998b. Special issue dedicated to the International Workshop on the Geology of Northwestern Sonora. Foreword. *Rev. Mex. Cien. Geol.,* 14(2):iii-viii.

Stanley, G. D., Jr., and C. M. González León. 1998c. New Late Triassic scleractinian corals from the Antimonio Formation, northwestern Sonora, Mexico. *Rev. Mex. Cien. Geol.*, 14(2):202-207.

Stewart, J. H., R. Amaya-Martínez, R. G. Stamm, B. R. Wardlaw, G. D. Stanley, Jr., and C. H. Stevens. 1998. Stratigraphy and regional significance of Mississippian to Jurassic rocks in Sierra Santa Teresa, Sonora, Mexico. *Rev. Mex. Cien. Geol.*, 14(2):115-135.

Urquiza, M. 1883. Exploracion del Distrito de Coalcoman, Estado de Michoacán. Anales del Ministerio de Fomento, Colonizacion, Industria y Comercio, México, 84 p., 24 figs., 1 map.

Wells, J. W. 1942. Jurassic corals from the Smackover Limestone, Arkansas. *J. Paleont.,* 16(1):126-129, pl. 21.

Wells, J. W. 1946. Some Jurassic and Cretaceous corals from northern Mexico. *J. Paleont.*, 20(1):1-7, pls. 1-2.

Winkler, G. G. 1861. Der Oberkeuper, nach Studien in den bayrischen Alpen. Zeitschrift der Deutschen Geologischen Gesellschaft, 13:459-521, pls. 5-9.

Chapter 4

Rudists of Mexico: Historical Aspects and Selected Areas of Research

PEDRO GARCÍA-BARRERA

Museo de Paleontología, Departamento de Biología Evolutiva, Facultad de Ciencias, UNAM. Ciudad Universitaria 04510, Coyoacán, Mexico, D. F. pgb@hp.fciencias.unam.mx.

1. Abstract

Mexico, as a country, is remarkable for its abundant Cretaceous fossil invertebrate fauna. During the Cretaceous, what now is Mexico, was almost entirely covered by large shallow seas, which record a number of rudist deposits. The distribution of land masses, seas, climate and low ocean water depth were very suitable for rudist growth. As time passed, the geography changed and rudist abundance, distribution and faunal associations declined. This chapter reviews the close relationship between rudist assemblages and Mexican paleogeography. The data offered herein, concerning the number of genera and species of rudists, is based on what is recorded in the literature. This is not a systematic review because we never take into account synonyms or particular opinions about the validity of one or another genera or species of rudists. Finally we will examine some selected areas with interesting fossil rudist faunas.

F. Vega et al. (eds.), Studies on Mexican Paleontology, 61–78.

2. Introduction

Rudists are extinct bivalve mollusks, which possess completely different external morphology when compared with extant bivalves (Fig. 1). Rudists are large, thick walled and exclusively marine bivalves that had distinctly unequal shaped valves. Rudists were gregarious animals that lived fixed to the substrate and grew constructing patches, banks and true reefs.

Rudists were widespread during the Cretaceous Period in shallow tropical seas within the Tethys Domain, which covered southern Europe, northern Africa, the Middle East, southern United States, the Caribbean region and all of the Mexican territory. As a biological group, rudists span almost 100 million years, first appearing in Europe at the end of the Jurassic (160 my ago) and disappearing at the end of the Cretaceous (65 my ago).

Rudist species are important stratigraphic tools because of their recognizable morphological changes through time and their habitats. Rudists lived in association with calcareous algae, foraminifera, sponges, corals, bryozoa, gastropods, pelecypods, echinoids and other groups, in a similar way as corals do today. The rapid growth of these organisms, the superposition of many generations of their skeletal remains and the construction of large and thick shell walls, made possible the accumulation of limestone sediments which reached up to 3000 m in thickness (Guzmán, 1967).

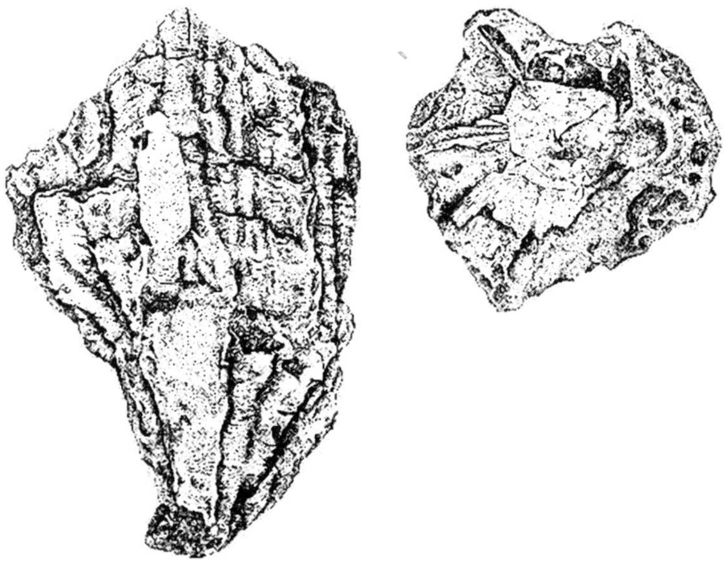

FIGURE 1. *Durania ojanchalensis,* Myers, 0.5 X, from Cárdenas Formation (Maastrichtian) North Central Mexico. Art by Edgar Gaytán.

Limestone, that often contain rudists remains, are exploited industrially as "marble" for making floors that decorate malls, airports, theaters, many public places and even in homes.

In some regions of Mexico, southern United States, Irak and Iran, rudist reefs often were surrounded by impermeable rocks constituting the porous rocks that trap and store gas and oil (Coogan, 1977; Scott *et al.*, 1993). For example, rich oil deposits found in limestone with rudists are reported at El Abra Formation along the Gulf of Mexico (Scott *et al.*, 1993).

3. Paleogeography and Mexican Rudists

During the Paleozoic Era there were restricted and apparently isolated seas in several different regions of the Mexican territory (López-Ramos, 1979). In the Mesozoic Era, during Triassic and most of the Jurassic, Mexico was an emerged territory with at least three narrow bays, one north of Sonora, a second one north of Veracruz and Puebla, and the third between Oaxaca and Guerrero in southern Mexico (Burckhardt, 1930; Erben, 1957; Imlay, 1980).

The Upper Jurassic is tectonically important because the opening of the Mexican Gulf began to occur at this time (Viniegra-Osorio, 1971, 1980). As a result of this phenomena, nearly 160 my ago during the Oxfordian, sediments started to accumulate within wide marine platforms of shallow seas (Fig. 2).

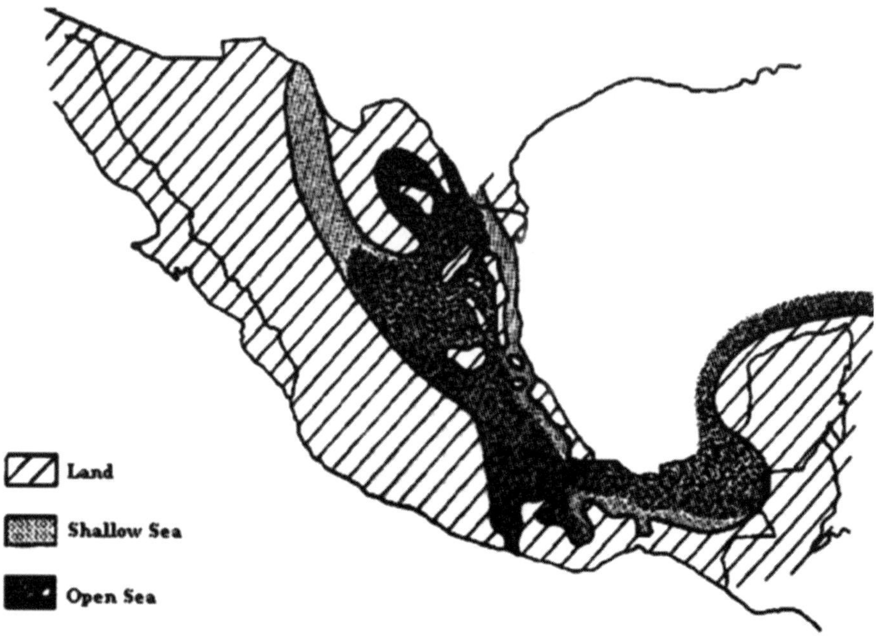

FIGURE 2. The Upper Jurassic transgression (160 my ago) covered almost fifty percent of the Mexican territory. After Alencáster, 1995.

An Upper Jurassic transgression advanced from the margin of the basin through the west, north and south and gradually covered large areas of the Mexican territory. Transgression was controlled by ancient topography whereas the high lands stood as peninsulas or as islands.

In Lower Cretaceous, during the Barremian and Aptian (120 my ago), there was a marine transgression, which covered all of the Mexican territory except in the west where a wide fringe extending from southern to northwestern Mexico existed (Cserna, 1976; Young, 1983). An ancient belt of high lands composed mainly of Paleozoic metamorphic and volcanic rocks along Guerrero, Oaxaca and Chiapas States also remained exposed (López-Ramos, 1981) (Fig. 3). In many areas, because of folding of the strata, shallow seas became much shallower. Emergence of very shallow seas, which deposited thick sequences of evaporites accumulated, which also allowed rudists to settle and form patches and reefs. Young (1983, 1984) believed that most of the evaporite deposits were destroyed by tectonic activity in the subduction zone along Mexico's southwestern coast, at the end of the Aptian. However in Jalisco, Colima, Michoacán, Guerrero and Oaxaca there are large fossiliferous deposits with rudists (Pantoja-Alor, 1993; García-Barrera, 1997).

During the Albian and Cenomanian (100 my ago) the sea reached its maximum extension covering almost completely the western land fringe and the high lands of southwestern Mexico. Some regions became wide platforms at this time.

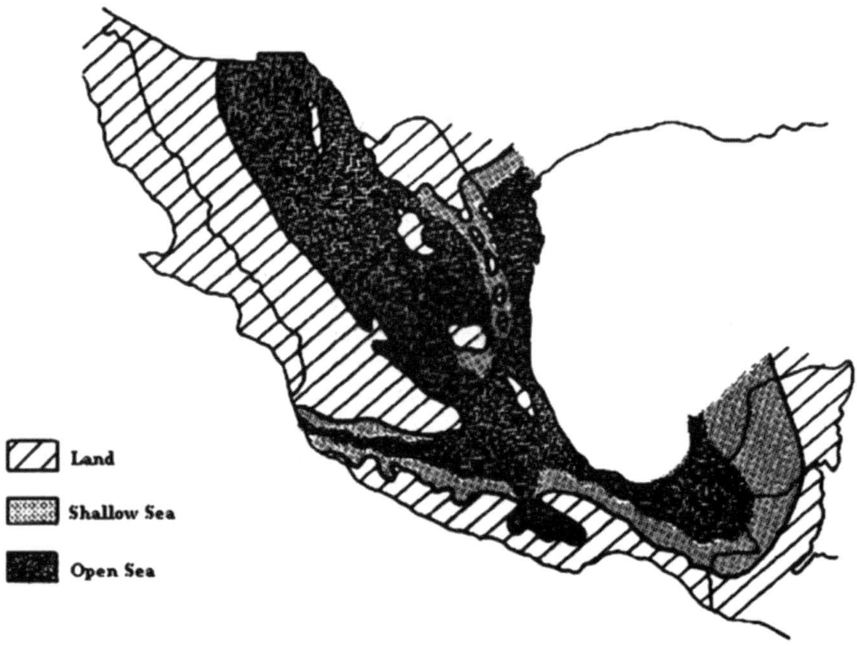

FIGURE 3. Mexico in the Lower Cretaceous, Barremian-Aptian, 120 my ago. After Alencáster, 1995.

Separated platforms eventually formed a mega-carbonated platform, which surrounded the Gulf of Mexico (Wilson, 1975; Michaud and Fourcade, 1989). The names of these platforms are: Bahamas, Florida, Texas, Coahuila, Valles-San Luis Potosí (Carrillo-Bravo, 1971) and Córdoba (González-Alvarado, 1976). Another large region with platform deposits and rudist bearing frameworks from the Albian-Cenomanian, was developed from Jalisco in central-west Mexico to southern Mexico in the region of Chiapas (Alencáster, 1987) (Fig. 4). During this time rudists reached their maximum diversity, abundance and geographic distribution in Mexico from Baja California to Yucatán (López-Ramos, 1973; Alencáster, 1978). Species of Caprinidae were the dominant taxa. A shell wall that has canals that run along the shell, called palial canals, which can be single or bifurcated once or twice and may become polygonal, characterizes the Caprinidae family.

The first endemic rudist genus in America, *Amphitriscoelus*, appeared either during early Aptian times (130 my ago) or in upper Barremian.

During Albian and Cenomanian there were other endemic genera such as *Coalcomana, Caprinuloidea, Jalpania, Mexicaprina, Texicaprina, Kimbleia* and others that are common in Texas, all Mexican territory and the Caribbean region. These endemic genera were distributed within a homogeneous faunal province which extended its boundaries from southern United States in the north to the Caribbean region reaching as far south as Venezuela. Due to the paleobiogeographic distribution, and according with Alencáster, 1987, this area should be called the Mexican Province.

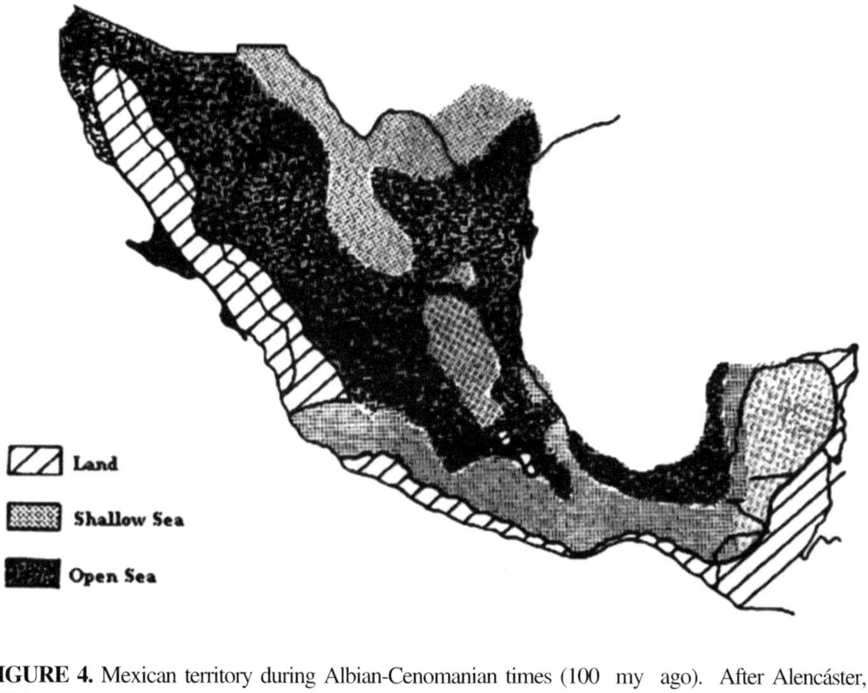

FIGURE 4. Mexican territory during Albian-Cenomanian times (100 my ago). After Alencáster, 1995.

During middle Cenomanian and because of intense volcanic activity both Albian and early Cenomanian rudists disappeared. Soon after, new species of rudists appeared. Some 90 my ago, during Turonian, the sea retreated from part of the Mexican territory leaving a wide fringe of land above the sea level to the west; this part of the territory was elevated by folding and large volumes of intrusive igneous rocks. The tectonic activity was the starting point of the elevation of the Sierra Madre Occidental, which by its posterior erosion produced great amounts of terrigenous deposits and along with volcanic rocks, displaced the sea progressively from west to east. Moreover, the continuous regression of the sea also was controlled by the cooling of the climate and the emergence of polar regions which captured water as permanent ice (Fig. 5). The regression of the sea was slow but continuous during upper Cretaceous up until the end of the Tertiary, when the entire continent was exposed as seen today.

When compared to fossil rudists in older deposits, Turonian rudists are scarce. The family Radiolitidae dominates and their genera are abundant. Also during the Turonian the last group of rudists, the Hippuritidae appears. Both groups had large shells and thick walls, vertical growth, large cylindrical attached valves and a reduced upper valve (opercular).

From Coniacian to Campanian (85-75 my ago) there are no longer large rudist banks or reefs; there are only isolated patches with sparse individuals. At the end of the Cretaceous (Campanian-Maastritchian, 75-65 my ago) the sea occupied a long fringe of the Mexican Gulf, which extended north to the Coahuila region, to the south up to northeast Guerrero and to the southeast to the Peninsula of Yucatán (Fig. 6).

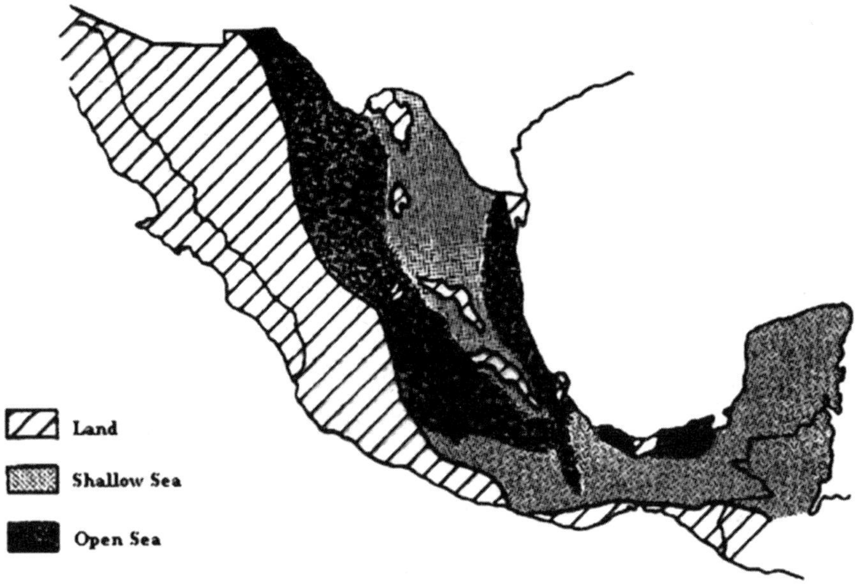

FIGURE 5. Mexican territory in the Turonian stage (90 my ago). After Alencáster, 1995.

The Cárdenas locality in San Luis Potosí is a septentrional area with an abundant and well preserved rudists fauna (Böse, 1906; Myers, 1968). The Cárdenas locality is the second most important Maastrichtian locality after the Central Chiapas region. In between the two areas, there are some fossil outcrops containing common species of rudists, which indicate the continuity of the platform from Cárdenas to Chiapas, passing through southern San Luis Potosí, Tamaulipas (Stephenson, 1922), southern Veracruz and Tabasco (Young, 1983) (Fig. 6).

4. Rudists from Southern Mexico

One of the most studied areas in southern Mexico is the Huetamo region in Michoacán and Guerrero. The area situated to the southwestern part of Mexico and to the south of the Mexican volcanic belt, is considered to be within the Guerrero Terrain (Campa and Coney, 1983). This tectonostratigraphic unit is characterized by magmatic, volcaniclastic, and sedimentary rocks typical of island arcs. The Guerrero Terrain was subdivided into the Huetamo, Ixtapa-Zihuatanejo and Teloloapan sub Terrains (Campa and Coney, 1983). Both, the eastern and western part of Huetamo region are in the Huetamo sub Terrain. The stratigraphic Mesozoic sequence and its geological units were studied for the first time by Pantoja-Alor in 1959, who divided the stratigraphic column as follows: at the base is the Angao Formation (Late Jurassic);

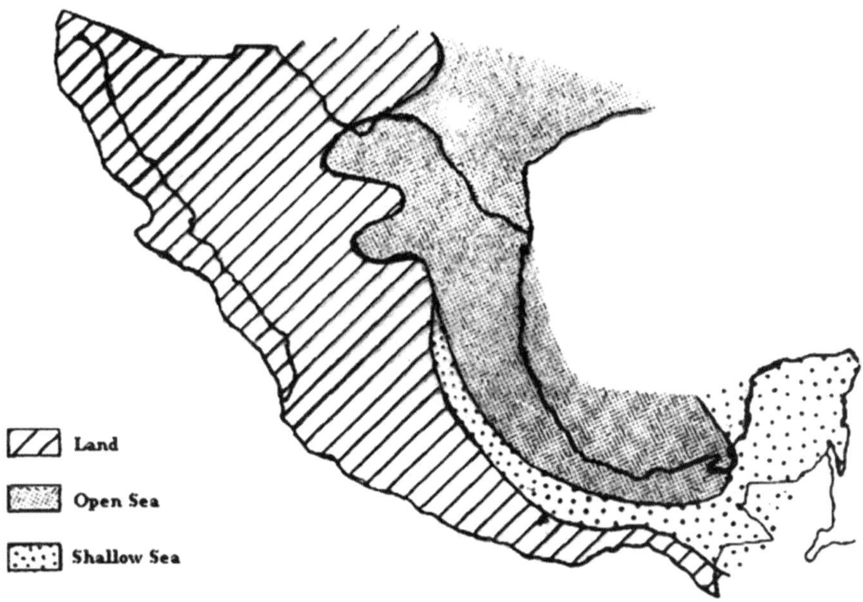

FIGURE 6. Mexico at the end of the Cretaceous, 70 my ago. After Alencáster, 1995.

San Lucas Formation, containing strata from Hauterivian to Aptian; Morelos Formation of Albian age and Mal Paso Formation from Cenomanian to Turonian. Nevertheless, more recent studies carried out by Pantoja-Alor (1990, 1992, 1993) offered new data about fossil assemblages, sub division of the extant geologic formations and description of new lithological units and formations. Paleontological studies focused on marine invertebrates, including rudists, were first done by Alencáster and Pantoja-Alor (1995, 1996, 1998) and by García-Barrera (1997). Recent contributions on paleontogical aspects of rudists can be found in Pantoja-Alor, 2000; Pantoja-Alor and Skelton, 2000; Filkorn, 2002).

At present, the sequence of Mesozoic rocks in Huetamo region, has been divided by Pantoja-Alor (1993) into two areas; east Huetamo and west Huetamo. Within the Eastern region, the stratigraphic sequence includes the Angao Formation, San Lucas Formation, El Cajón Formation, Mal Paso Formation and Cutzamala Formation.

At the base, the Angao Formation is considered a flysch turbiditic distal sequence. Conformably overlying the Jurassic sediments, are the strata of San Lucas anticline, which is a flysch sequence containing shale, sandstone and conglomerate (turbiditic) that corresponds to San Lucas Formation (Valanginian-Barremian-Aptian).

The San Lucas Formation was divided into two members: the lower, Terrero Prieto member and the upper Las Fraguas member. The Terrero Prieto member consists of shale, calcareous sandstone and conglomerates which contain rudists banks at the upper part of the sequence and Upper Valanginian and Barremian ammonoids (Gómez Luna *et al.*, 1993) and lower Barremian ammonoids (González-Arreola *et al.*, 1996). The Las Fraguas member is almost 400 meters thick, consisting of shale, feldespathic calcareous sandstone, mudstone and argillite. Dominating the sequence there are volcaniclastic sediments that contain abundant and diverse brachiopods, gastropods, echinoids and ammonoids of Aptian age (Sour-Tovar and García-Barrera, 1995).

The late Aptian El Cajón Formation conformably overlies the San Lucas Formation. The El Cajón Formation consists of a thick limestone strata (biogenic) interbedded with coquina composed of *Orbitolina* and argillaceous sediments. The exposed rocks represent lagoonal facies of a shallow platform, developed among rudists banks and ostreids in an intertidal zone.

Strata of the Mal Paso Formation overlay, with a slight angular unconformity, the strata of El Cajón Formation. The El Cajón Formation is a clastic marine sequence consisting of shale, sandstone, marl, limestone and interbedded conglomerate. The Mal Paso Formation is between 750-1000 meters thick. The Mal Paso Formation is considered mainly Albian, however the upper portion may have Cenomanian strata.

The Mal Paso Formation is divided into two members, the lower (clastic) member represents deltaic facies, whereas the upper member, which is mainly calcareous, represents lacunar environments. The upper member is more fossiliferous than the lower; with a great quantity and variety of rudist species (Table 1); the lower member contains invertebrate remains and rudist banks (García-Barrera, 1997).

Strata of the Mal Paso Formation overlay, with a slight angular unconformity, the Cutzamala Formation, proposed by Campa (1978). This unit is composed by argillite, mudstone, sandstone and abundant conglomerate. The Cutzamala Formation represents distal deltaic environments, fluvial continental and alluvial deposits. Polymictic and oligomictic conglomerates are composed mainly of limestone fragments from older Mesozoic formations.

At the western portion of the Huetamo region, the sequence, described by Pantoja-Alor, 1993 include the Zirándaro and Cumburindio formations.

The upper Jurassic-Lower Cretaceous strata in Huetamo, are included in the Zirándaro and Cumburindio Formations. The Zirándaro Formation was proposed by Pantoja-Alor (1993) for the lowest lithostratigraphic unit mapped in and around the town of Zirandaro, located 10 km south of the town of Turitzio. The Zirándaro Formation is composed of a thick succession of marine conglomeratic red beds, calcareous siltstone, sandstone and limestone intercalated by andesitic-dacitic lava flows, breccia and tuff. The succession is interpreted as marine deltaic clastic sediments with some flysch-like facies deposited in a small subsiding shelf affected by volcanic events. Eastwards this unit gradually passes into distal flysch, turbiditic

TABLE 1. Rudist species found at Huetamo region.

Formation	Species	Locality
Mal Paso	*Caprinuloidea perfecta* Palmer	El Encino
Upper Member	*Caprinuloidea gracilis* Palmer	
	Coalcomana cf ramosa (Boehm)	
	Kimbleia albrittoni (Perkins)	
Mal Paso	*Caprinuloidea* Palmer	Las Latas
Upper Member	*Coalcomana sp.*	
	Mexicaprina cornuta Coogan	
	Mexicaprina alata Filkorn	
	Tepeyacia corrugata Palmer	
Mal Paso	*Toucasia hancockensis* Whitney	La Piñuela
Upper Member	*Toucasia sp.*	
San Lucas	*Amphitriscoelus waringi* Harris & Hodson	El Cuachalalate
Lower Member		
	A. pluriloculata Alencáster & Pantoja	
	Caprina massei Alencáster	
Upper Member	*Amphitriscoelus waringi* Harris & Hodson	El Martillo
El Cajón	*Amphitriscoelus waringi* Harris & Hodson	Rancho El Paraís
	A. pluriloculata Alencáster & Pantoja	
	Coalcomana ramosa (Boehm)	
	Toucasia. Munier-Chalmas	
	Caprina d'Orbigny	
Cumburindio	*Amphitriscoelus waringi* Harris & Hodson	Los Hornos
	A. pluriloculata Alencáster & Pantoja	
	Caprina massei Alencáster	
	Huetamia buitronae Alencáster & Pantoja	
	Pantojaloria sphaerica Alencáster	
	Douvillelia skeltoni Alencáster & Pantoja	

and deep water basin deposits of the upper Jurassic Angao Formation (Pantoja-Alor, 1959, 1990).

The Cumburindio Formation was proposed also by Pantoja-Alor in 1993 for a Barremian?-Aptian succession more than 800 m thick composed of sandstone, argillaceous siltstone, and biostromic rudist limestone. The succession progressively changes to delta facies composed of red shale, siltstone, sandstone, conglomerate and volcanic breccia with interbedded lava flows and tuff of andesitic-dacitic composition. According to Pantoja-Alor (1993), the Cumburindio Formation can be subdivided into two members: a lower calcareous-sandy feldespathic member, which is very fossiliferous, having at least four rudist biostromes; and an upper deltaic-volcaniclastic and prograding member, made of red beds with claystone, siltstone, sandstone, conglomerate and intercalations of volcanic flows, breccia and tuff of andesitic to rhyolitic composition. The base of the lower calcareous sandy-feldespathic member of the Cumburindio Formation consists of 15 to 50 m of fine conglomerate, marl and limestone which unconformably overlay the volcaniclastic, lava flows, green tuff and reddish limestone of the Zirándaro Formation.

5. Rudists from Central Mexico

Within central Mexico the El Abra Formation of Querétaro State has an extraordinary fossiliferous locality called El Madroño, where excellent preservation of the invertebrate fauna exists including rudist species, especially caprinids (Kellum, 1930; Bonet, 1952, 1963) (Table 2). The outcrops of El Madroño, are located on both

TABLE 2. Rudist species found at El Madroño Region.

Formation	Species	Locality
El Abra	*Jalpania queretana* Alencáster & Aguilar-Pérez	El Madroño
	Texicaprina kugleri (Bouwman)	
	Texicaprina vivari (Palmer)	
	Mexicaprina quadrata Alencáster & Oviedo-García	
	Mexicaprina cornuta Coogan	
	Mexicaprina minuta Coogan	
	Kimbleia albrittoni (Perkins)	
	Guzzyella bisulcata Alencáster	
	Guzzyella acuminata Alencáster	
	Mullerriedia boesei Alencáster	
	Radiolites costata Scott	
	Eoradiolites davidsoni Hill	
	Caprinuloidea perfecta Palmer	
	Pacificaprina pyriformis Chartousse & Masse	
	Requienia Matheron.	
	Monopleura marcida White	
	Monopleura pinguiscula White	
	Monopleura ayalai Alencáster & Pichardo-Barrón	
	Monopleura bicornuta Alencáster & Pichardo-Barrón	

sides of federal road 120 between Querétaro city and Xilitla, San Luis Potosí (Guzzy, 1990). El Madroño is also known as Laguna Colorada (Red Lagoon in English).

Based upon the fossil assemblage of foraminifers and rudists the El Abra Formation is generally considered Middle Albian-Cenomanian (Enos, 1974). However portions of the El Abra Formation are considered Albian-Turonian (Aguayo, 1978), and even Barremian-Coniacian (Suter, 1987). Some authors include El Madroño region within a stratigraphic unit called El Doctor Formation (Wilson *et al.* 1955). The lithology and faunal content of El Abra and El Doctor formations are very similar and some researches are using both names based on geographic positioning of the formations. Among most geologists and paleontologists it is accepted that both names represent the same unit, El Doctor should be suppressed, as El Abra is the older name (Carrasco, 1970). El Abra Formation includes the Taninul facies and El Abra facies. The first corresponds to the reefal carbonated environment developed at the margin of the large Valles-San Luis Potosí Platform and the second to the lagoonal or back reef environment (Bonet, 1952; Aguayo, 1978).

The age of El Madroño deposits was considered late Albian-early Cenomanian (Alencáster and Aguilar-Pérez, 1995), but considering the species of rudists and foraminirers found therein, the age should be restricted to late Albian (Alencáster, 1998; Alencáster and Oviedo-García, 1998).

Recent paleontological publications (Aguilar-Pérez and Pichardo-Barrón, 2000; Pichardo-Barrón and Aguilar-Pérez, 2000; Alencáster and Pichardo-Barrón, 2003) have contributed to the rudists species known from the El Madroño deposits.

6. Rudists from North Central Mexico

To the north, Mexico also has good exposures of Upper Cretaceous rocks, containing a diverse and very well preserved invertebrate marine fauna, including rudists. As an example, we will take a look at the Cárdenas Formation and its rudist fauna (Alencáster *et al.* 1999; Schafhauser *et al.* 2003) (Table 3). In addition, other

TABLE 3. Rudist species found at Cárdenas Region.

Formation	Species	Locality
Cárdenas	*Coralliochama gboehmi* Böse *Hippurites muellerriedi* (Vermunt) *Hippurites perkinsi* Myers *Hippurites ceibarum* Muellerried *Biradiolites aguilerae* Böse *Biradiolites cardenasensis* Böse *Tampsia* sp. *Tampsia floriformis* Myers *Tampsia poculiformis* Myers *Durania ojanchalensis* Myers *Praebarrettia* sp. *Biradiolites rudissimus* Trechmann	Arroyo Atarjea

localities such as Charco Blanco, Guadalcazar, Llanos del Carmen and Santa Isabel will be mentioned (Sierra de la Ruda).

The Cárdenas Formation crops out in San Luis Potosí in north central Mexico, near the town of Cárdenas, some 130 km east of San Luis Potosí City. The town of Cárdenas is on the western limb of an asymmetric syncline of upper Cretaceous rocks, mainly of Maastrichtian age. East of the syncline is the eastern slope of the Sierra Madre Oriental, which consists of middle Cretaceous limestone, and the Tamasopo, San Felipe, and El Abra Formations. West of Cárdenas there are anticlinal hills of Middle Cretaceous limestone separated by Cenozoic alluvium (Schafhauser *et al.*, 2003).

According to Myers (1968), the Cárdenas Formation is a very fossiliferous unit more than 1000 meters thick, composed of sedimentary rocks that crop out in an asymmetrical syncline of the Sierra Madre Oriental. The Cárdenas Formation was originally divided into three members: the lower member consists of 180 meter thick sequence composed of alternating shale, sandstone and biosparite; the middle member consists of 445 m of shale and siltstone; the upper member consists of 430 m of thick siltstone, sandstone and biosparudite. The Cárdenas Formation is unconformably overlain by unfossiliferous siltstone, sandstone, and conglomerate of the Tabaco Formation. Laterally, the sedimentary rocks of the Cárdenas Formation pass into the marls of the Méndez Formation, which is equivalent, but deposited in a deeper water environment than the Cárdenas Formation. The Cárdenas Formation conformably overlies the uppermost limestone units of the Tamasopo Formation, (Cenomanian to Santonian age) which is part of the Valles-San Luis Potosí platform.

Some other important fossiliferous localities in San Luis Potosí is the mainly Albian Santa Isabel located at Sierra de la Ruda (north east), where there are *Coalcomana* sp., *Caprinuloidea* sp., *Kimbleia* sp. and other species. Turonian hippuritids can be found at Llanos del Cármen (northwest), where there are *Hippurites* sp., *Vaccinites* sp., and radiolitids like *Eoradiolites* sp. *Durania* sp.

Two excellent localities Charco Blanco and Guadalcazar are situated 80 km north of San Luis Potosí on both sides of the road to Matehuala, where there are abundant Upper Cretaceous rudists such as *Durania ojanchalensis, Biradiolites rudissimus, Vaccinites vermunti* and other species, including probable new genera.

7. Rudists from Chiapas, SE Mexico

Rudists from Albian and Cenomanian stages are found in the Sierra Madre Formation, which is exposed in almost a third of Chiapas. It extends from NE Tuxtla Gutiérrez city to Guatemala, reaching more than 3000 meters thickness. The Sierra Madre Formation consists of both inner and outer shelf deposits. The unit of limestone where rudists are found, was described for the first time by Sapper (1894). The same limestone unit was named by Böse (1906) as "Cretaceous limestone with Rudists". Very important localities containing rudists from Albian and Cenomanian stages are found at Cintalapa, Ocozocoautla, Berriozabal, El Coyol, Venustiano Carranza, La Concordia and Comitán (Alencáster, 1971; Steele, 1986; Waite, 1986) (Table 4).

Upper Cretaceous strata are exposed at a low topographic region called Central Depression in Chiapas, which extends throughout the State from NW-SE and also within the highlands of Chiapas (Müllerried, 1957). Most of the fossiliferous strata are assigned to the Angostura Formation (Sánchez-Montes de Oca, 1969). Rudists come from this unit, which is Upper Campanian-Maastrichtian age. The type locality is SE from Tuxtla Gutiérrez city, around Presa Angostura, where deposits reach 1330 meters consisting mainly of micrite, biomicrite and biospatite and containing large quantities of foraminifera, gastropods, echinoderms, corals, calcareous algae and rudists.

TABLE 4. Rudist species found in Chiapas, SE Mexico.

Formation	Species	Locality
Sierra Madre	*Radiolites* sp. *Sauvagesia* Choffat *Toucasia* Munier-Chalmas *Requienia* sp. *Coalcomana* Harris & Hodson *Caprinuloidea* Palmer *Kimbleia* Coogan *Texicaprina* Coogan	Ocozocoautla, Cintalapa
Angostura	*Plagioptychus jamaicensis* (Whitfield) *Plagioptychus muellerriedi* Alencáster *Plagioptychus agariciformis* Alencáster *Titanosarcolites giganteus* (Whitfield) *Titanosarcolites macgillavryi* Alencáster *Biradiolites cancellatus* (Whitfield) *Biradiolites rudissimus* Trechmann *Bournonia cardenasensis* (Böse) *Thirastylon chubby* Alencáster *Durania nicholasi* (Whitfield) *Durania ojanchalensis* Myers *Sauvagesia macroplicata* (Whitfield) *Tampsia floriformis* Myers *Barrettia monilifera* Woodward *Barrettia multilirata* Whitfield *Barrettia gigas* Chubb *Praebarrettia sparcilirata* (Whitfield) *Chiapasella radiolitiformis* (Trechmann) *Coralliochama sp.* *Mitrocaprina sp.* *Antillocaprina trilobata* García-Barrera *et al.* *Vaccinites inaequicostatus vermunti* MacGillavry *Parastroma sp.*	Tuxtla, Berriozabal, Ocozocoautla

At the end of Cretaceous, rudists were more diverse than in any other geologic age. During Campanian and Maastrichtian rudists reached their maximum number of genera and species, consisting mainly of caprinids, radiolitids and hippuritids. In addition, there are several genera whose complicated structure do not allow assignment to any recognized rudists families at this time.

Upper Cretaceous rudists from Chiapas belong to a unique fauna because most of the taxa studied are endemic, and also because their populations developed within a paleobiogeographic sub province which was designated the Antillean sub province by Kauffman (1973). This region comprises the northern part of South America, Central America, Cuba, Jamaica, Puerto Rico, all of the Caribbean islands, southern Mexico and north central Mexico (Cárdenas region) in which there are many common rudists species with Chiapas. Although there are some different species within the association as a whole, many consider it as the most septentrional part of the Antillean sub province (García-Barrera *et al.* 1998).

Provincialism also took place in other regions within the Tethys Realm where endemic faunas developed: such as western and central Europe, eastern Africa and the Middle East. As in Chiapas, the other tethys endemic rudist faunas were complex and reached large sizes. Among the largest organisms are *Titanosarcolites*, which reached two meters long; *Barrettia* and *Parastroma* could reach up to 70 cm in diameter and 50 cm high; and *Durania nicholasi* could reach 80 cm in diameter. Endemic rudists genera are represented by *Barrettia, Praebarrettia, Titanosarcolites, Chiapasella, Tampsia, Thyrastylon* and *Antillocaprina* (Alencáster, 1971).

Just as rudists reached their greatest diversity and development throughout the world they quickly declined and disappeared by the end of the Maastrichtian. Rudist extinction can be explained because they were part of a tropical fauna adapted to very special environmental conditions, therefore they could not evolve to the changes produced by tectonism, volcanism and oceanic regressions that took place at the end of the Maastrichtian and continued through early and middle Cenozoic times.

Another possible explanation of rudists extinction is that the group disappeared when a meteorite impacted the Earth 65 my ago (Steuber *et al.* 2002).

Finally, not only rudists disappeared, but many land and marine faunas and floras also went extinct at the end of the Maastrichtian. Because rudists do not have descendants, their way of life, physiology and tissue structure must be deduced from recent mollusks, particularly bivalves. Fortunately the Cretaceous fossil record the opportunity to study and interpret rudists as part of the living species that populated the world almost 80 million years ago.

8. Final Comments

Studies of Mexican rudists began in the nineteen century. One of the earliest formal publications concerning rudists was written by Felix in 1891, after that, there are many publications made by diverse authors. In Mexico only two researchers have dedicated efforts to the study of rudists: the late Dr. Federico K. Müllerried (1891-1952) who described many species of rudists from several regions of the country; and Dr. Gloria Alencáster, who is still working and publishing important papers on Mexican rudists fauna.

The two researchers mentioned above, have made extremely valuable contributions to Mexican Paleontology, however new discoveries of rudists emphasize the need to continue rudists studies. It is difficult to report how many rudists species are registered in Mexico; however taking into account what appears in the literature, there are around 57 genera and 250 species, but again, without considering synonyms, *affinis*, doubtful assignments, varieties and other problematic taxa these numbers are somewhat variable.

References

Aguayo, E., 1978, Sedimentary environments and diagenesis of a Cretaceous reef complex, eastern Mexico, *An. Inst. Cien. Mar Limnol. UNAM,* 5:83-140.

Aguilar-Pérez and Pichardo-Barrón, Y., 2000, Los Rudistas de la Formación Cupido en la zona del cerro de Labradores, Municipio de Galeana, Nuevo León, VII Congreso Nacional de Paleontología, Libro de resúmenes, *Facultad de Ciencias de la Tierra, U.A.N.L. and Soc. Mex. Paleont.* p. 4.

Alencáster, G., 1971, Rudistas del Cretácico Superior de Chiapas, UNAM, Instituto de Geología, *Paleontología Mexicana 34.*

Alencáster, G., 1978, Distribución de faunas marinas del sur de Mexico y del Norte de América Central durante el Cretácico, UNAM, Instituto de Geología, *in*: Ferrusquía, V. I,. (Ed.), Conexiones terrestres entre Norte y Sudamérica, UNAM, Instituto de Geología., 101:47-64.

Alencáster, G., 1987, Moluscos bentónicos de la mitad meridional de Mexico durante el Cretácico. *Rev. Soc. Mex. Paleontol.* 1:1-23.

Alencáster, G., 1995, Moluscos extintos (Rudistas) del Cretácico de Chiapas, *Rev. ICACH* 1 1:68-94.

Alencáster, G., 1998, New Caprinid rudist genera *Guzzyella* and *Muellerriedia* (Bivalvia-Hippuritacea) from the Albian (Cretaceous) of Central Mexico, *Rev. Mex. Cien. Geol.* 15:91-105.

Alencáster, G., and Aguilar-Pérez, J., 1995, *Jalpania queretana*, a new rudist (Bivalvia-Hippuritacea) from the Mid-Cretaceous of Central Mexico, *Rev. Mex. Cien. Geol.* 12:117-122.

Alencáster, G., and Oviedo-García, A., 1998, Re-examination of the genera *Texicaprina* Coogan, *Mexicaprina* Coogan and *Kimbleia* Coogan Caprinid Rudists from the Albian of Central Mexico, *Rev. Soc. Mex. Paleontol.* ,8:163-179.

Alencáster, G., and Pantoja-Alor, J., 1995, New Early Aptian rudists (Bivalvia-Hippuritacea) from the Huetamo area of southwestern Mexico, *Rev. Mex. Cien. Geol.* 22:123-134.

Alencáster, G., and Pantoja-Alor, J., 1996, The rudist *Amphitriscoelus* (Bivalvia-Hippuritacea) in the Lower Cretaceous of southwestern Mexico, *J. Paleontol.* 70:399-407.

Alencáster, G., and Pantoja-Alor, J. 1998, Two new Lower Cretaceous Rudists (Bivalvia-Hippuritacea) in the Huetamo region of southwestern Mexico, *GEOBIOS, Mem. Spec.*, 22:15-28.

Alencáster, G., Omaña, L., Pons, J. M., Torres, J. M., Tristan, M., and Vázquez, N., 1999, Upper Cretaceous rudists and larger foraminifera from the Cárdenas region, San Luis Potosí, Mexico, *Erl. Geol. Abh.,* 3:6-7.

Alencáster, G., and Pichardo-Barrón, Y., 2003, El género *Monopleura* (Bivalvia-Hippuritoidea) del Albiano de la Paltaforma Valles-San Luis Potosí, *in:Agustín Ayala-Castañares: universitario, impulsor de la investigación científica*, Soto, L. A., (Editor), *Inst. Cien. Mar Limnol.,* UNAM, Mexico, 167-176.

Bonet, F., 1952, La facies Urgoniana del Cretácico medio de la región de Tampico, *.Bol. Asoc. Mex. Geól. Petrol.* 4:153-262.

Bonet, F., 1963, Biostratigraphic notes on the Cretaceous of eastern Mexico, in: Boyd R. D., (Editor),

Geology of Peregrina Canyon and Sierra El Abra, Mexico, Corpus Christi, *Geol. Soc. Annual trip guide book,* 15: 23-48.

Böse, E., 1906, La fauna de moluscos del Senoniano de Cárdenas, San Luis Potosí, *Bol. Inst. Geol. Méx.* 24:1-92.

Burckhardt, C., 1930, Etude sur le Mésozoique Mexicain, *Mem. Soc. Paléontol. Suisse,* 49 and 50.

Campa, M. F., 1978, La evolución tectónica de Tierra Caliente, Guerrero, *Bol. Soc. Geol. Mex.* 39:52-64.

Campa, M. F., and Coney, J. P., 1983, Tectonostratigraphic Terrains and mineral resources distribution in Mexico, *Can. J. Earth Sci.* 26:1040-1050.

Carrasco, V. B., 1970, La Formación El Abra (Formación El Doctor) en la Plataforma Valles-San Luis Potosí, *Rev. Inst. Mex. Petról.* 2:97-100.

Carrillo-Bravo, J., 1971, La Plataforma Valles-San Luis Potosí, *Bol. Asoc. Mex. Geol. Petrol.* 23:1-102.

Coogan, A. H., 1977, Early and Middle Cretaceous Hippuritacea (Rudists) of the Gulf Coast, *Univ. Texas, Austin, Bur. Econ. Geol., Rept. Inv.*32-70.

Cserna, Z., 1976, Mexico Geotectonics and Mineral deposits. *New Mexico Geol. Soc, Spec. Pub.* 18-25.

Enos, P., 1974, Reefs, platforms and basins of Middle Cretaceous in northeast Mexico, *Am. Assoc. Petrol. Geol. B.* 58:800-809.

Erben, H. K., 1957, Paleogeogrphic reconstruction of the Lower and Middle Jurassic and for the Callovian of Mexico, in: El Mesozoico del hemisferio Occidental y sus correlaciones mundiales, XX Congr. Geol. Internal. Mexico D. F. Sección 2, 35-41.

Felix, J., 1891, Versteinerungen aus der mexicanischen Jura und Kreide Formation, *Palaeontographica,* 37:140-199.

Filkorn, F. H., 2002, A new species of *Mexicaprina* (Caprinidae, Coalcomaninae) and review of the age and paleogeography of the genus, *J. Paleontol.*, 76:672-691.

García-Barrera, P., 1997, Bioestratigrafía y biogeografía local de los rudistas cretácicos (Hippuritacea) de la región de Huetamo, Michoacán, Tesis doctoral, Facultad de Ciencias, UNAM, Mexico, unpublished.

García-Barrera, P., Avendaño-Gil, J., Omaña, L., and Alencáster, G., 1998, *Antillocaprina trilobata* nov. sp. and Upper Cretaceous associated fauna from Chiapas, southeast Mexico, *GEOBIOS, Mém. Spéc.* 22:125-135.

Gómez-Luna, M. E., Contreras y Montero B., Suastegui M., and Ramírez E. J., 1993, Amonitas del Valanginiano Superior-Barremiano de la Formación San Lucas en el área de Huetamo, Michoacán, *Rev. Soc. Mex. Paleontol.*, 6:57-65.

González-Alvarado, J., 1976, Resultados obtenidos de la exploración de la Plataforma Córdoba y principales campos productores, *Soc. Geol. Mex.* 37:53-59.

González-Arreola, C., Pantoja-Alor, J., Oloriz, F., Villaseñor, A. B., and García-Barrera, P., 1996, Lower Aptian Ammonitina, *Pseudohaploceras liptoviense* (Zeuschner) in Cumburindio Formation (southwestern Mexico), *GEOBIOS*, 29:35-43.

Guzmán, E. J., 1967, Reef type stratigraphic traps in Mexico, *in*: Origin of oil, Geology and Geophysics, The World Petroleum Congress 2:461-470.

Guzzy, A. G., 1990, Radiolitidos (Mollusca-Bivalvia Extinta) del Cretácico Medio de la región de Jalpan, Querétaro, Tesis de Licenciatura, Facultad de Ciencias, UNAM, Mexico, unpublished,

Imlay, R. W., 1980, Jurassic paleobiogeography of the conterminous United States in its continental setting, *U. S. Geol. Surv. Prof. Paper* 1062.

Kauffman, E. G., 1973, Cretaceous Bivalvia in: Hallam, A., (Editor) Atlas of Paleogeography, Elsevier Sci. Publ., Amsterdam, 353-383.

Kellum, L. B., 1930, Similarity of surface geology *in* front ranges of Sierra Madre Oriental to subsurface in Mexican south fields, *Am. Assoc. Petrol. Geol.* B. 16:73-91.

López-Ramos, E., 1973, Estudio geológico de la Península de Yucatán, Mexico, *Bol. Asoc. Mex. Geól. Petrol.* 35:24-76.

López-Ramos, E., 1979 Geología de Mexico, Tomos I, II, III, Author`s Edition, Mexico D.F.

López-Ramos, E., 1981, Paleogeografía y Tectónica del Mesozoico de Mexico, UNAM, *Rev. Inst. Geol.* 5:158-177.

Michaud, F., and Fourcade, F., 1989, Stratigraphie et paléogeógraphie du Jurassique et du Crétacé du Chiapas (Sud-Est du Méxique), *B. Soc. Géol. Fr.* 5:638-650.

Müllerried, F. K. G., 1957, La Geología de Chiapas, Gobierno Constitucional del Estado de Chiapas.

Myers, R. L., 1968, Biostratigraphy of the Cárdenas Formation (Upper Cretaceous), San Luis Potosí, Mexico. UNAM, Instituto de Geología, *Paleontología Mexicana 24:1-89.*

Pantoja-Alor, J., 1959, Estudio geológico de reconocimiento de la región de Huetamo, Estado de Michoacán, *Bol. Cons.Rec. Nat. no Renov.*, 50:1-54.

Pantoja-Alor, J., 1990, Redefinición de las unidades estratigráficas de la secuencia mesozoica de la región de Huetamo-Altamirano, estados de Michoacán y Guerrero, Libro de resúmenes, X Convención Geológica Nacional, *Soc. Geol. Mex.* p. 66.

Pantoja-Alor, J., 1992, La formación Mal Paso y su importancia en la estratigrafía del sur de Mexico, Memoria, XI Convención Geológica Nacional, *Soc. Geol. Mex.* p.121.

Pantoja-Alor, J., 1993, Description of the localities visited *in* the field trip B, in: Pantoja-Alor, J., Alencáster, G., and Gómez-Caballero (Editors), Geology and rudist communities of the Huetamo region, state of Michoacán, Mexico, UNAM, Instituto de Geología, 3rd International Conference on Rudists, Mexico, *Guide Book field Trip B,* 11-40.

Pantoja-Alor, J., 2000, Contributions to the biostratigraphy of the Mal Paso Formation, SW Mexico, VII Congreso Nacional de Paleontología, Libro de resúmenes, *Facultad de Ciencias de la Tierra, U.A.N.L. y Sociedad Mexicana de Paleontología,* 113-114.

Pantoja-Alor, J., and Skelton, P. W., 2000, Tepeyacia corrugata Palmer: Rudist of the Polyconitidae Family, VII Congreso Nacional de Paleontología, Libro de resúmenes, *Facultad de Ciencias de la Tierra, U.A.N.L. y Soc. Mex. Pal.*, p. 57.

Pichardo-Barrón, Y., and Aguilar-Pérez, J., 2000, Rudistas Caprinidos de la Formación el Abra en el Municipio de Landa de Matamoros, Querétaro, VII Congreso Nacional de Paleontología, Libro de resúmenes, *Facultad de Ciencias de la Tierra, U.A.N.L. y Soc. Mex. Pal.*, p.61.

Sánchez-Montes de Oca, R.,1969, Estratigrafía y Paleontología del Mesozoico de Chiapas, Seminario sobre exploración petrolera, *Rev. Inst. Mex. Petról.*, Mesa redonda 1-31.

Sapper, K., 1894, Informe sobre la geografía física y la geología de los estados de Chiapas y Tabasco, *Bol. Agr., Min. Ind.,* 3:187-211.

Schafhauser, A., Götz, S., Baron-Szabo, R., and Stinnesbeck, W., 2003, Depositional environment of Coral-Rudist associations in the Upper Cretaceous Cárdenas Formation (Central Mexico), *Geol. Croatica,* 56: 87-198.

Scott, R. W., Toni-Simo J. A., and Masee, J. P., 1993, Economic resources in Cretaceous carbonate platforms: an overview, *in*: Scott, R. W., Toni-Simo J. A., and Masse J. P., (Editors), *Cretaceous Carbonate Platforms*, AAPG Memoir 56:15-23.

Sour-Tovar, F., and García-Barrera, P., 1995, Braquiópodos fósiles de la Formación San Lucas (Neocomiano-Aptiano) en el estado de Michoacán, significado paleobiogeográfico, V Congreso Nacional de Paleontología, Memoria, *Soc. Mex. Pal.*, p. 40-41.

Steele, D. R., 1986, Physical stratigraphy and petrology of the Cretaceous Sierra Madre Limestone, west-central Chiapas, *in*: Cserna, Z., and F. Ortega (Editors), Contributions to the stratigraphy of the Sierra Madre Limestone (Cretaceous) of Chiapas, UNAM, *Bol. Inst. Geol. Mex.* 102:1-101.

Stephenson, L. W., 1922, Some Upper Cretaceous shells of the rudistid group from Tamaulipas, Mexico, *U. S. Nat. Mus. Proc.*, 61:1-28.

Steuber, T., Mitchell, S. F., Buhl, D., Gunter, G., and Kasper H. U., 2002, Catastrophic extinction of Caribbean rudist bivalves at the Cretaceous-Tertiary boundary, *Geology*, 30:999-1002.

Suter, M., 1987, Structural transverse across the Sierra Madre Oriental fold-thrust belt in east-central Mexico, *Geol. Soc. Am. Bull.* 98:249-264.

Viniegra-Osorio, F., 1971, Age and evolution of salt basins of southeastern Mexico, *Am. Assoc. Petrol. Geol. B.* 55:478-494.

Viniegra-Osorio, F., 1980, Great carbonate bank of Yucatán, southern Mexico, *J. Petrol. Geol.,* 3:247-278.

Waite, L. E., 1986, Biostratigraphy and paleoenvironmental analisys of the Sierra Madre Limestone (Cretaceous), Chiapas, *in*: Cserna, Z., and F. Ortega, (Editors), Contributions to the stratigraphy of the Sierra Madre Limestone (Cretaceous) of Chiapas, UNAM, *Bol. Inst. Geol. Mex.* 103:103-245.

Wilson, B. W., Hernández, J. P., and Meave, E., 1955, Un banco calizo del Cretácico en la parte oriental del estado de Querétaro, Mexico, *Bol. Soc. Geol. Mex.* 18:1-10.

Wilson, J. L., 1975, Carbonate facies in Geologic History, Springer-Verlag.

Young, K., 1983, Mexico, Chapter 3, *in*: Moullade, M., and A. E., Nairn (Editors), The phanerozoic geology of the world, II The Mesozoic, B:61-68, Elsevier Sci. Publishing Co. Amsterdam.

Young, K., 1984, Biogeography and stratigraphy of selected middle Cretaceous rudists in southwestern North America, *Memoria Tercer Congreso Latinoamericano de Paleontología, Mexico*, 341-359.

Chapter 5

Mesozoic and Tertiary Decapod Crustacea from Mexico

FRANCISCO J. VEGA[1], TORREY G. NYBORG[2] and MARÍA DEL CARMEN PERRILLIAT[1]

[1] · Instituto de Geología, Universidad Nacional Autónoma de México, Ciudad Universitaria, Delegación Coyoacán, 04510 – México, DF, vegver@servidor.unam.mx; mariacp@servidor.unam.mx.

[2] · Department of Natural Sciences, Loma Linda University, Loma Linda, California, 92350 USA, tnyborg06g@ns.llu.edu.

---

1. Abstract

In recent years, the number of described species of fossil crustaceans from Mexico has increased noticeably. Numerous new occurrences and new taxa were described from lithostratigraphic units that range in age from Early Cretaceous to Miocene. These reports included not only the systematic description of the specimens, but also information on paleoenvironment, paleoecology and evolution. The present work is a review of the fossil record of fossil crustaceans from Mexico, including comments on new localities and fauna that are under study. Based on the ongoing research in this field, it is possible that the number of fossil species of brachyurans will reach twice as much as the ones reported nowadays in a few years.

F. Vega et al. (eds.), Studies on Mexican Paleontology, 79–100.

2. Introduction

Fossil decapod crustaceans, when compared to other arthropod groups such as trilobites and insects, are rare in the fossil record. At least, in part, this rarity may actually reflect a lack of collecting and attention by the scientific community, especially in countries outside of Europe and the United States. This is exemplified in the fact that approximately 70% of the known fossil species of decapod crustaceans are reported from Europe and the United States. Therefore, systematic paleontology and paleobiogeographic patterns, based upon decapod crustaceans, maybe biased to those areas where most scientific work is focused.

Historically, Mexico has had very little systematic paleontology work concerning fossil decapod crustaceans. The only previous work was by Mary J. Rathbun of the Smithsonian (Rathbun, 1930). In the past twenty years, the fossil decapod crustaceans of Mexico have received renewed interest (Vega and Perrilliat, 1989; Vega and Feldmann, 1991; 1992; Vega *et al.* 1994; 1995; 1997; 1999; 2003; Feldmann *et al.* 1995b; 1996; 1998).

To date, only a small amount of potential species of fossil decapod crustaceans have been reported. Mexican fossil decapod crustaceans are significant because they represent an important crossroads for decapod crustacean originations and dispersals.

The objective of this contribution is to report the work that has been previously done and highlight the ongoing studies of fossil decapod crustaceans of Mexico, including a list of formally reported species, and preliminary information on those assemblages that are being studied, including locality information, age and paleoenvironment.

3. Occurrence, Curation and Identification of Fossil Decapod Crustaceans of Mexico

In Mexico fossil decapod crustaceans primarily occur within shallow marine sedimentary rocks, such as deltas and lagoons. Furthermore, the occurrence of fossil decapod crustaceans is selectively associated with storm deposits or episodic opportunistic population growth (Bishop, 1986). Subsequently one episodic deposit that contains fossil decapod crustaceans can be traced laterally along a sedimentary layer usually revealing additional specimens. Quality of fossil decapod crustacean preservation is dependant on various factors, such as the level of cuticle calcification, nature of the remains (corps or molts), sediment grain size, depositional environment, type of fossilization (phosphatic, calcitic, concretions), rate of burial, and other depositional factors. For example, within the families Callianassidae and Paguridae only the claw parts are usually fossilized because it is the only portion of the crustacean that is calcified. In addition, decapod crustaceans go through several molts within their life spans, each molt maybe preserved in the fossil record such that a decapod crustacean may have left several carapace molts within the fossil record.

Similar geologic tools used for the extraction of fossil crustaceans represent no less effort than the ones needed for any other fossil invertebrate group. Geologic hammers, chisels, pneumatic air tools, and air abrasives are used to reveal the morphology of the fossil decapod crustacean specimens. Fossil decapod Crustacea occur in all rock types including limestone, coarse sandstone, or fine mudstone. In addition, fossil decapod crustaceans are often times contained within concretions. Morphology differences amongst the many families of decapod crustaceans and morphological differences of the dorsal and ventral carapace surfaces make cleaning of specimens especially difficult. The final result is only a portion of morphology such as a dorsal carapace view or a ventral carapace view. Appendages, if preserved, are valuable clues to the systematic placement and habitat type of decapod crustaceans.

Identification of fossil decapod crustaceans is primarily based upon dorsal carapace morphology, which is the most readily fossilized portion of the crustacean and in the cases of molts is the only preserved portion. Many workers have stressed the fact that systematic identifications can be based solely upon the dorsal carapace morphology however mistakes arise when systematic placement is based solely upon dorsal carapace morphology. This is exemplified in the number of walking legs that, in part, determines the differences between the families Pinnotheridae and Hexapodidae. This is also true in decapod crustacean taxa that have very similar dorsal carapace morphologies throughout the genus such as in the genus *Cancer*. In these cases, further collecting of fossil decapod crustaceans are needed before systematic placement can be confidentially achieved.

The main problem however is the fact that neontologists use portions of the crustacean that are not usually preserved in the fossil record such as features of the ventral surface and mouth parts. Indeed the fossil process does not normally preserve all the morphology parts of the species however it is very important to stress that inclusion of all morphological features of fossil decapod crustaceans beyond the dorsal carapace is essential if neontologists are going to accept paleontological classification of fossil decapod crustaceans.

4. Systematics, Stratigraphy and Distribution

We offer a list of species formally reported from Mexico and fossil decapod crustaceans currently under study or in the process of being published. In addition, fossil occurrences of the orders Tanaidacea and Isopoda are included.

Systematic Paleontology

Order Tanaidacea Hansen, 1895

Ophthalmapseudes sp.

Figure 1.1

Locality/Formation: Chiapas, El Espinal Quarry. Sierra Madre Formation. Age: Aptian-Albian. Discussion: Hundreds of specimens in mass mortality horizons. Lagoon deposits. First record of fossil Tanaidacea in Mexico. Reference: Vega *et al.* 2003.

Order Isopoda Latreille, 1817

Family Sphaeromidae White, 1847

Sphaeroma burkartii Bárcena, 1875

Locality/Formation: Jalisco. Age: Tertiary. Discussion: Continental deposits. Reference: Bárcena, 1875; Hessler, 1969.

Family Archaeoniscidae Haack, 1918

Archaeoniscus aranguthyorum Feldmann, Vega, Applegate and Bishop, 1998

Figure 1.2

Locality/Formation: Puebla, Tepexi de Rodríguez. Tlayúa Formation. Age: Albian. Discussion: Shallow marine deposits. Reference: Feldmann *et al.* 1998.

Suborder Flabellifera Sars, 1882

Family Sphaeromatidae Milne Edwards, 1840

Subfamily Sphaeromatinae Milne Edwards, 1840

Figure 1.3

Locality/Formation: Chiapas, El Espinal Quarry. Sierra Madre Formation. Age: Aptian-Albian. Discussion: At least two species represented by only a few specimens. Brackishwater deposits. Reference: Vega *et al.* 2003.

Order Decapoda Latreille, 1802

Suborder Pleocyemata Burkenroad, 1963

Infraorder Astacidea Latreille, 1803

Family Erymidae Van Straelen, 1925

Enoploclytia tumimanus Rathbun, 1935

Locality/Formation: Coahuila, Escondido Formation. Age: Maastrichtian. Discussion: Marine platform. Reference: Rathbun, 1935.

Infraorder Palinura Latreille, 1803

Family Mecochiridae Van Straelen, 1925

Mecochirus sp.

Locality/Formation: Hidalgo, Zimapán. Trancas Formation. Age: Cenomanian. Discussion: Marine carbonate platform. Reference: Vega *et al.* 2003.

Meyeria mexicana Rathbun, 1935

Locality/Formation: Chihuahua. Cuchillo Parado Formation. Age: Aptian. Discussion: Marine platform. Reference: Rathbun, 1935.

Meyeria pueblaensis Feldmann, Vega, García-Barrera, Rico-Montiel and Martínez-López, 1995

Locality/Formation: Puebla, San Juan Raya. San Juan Raya Formation. Age: Aptian. Discussion: Shallow marine deposits. Reference: Feldmann *et al.* 1995b.

Family Palinuridae Latreille, 1802

Palinurus sp.

Figure 1.4

Locality/Formation: Chiapas, El Espinal Quarry. Sierra Madre Formation. Age: Aptian-Albian. Discussion: One specimen represented by an exuvia. Brackish water deposits. First record for the genus as a fossil in Mexico. Reference: Vega *et al.* 2003.

Astacodes sp.

Figure 1.5

Locality/Formation: Guerrero. San Lucas Formation. Age: Hauterivian-Aptian. Discussion: Marine calcareous platform. Reference: Alencáster, 1977.

Infraorder Anomura MacLeay, 1838

Superfamily Thalassinoidea Latreille, 1831

Family Callianassidae Dana, 1852

Callianassa tepetatensis Rathbun, 1930

Locality/Formation: Baja California Sur. Tepetate Formation. Age: Upper Eocene. Discussion: Marine platform. Reference: Rathbun, 1930.

Callianassa pustulata Withers, 1926

Locality/Formation: Veracruz. Age: Eocene. Discussion: Marine platform. Reference: Rathbun, 1930.

Callianassa vaughani Rathbun, 1935

Locality/Formation: Tamaulipas, San Fernando. Age: Oligocene. Discussion: Marine platform. Reference: Rathbun, 1935.

Callianassa pellucida Rathbun, 1930

Locality/Formation: Veracruz. Age: Miocene. Discussion: Marine platform. Reference: Rathbun, 1930.

Callianassa sp.

Locality/Formation: Chiapas, Veinte de Noviembre. El Bosque Formation. Age: Lower Eocene. Discussion: Shallow marine deposits. Reference: Vega *et al.* 2001b.

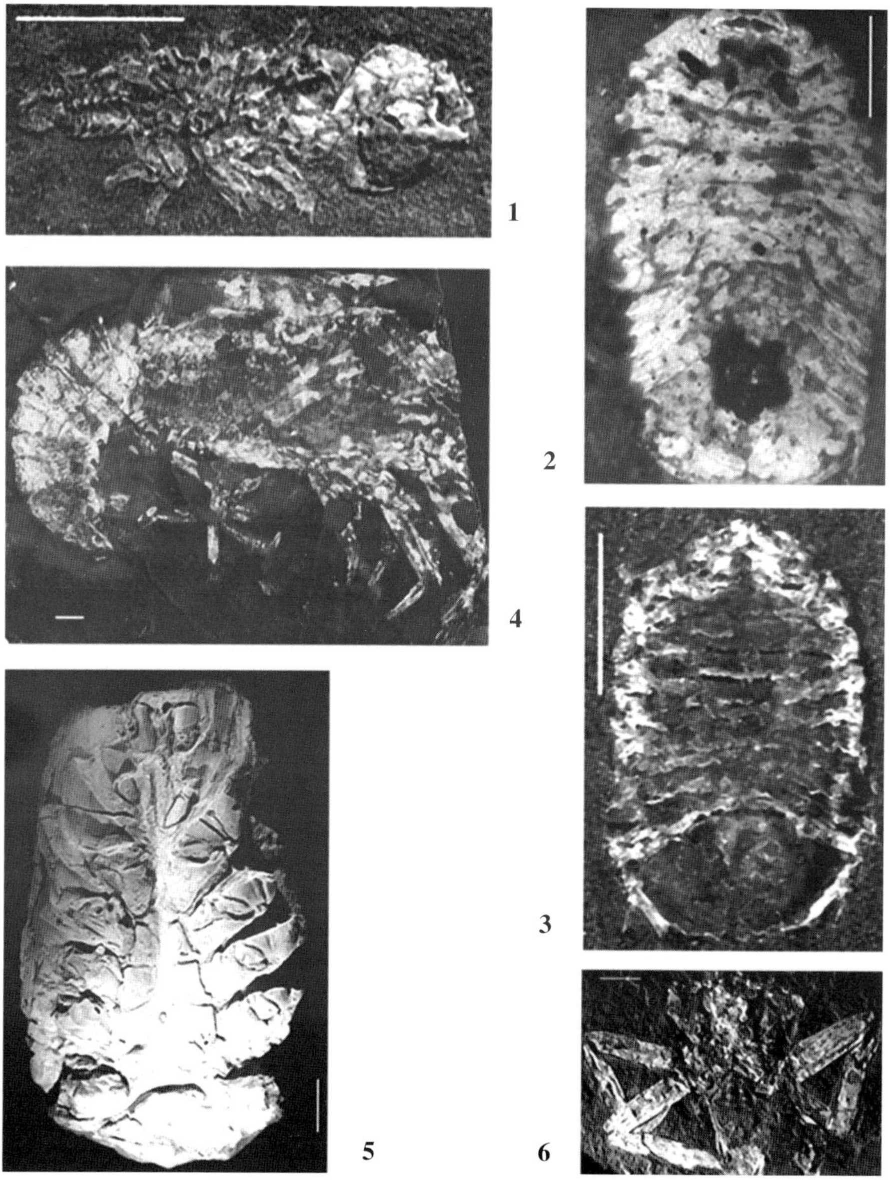

FIGURE 1. 1, *Ophthalmapseudes*? sp. El Espinal Quarry, Sierra Madre Formation, Chiapas, Aptian-Albian. 2, *Archaeoniscus aranguthyorum* Feldmann, Vega, Applegate and Bishop. Tepexi de Rodríguez, Tlayúa Formation, Puebla, Albian. 3, Flabellifera. El Espinal Quarry, Sierra Madre Formation, Chiapas, Aptian-Albian. 4, *Palinurus* sp. El Espinal Quarry, Sierra Madre Formation, Chiapas, Aptian-Albian. 5, Astacodes sp. San Lucas Formation, Guerrero, Hauterivian-Aptian. 6, *Tepexicarcinus tlayuaensis* Feldmann, Vega, Applegate and Bishop. Tepexi de Rodríguez, Tlayúa Formation, Puebla, Albian. Scale bars = 0.5 cm.

?*Callianassa* sp.
Locality/Formation: Nuevo León, El Antrisco. Potrerillos Formation. Age: Maastrichtian. Discussion: Deltaic front. Reference: Vega *et al.* 1995.

Protocallianassa sp.
Locality/Formation: San Luis Potosí, Cárdenas. Cárdenas Formation. Age: Maastrichtian. Discussion: Deltaic platform. Reference: Vega *et al.* 1995.

Cheramus sp.
Locality/Formation: San Luis Potosí, Cárdenas. Cárdenas Formation. Age: Maastrichtian. Discussion: Deltaic platform. Reference: Vega *et al.* 1995.

Infraorder Brachyura Latreille, 1802
Family Paguridae Latreille, 1802
Paleopagurus cf. *P. pilsbryi* Roberts, 1962
Locality/Formation: Nuevo León, El Antrisco. Potrerillos Formation. Age: Maastrichtian. Discussion: Deltaic front. Reference: Vega *et al.* 1995.

Family Diogenidae Ortmann, 1892
Dardanus mexicanus Vega, Cosma, Coutiño, Feldmann, Nyborg, Schweitzer and Waugh, 2001
Locality/Formation: Chiapas, Copoya. San Juan Formation. Age: Middle Eocene. Discussion: Shallow marine deposits. Reference: Vega *et al.* 2001b.

Dardanus cf. D. *mexicanus* Vega, Cosma, Coutiño, Feldmann, Nyborg, Schweitzer and Waugh, 2001
Locality/Formation: Baja California Sur, Arroyo Yesenia. Bateque Formation. Age: Eocene. Discussion: Marine platform. Reference: Schweitzer *et al.* 2002.

Family Aeglidae Dana, 1852
Protaegla minuscula Feldmann, Vega, Applegate and Bishop, 1998
Locality/Formation: Puebla, Tepexi de Rodríguez. Tlayúa Formation. Age: Albian. Discussion: Marine platform. Reference: Feldmann *et al.* 1998.

Family ?Homolidae White, 1847
Tepexicarcinus tlayuaensis Feldmann, Vega, Applegate and Bishop, 1998
Figure 1.6
Locality/Formation: Puebla, Tepexi de Rodríguez. Tlayúa Formation. Age: Albian. Discussion: Shallow marine deposits. Reference: Feldmann *et al.* 1998.

Family Dynomenidae Ortmann, 1892
Graptocarcinus muiri Stenzel, 1944
Locality/Formation: San Luis Potosí. Taninul Limestone. Age: Aptian. Discussion:

Marine platform. Reference: Stenzel, 1944.

Family Dakoticancridae Rathbun, 1917

Dakoticancer australis Rathbun, 1935

Figure 2.1

Locality/Formation: Nuevo León, El Antrisco. Potrerillos Formation; Coahuila, Saltillo, Cerro del Pueblo Formation. Age: Maastrichtian. Discussion: Deltaic platform. Reference: Vega and Feldmann, 1991. Locality/Formation: San Luis Potosí, Cárdenas. Cárdenas Formation. Age: Maastrichtian. Discussion: Deltaic platform. Reference: Vega *et al.* 1995.

Family Dorippidae MacLeay, 1838

Sodakus mexicanus Vega, Feldmann and Villalobos-Hiriart, 1995

Locality/Formation: Nuevo León, El Antrisco. Potrerillos Formation. Age: Maastrichtian. Discussion: Deltaic platform. Reference: Vega *et al.* 1995.

Family Calappidae de Haan, 1833

Calappilia hondoensis Rathbun, 1930

Locality/Formation: Baja California Sur. Tepetate Formation. Age: Upper Eocene. Discussion: Marine platform. Reference: Rathbun, 1930.

Calappilia cf. *C. hondoensis* Rathbun, 1930

Locality/Formation: Chiapas, El Jobo. San Juan Formation. Age: Middle Eocene. Discussion: Shallow marine deposits. Reference: Vega *et al.* 2001b.

Calappa zurcheri Bouvier, 1899

Locality/Formation: Veracruz. Age: Oligocene. Discussion: Marine platform. Reference: Rathbun, 1930.

Calappa flammea (Herbst, 1794)

Locality/Formation: Baja California Sur, Tepetate. Age: Oligocene. Discussion: Marine platform. Reference: Rathbun, 1930.

Prehepatus harrisi Bishop, 1985

Locality/Formation: Nuevo León, El Antrisco. Potrerillos Formation. Age: Maastrichtian. Discussion: Deltaic front. Reference: Vega *et al.* 1995.

Family Hepatidae Stimpson, 1871

Eriosachila bajensis Schweitzer, Feldmann, González-Barba and Vega, 2002

Locality/Formation: Baja California Sur, Cañada El Mangle. Age: Eocene. Discussion: Marine platform. Reference: Schweitzer *et al.* 2002.

Eriosachila sp.
Locality/Formation: Chiapas, El Jobo. San Juan Formation. Age: Middle Eocene. Discussion: Shallow marine deposits. Reference: Vega *et al.* 2001b.

Family Raninidae de Haan, 1839
Ranina berglundi Squires and Demetrion, 1992
Locality/Formation: Baja California Sur, Mesa La Salina. Bateque Formation. Age: Eocene. Discussion: Marine platform. Reference: Squires and Demetrion, 1992.

Lophoranina precocious Feldmann, Vega, García-Barrera, Tucker and Avendaño, 1996
Figure 2.2
Locality/Formation: Chiapas, Tuxtla Gutiérrez, Ocozocoautla. Ocozocoautla Formation. Age: Maastrichtian. Discussion: Shallow carbonate platform. Reference: Feldmann *et al.* 1996.

Lophoranina cristaspina Vega, Cosma, Coutiño, Feldmann, Nyborg, Schweitzer and Waugh, 2001
Figure 2.3
Locality/Formation: Chiapas, El Jobo. San Juan Formation. Age: Middle Eocene. Discussion: Shallow marine deposits. Reference: Vega *et al.* 2001b.

Lophoranina bishopi Squires and Demetrion, 1992
Locality/Formation: Baja California Sur, Mesa La Salina. Bateque Formation. Age: Eocene. Discussion: Marine platform. Reference: Squires and Demetrion, 1992.

Laeviranina sp.
Locality/Formation: Chiapas, Veinte de Noviembre. El Bosque Formation. Age: Lower Eocene. Discussion: Shallow marine deposits. Reference: Vega *et al.* 2001b.

Notopus minutus Vega, Cosma, Coutiño, Feldmann, Nyborg, Schweitzer and Waugh, 2001
Locality/Formation: Chiapas, Veinte de Noviembre. El Bosque Formation. Age: Lower Eocene. Discussion: Shallow marine deposits. Reference: Vega *et al.* 2001b.

Raninoides mexicanus Rathbun, 1930
Locality/Formation: Veracruz. Age: Miocene. Discussion: Marine platform. Reference: Rathbun, 1930.

Family ? Majidae Samouelle, 1819
Uncertain species
Locality/Formation: San Luis Potosí, Cárdenas. Cárdenas Formation. Age:

Maastrichtian. Discussion: Marine platform. Reference: Vega *et al.* 1995.

Family Inachidae MacLeay, 1838

Oregonia spinifera Schweitzer, Feldmann, González-Barba and Vega, 2002

Locality/Formation: Baja California Sur, San Juan de la Costa. El Cien Formation. Age: Oligocene. Discussion: Marine platform. Reference: Schweitzer *et al.* 2002.

Family Atelecyclidae Ortmann, 1893

Levicyclus tepetate Schweitzer, Feldmann, González-Barba and Vega, 2002

Locality/Formation: Baja California Sur, Cañada Palo Alto. Tepetate Formation. Age: Eocene. Discussion: Marine platform. Reference: Schweitzer *et al.* 2002.

Montezumella tubulata Rathbun, 1930

Locality/Formation: Baja California Sur, Arroyo Colorado. Tepetate Formation. Age: Upper Eocene. Discussion: Marine platform. Reference: Rathbun, 1930.

Family Carcineretidae Withers, 1922

Carcineretes planetarius Vega, Feldmann, Ocampo and Pope, 1997

Figures 2.4

Locality/Formation: Chiapas, Ocozocoautla. Ocozocoautla Formation and Belize, Albion Island, Barton Creek Formation. Age: Maastrichtian. Discussion: Lagoon deposits. Reference: Vega *et al.* 1997; Vega *et al.* 2001a.

Ophthalmoplax difuntaensis (Vega and Feldmann, 1991)

Figure 2.5

Locality/Formation: Nuevo León, El Antrisco. Potrerillos Formation; Coahuila, Paredón. Las Encinas Formation. Age: Maastrichtian. Discussion: Deltaic platform. Reference: Vega and Feldmann, 1991. Originally described as *Mascaranada difuntaensis.* More complete specimens allow its generic placement as Ophthalmoplax. Under study.

Branchiocarcinus cornatus Vega, Feldmann and Sour-Tovar, 1995

Locality/Formation: San Luis Potosí, Cárdenas. Cárdenas Formation. Age: Maastrichtian. Discussion: Deltaic platform. Reference: Vega *et al.* 1995.

Icriocarcinus xestos Bishop, 1988

Figure 2.6

Locality/Formation: Baja California. Rosario Formation. Age: Maastrichtian. Discussion: Marine platform. Reference: Bishop, 1988; Schweitzer *et al.* 2002.

Family Portunidae Rafinesque, 1815

Portunus atecuicitlis Vega, Feldmann, Villalobos-Hiriart and Gío-Argáez,1999

Locality/Formation: Veracruz, San Pablo. Tuxpan Formation. Age: Miocene. Discussion: Marine platform. Reference: Vega *et al.* 1999.

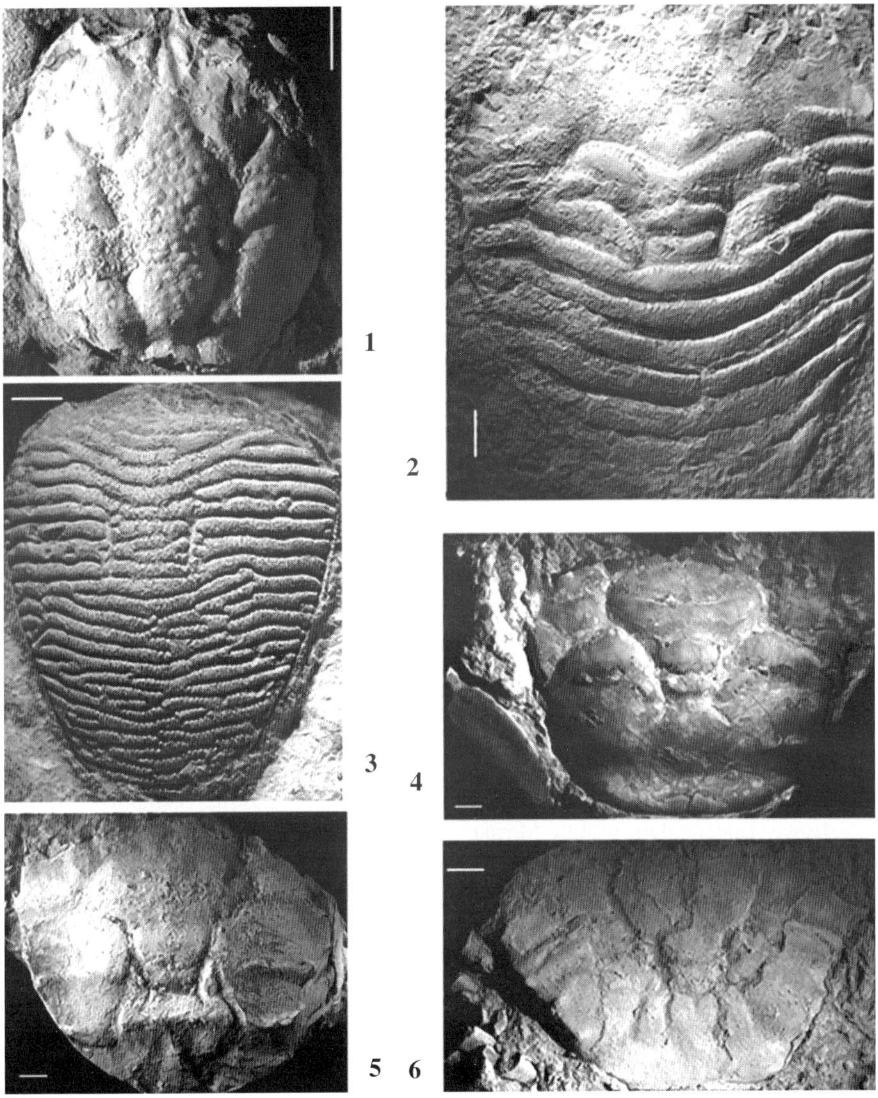

FIGURE 2. 1, *Dakoticancer australis* Rathbun (juvenile specimen). Sierra El Antrisco, Nuevo León, Potrerillos Formation, Maastrichtian. 2, *Lophoranina precocious* Feldmann, Vega, García-Barrera, Tucker, and Avendaño. Tuxtla Gutiérrez, Chiapas, Ocozocoautla Formation, Maastrichtian. 3, *L. cristaspina* Vega, Cosma, Coutiño, Feldmann, Schweitzer, Nyborg and Waugh. El Jobo, Chiapas, San Juan Formation, Middle Eocene. 4, *Carcineretes planetarius* Vega, Feldmann, Ocampo and Pope. Ocozocoautla, Chiapas, Ocozocoautla Formation, Maastrichtian. 5, *Ophthalmoplax difuntaensis* (Vega and Feldmann). Sierra El Antrisco, Nuevo León, Potrerillos Formation, Maastrichtian. 6, *Icriocarcinus xestos* Bishop. Baja California, Rosario Formation, Maastrichtian. Scale bars = 0.5 cm.

Necronectes nodosa Schweitzer, Feldmann, González-Barba and Vega, 2002
Locality/Formation: Baja California Sur, San Juan de la Costa. El Cien Formation. Age: Oligocene. Discussion: Marine platform. Reference: Schweitzer *et al.* 2002.

Necronectes tajinensis Vega, Feldmann, Villalobos-Hiriart and Gío-Argáez, 1999
Locality/Formation: Veracruz, San Pablo. Tuxpan Formation. Age: Miocene. Discussion: Shallow marine deposits. Reference: Vega *et al.* 1999.

Scylla costata Rathbun, 1930
Locality/Formation: Unknown locality. Age: Oligocene. Reference: Rathbun, 1930.

Podophthalmus(?) sp.
Locality/Formation: Tamaulipas, Villagrán. Age: Upper Cretaceous. Reference: Rathbun, 1930.

Family Xanthidae MacLeay, 1838
Amydrocarcinus dantei Schweitzer, Feldmann, González-Barba and Vega, 2002
Locality/Formation: Baja California Sur. Bateque Formation. Age: Eocene. Discussion: Marine platform. Reference: Schweitzer *et al.* 2002.

Verrucoides stenohedra Vega, Cosma, Coutiño, Feldmann, Nyborg, Schweitzer and Waugh, 2001
Figure 3.1
Locality/Formation: Chiapas, Veinte de Noviembre. El Bosque Formation. Age: Lower Eocene. Discussion: Marine platform. Reference: Vega *et al.* 2001b.

Xanthosia americana Rathbun, 1930
Locality/Formation: Baja California, Tepetate. Age: Oligocene. Discussion: Marine platform. Reference: Rathbun, 1930.

Megaxantho zoque Vega, Feldmann, García-Barrera, Filkorn, Pimentel and Avendaño, 2001
Figure 3.2
Locality/Formation: Chiapas, Ocozocoautla. Ocozocoautla Formation. Age: Maastrichtian. Discussion: Lagoon deposits. Reference: Vega *et al.* 2001a.

Parazanthopsis meyapaquensis Vega, Feldmann, García-Barrera, Filkorn, Pimentel and Avendaño, 2001
Figure 3.3
Locality/Formation: Chiapas, Ocozocoautla. Ocozocoautla Formation. Age: Maastrichtian. Lagoon deposits. Reference: Vega *et al.* 2001a.

Lobonotus mexicanus Rathbun, 1930
Locality/Formation: Baja California Sur, Arroyo Colorado. Tepetate Formation. Age: Upper Eocene. Discussion: Marine platform. Reference: Rathbun, 1930; Schweitzer *et al.* 2002.

Harpactocarcinus americanus Rathbun, 1930
Locality/Formation: Tamaulipas. Age: Eocene. Discussion: Marine platform. Reference: Rathbun, 1930.

Xandaros sternbergi (Rathbun, 1926)
Locality/Formation: Baja California. Rosario Formation. Age: Maastrichtian. Discussion: Marine platform. Reference: Rathbun, 1926; Bishop, 1988; Schweitzer *et al.* 2002.

Family Goneplacidae MacLeay, 1838
Stoaplax nandachare Vega, Cosma, Coutiño, Feldmann, Nyborg, Schweitzer and Waugh, 2001
Locality/Formation: Chiapas, Veinte de Noviembre. El Bosque Formation. Age: Lower Eocene. Discussion: Shallow marine deposits. Reference: Vega *et al.* 2001b.

Family Pinnotheridae de Haan, 1833
Viapinnixa alvarezi Vega, Cosma, Coutiño, Feldmann, Nyborg, Schweitzer and Waugh, 2001
Figure 3.4
Locality/Formation: Chiapas, Veinte de Noviembre. El Bosque Formation. Age: Lower Eocene. Discussion: Shallow marine deposits. Reference: Vega *et al.* 2001b.

Family Retroplumidae Gill, 1894
Archaeopus mexicanus Schweitzer, Feldmann, González-Barba and Vega, 2002
Locality/Formation: Baja California, Sal Sipuedes. Rosario Formation. Age: Maastrichtian. Discussion: Marine platform. Reference: Schweitzer *et al.* 2002.

Costacopluma mexicana Vega and Perrilliat, 1989
Figure 3.5
Locality/Formation: Nuevo León, El Antrisco; Coahuila, General Cepeda. Age: Maastrichtian. Discussion: Hundreds of specimens preserved in different styles. Deltaic platform. Reference: Vega and Perrilliat, 1989.

Costacopluma bishopi Vega and Feldmann, 1992
Figure 3.6
Locality/Formation: Guerrero, Temalac. Mexcala Formation. Age: Turonian. Discussion: Small specimens, width of carapace = less than 4 mm. Deltaic platform.

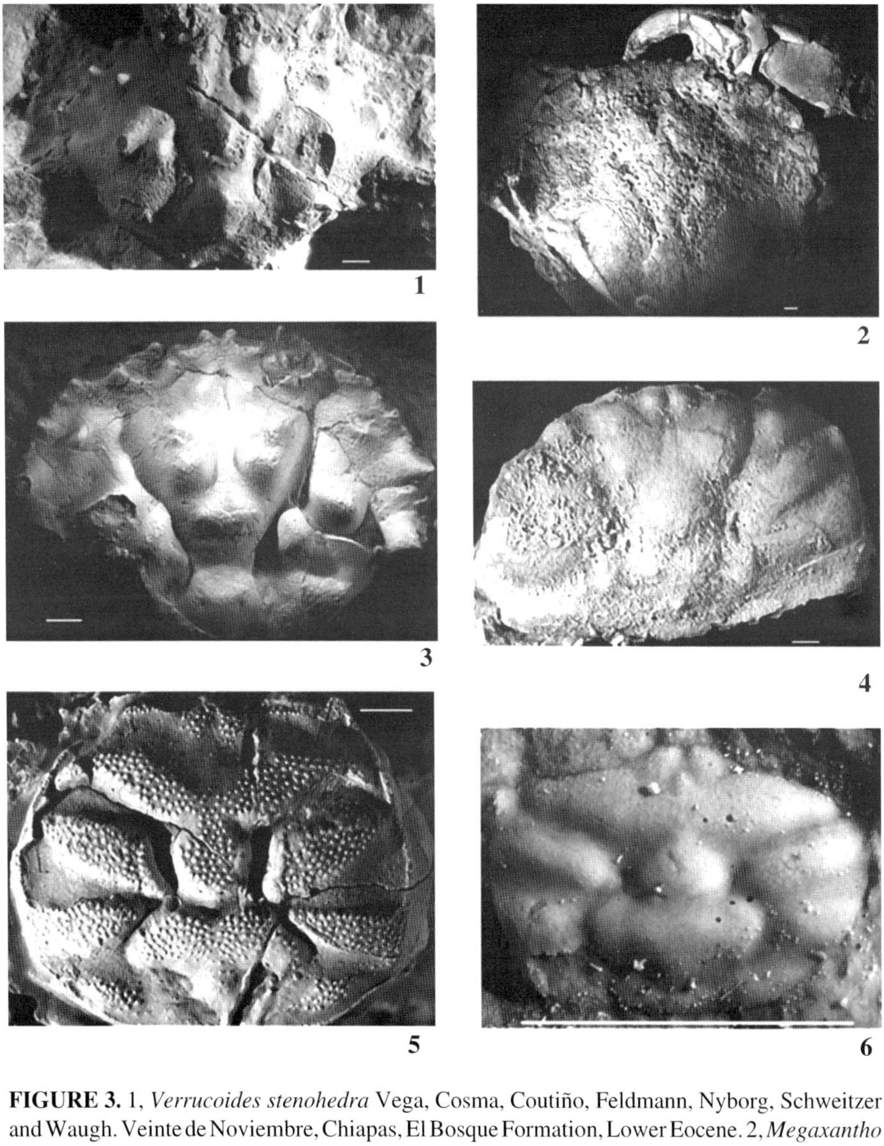

FIGURE 3. 1, *Verrucoides stenohedra* Vega, Cosma, Coutiño, Feldmann, Nyborg, Schweitzer and Waugh. Veinte de Noviembre, Chiapas, El Bosque Formation, Lower Eocene. 2, *Megaxantho zoque* Vega, Feldmann, García-Barrera, Filkorn, Pimentel and Avendaño. Ocozocoautla, Chiapas, Ocozocoautla Formation, Maastrichtian. 3, *Parazanthopsis meyapaquensis* Vega, Feldmann, García-Barrera, Filkorn, Pimentel and Avendaño. Ocozocoautla, Chiapas, Ocozocozutla Formation, Maastrichtian. 4, *Viapinnixa alvarezi* Vega, Cosma, Coutiño, Feldmann, Nyborg, Schweitzer and Waugh. Veinte de Noviembre, Chiapas, El Bosque Formation, Lower Eocene. 5, *Costacopluma mexicana* Vega and Perrilliat. Sierra El Antrisco, Nuevo León, Potrerillos Formation, Maastrichtian. 6, *C. bishopi* Vega and Feldmann. Temalac, Guerrero, Mexcala Formation, Coniacian? Scale bars = 0.5 cm.

Reference: Vega and Feldmann, 1992. C. *bishopi* from the Cárdenas Formation, is *C. mexicana.* Reference: Vega *et al.* 1995.

5. Origin, Evolution and Extinction

Fossil decapod crustaceans first appear in the fossil record at the end of the Jurassic Period. Since the Jurassic, decapod crustaceans have had at least two episodes of radiation. The first occurred during the Late Cretaceous, with many species associated with deltaic environments. The second occurred during the Eocene, where diverse assemblages occurred in reef environments (Glaessner, 1969).

Notably the oldest records of origination for certain decapod crustacean genera occurred in Mexico with subsequent worldwide distribution thereafter. Some examples are given below.

Panulirus.- Recently exuvia of a species of the genus *Panulirus* from the El Espinal Quarry, Chiapas, within Aptian-Albian rocks of the Sierra Madre Formation was discovered (Vega *et al*., 2003). This specimen marks the oldest known occurrence of *Panulirus* in America. Previously the oldest record for this genus was from the Lower Cretaceous of Europe (Glaessner, 1969). Interestingly the carapace morphology between the Mexican and European genus has changed little, therefore this may be considered as a pancronic genus.

Lophoranina.- This raninid genus had a wide geographic distribution during early Tertiary times. The fossil record of *Lophoranina* extends from the lower Eocene to the Oligocene of southeastern United States, Baja California Sur, Chiapas, Spain, France, Hungary, Italy and the Indopacific region (Tucker, 1995). *Lophoranina* possesses a series of dorsal transverse terraces with spines and frontal spines that are directed forward. It has been suggested that the spines of the transverse terraces serve to anchor *Lophoranina* to algal mats for mimetism, or as a burial anchor (Savazzi, 1981). The oldest known species is from Maastrichtian rocks of Chiapas. Terrace pattern of the Chiapas *Lophoranina* is simplistic when compared to Tertiary species (Feldmann *et al.* 1996; Vega *et al.* 2001). It is also important to note that the mean size of *Lophoranina praecocious* from the Maastrichtian of Chiapas is slightly larger than that of the species that followed. Anterior surface of the *Lophoranina praecocious* carapace lacks any terraces, and the posterior two thirds are ornamented only with a few parallel terraces. In addition, anterolateral spines tend to become more branched and complex in Tertiary specimens. Via (1969) suggested the possibility of sexual dimorphism inferred from the different branching pattern of a same species of *Lophoranina* from the Eocene of Spain. Interestingly *Lophoranina bishopi* is found from the Middle Eocene Bateque Formation of Baja California Sur. In addition the oldest record of the genus *Ranina* also occurs at the Bateque Formation. It is believed by the authors that through time *Lophoranina* spp. lost its terraces and the dorsal carapace became scabrous (spiny), most likely reflecting a habitat change from carbonate rich reef complexes to more sandy environments. Today the only living species of this genus, *Ranina ranina,* is found exclusively on sandy substrates of the Indopacific region.

Family Dakoticancridae.- The Late Cretaceous family Dakoticancridae has been reported from rocks extending from South Dakota, USA to San Luis Potosí, Mexico. Species of this family possess a characteristic robust carapace. The Dakoticancridae were very abundant, as thousands of specimens have been reported from the Campanian-Maastrichtian of southeastern United States (Bishop, 1983). *Tetracarcinus, Dakoticancer, Seorsus,* and *Avithelmessus*, are the representative genera of this family. This family disappeared at the end of the Cretaceous Period (Bishop *et al.*, 1998). One possible explanation for the disappearance of this family is related to its restricted paleobiogeographic distribution north of the Chicxulub impact site. According to models from the impact trajectory, influence of secondary effects should had been worst northwest of the impact site (Sharpton *et al.* 1996). Along with the extinction of the Family Carcineretidae, also distributed relatively near the impact site, the Dakoticancridae seems to have been another victim of the mass extinction that occurred 65 million years ago.

Family Carcineretidae.- Regardless that there is ongoing debate on the genera that makeup this family, most authors agree that this group disappeared at the end of the Cretaceous. Some authors have reported the presence of members of this family in Europe within Maastrichtian rocks (Fraayé, 1996). However, the most important diagnostic features; the presence of fifth flattened pereiopods; a robust and spiny chelae; and a carapace with sharp, transverse ridges are to be found only in three genera, *Carcineretes, Ophthalmoplax* and *Branchiocarcinus. Carcineretes* has been reported from Maastrichtian rocks of Jamaica, Belize and Chiapas (Whitters, 1922; Morris, 1993; Vega *et al.* 1997; Vega *et al.* 2001a). *Ophthalmoplax* has been reported from Maastrichtian localities of southeastern United States, northeastern Mexico (Coahuila, Nuevo León), and Colombia (Rathbun, 1935; Vega and Feldmann, 1991; Feldmann *et al.* 1999; Feldmann and Villamil, 2002); *Branchiocarcinus* was found from the Cárdenas Formation (Maastrichtian) of San Luis Potosí. If we consider only these three genera as part of the Carcineretidae, then its paleobiogeographic distribution was restricted to the Gulf of Mexico and Caribbean region. Within the center of this distribution we find the impact site of the Chicxulub event, and this may be the second family of decapod crustaceans to become extinct at the end of the Cretaceous Period (Vega *et al.* 1997).

Costacopluma.- The genus *Costacopluma* has been reported from the Upper Cretaceous of Nigeria, India, Greenland and Mexico (Nuevo León, Coahuila, Guerrero) (Collins and Morris, 1975; Gaetani *et al.* 1983; Collins and Rasmussen, 1992; Vega and Perrilliat, 1989; Vega and Feldmann, 1992, Vega *et al.* 1995). Also the genus has been reported from Paleocene rocks of Senegal, Argentina, Brazil and Venezuela (Rémy, 1959; Feldmann *et al.* 1995a; Feldmann and Martins-Neto, 1995; Collins *et al.* 1994). In addition, there is a recent report of this genus from Paleocene rocks California, USA (Nyborg *et al.*, 2003). The distribution pattern of *Costacopluma* is distinctly different from the Cretaceous to the Paleocene. This difference may represent the extinction dynamics that occurred at the end of the Cretaceous Period. To date, the oldest record of *Costacopluma, Costacopluma bishopi*, is from the Coniacian of Guerrero, Mexico (Vega and Feldmann, 1992), suggesting an American origin for a group only found today in the Indopacific region.

In difference to other crustacean groups, retroplumid crabs seem to have been deposit feeders. Recent retroplumids live on the Indopacific Ocean bottom, more than 500 m deep, feeding on organic matter that is always available (Saint-Laurent, 1989).

6. Paleobiogeography

The fossil record for decapod crustacean species is not clearly understood. However it is clear that while some genera lasted tens of million years *(Costacopluma, Lophoranina, Panulirus)*, others lasted no more than 5 million years (*Carcineretes, Dakoticancer).* Apparently this duration is related to the paleobiogeographic distribution of the group. Those genera that had a larger geographic distribution tended to have a long and robust fossil record whereas those that had a restricted distribution tended to succumb to evolutionary extinction.

Due to the relative position of the Mexican territory during Cretaceous and Tertiary times, Mexico appears to have been a major area for fossil decapod crustacean originations and distribution patterns. The relatively scarce record of fossil decapod crustaceans in Mexico during the Lower Cretaceous is still relatively unknown. Aptian Mexican mecochirids have affinities to European mecochirid species. Upper Cretaceous crustaceans from Chiapas are related to crustaceans of the Caribbean region, while Maastrichtian crustaceans from northeastern Mexico are related to the Mississipi Embayment and Atlantic Coastal Plain crustaceans. Eocene species from Baja California Sur and Chiapas have relationships with North America west coast crustaceans, but also show some Tethyan influence. The Miocene species of Veracruz and Chiapas are related to Central America and Venezuela crustaceans.

7. Paleoecology

The type of rock in which decapod crustaceans are found, along with the associated fauna, represents the key to the kind of environment in which they lived. In many cases, there are few species of fossil decapod crustaceans found in one locality and there usually is only one or two dominant species. For example, in several Maastrichtian localities of Coahuila and Nuevo León, *Costacopluma mexicana* is found only with *Sodakus mexicanus*, while in other localities *Costacopluma mexicana* is only found with *Ophthalmoplax difuntaensis*, but it is never found along with *Dakoticancer australis* that appears to have excluded any other crustacean, except for *Prehepatus* sp.

Dakoticancrid crabs were a competitive group, possessing a large, robust carapace and strong chelae. In addition, short but resistant spines covered the carapace and pereiopods. Dakoticancrids inhabited shallow marine environments, and may have even been able to have sub aerial incursions. Galleries have been observed at localities where *Dakoticancer australis* is abundant.

Carcineretid crabs were either swimmer and/or burrower crabs, due to their typical flattened fifth pereiopods, and to their flat carapaces. Similarities between this extinct family and a similar decapod crustacean family, the Portunidae, are important, although no direct phylogenetic relationship has yet been accepted (Glaessner, 1969). Carcineretids have elongated chelae, with sharp teeth and acute, hook-shaped ends. It may be possible that, as recent portunids do, carcineretids were scavengers or even able to capture fish.

Retroplumid crabs feed on organic matter accumulated on their feather shaped fifth pair of pereiopods (Saint-Laurent, 1989). Their chelae are rather small probably not used for defense or attack. However, *Costacopluma mexicana,* from Maastrichtian localities of northeastern Mexico are the most abundant decapod crustacean. It appears that even with small chelae *Costacopluma mexicana* was able to share environments with other decapod crustacean species. One specimen of this species was recently found inside a gallery.

8. Cuticule Structure

An interesting fact of the study of crustaceans is to review and compare their cuticle structure. In many cases, this structure has been obscured by mineral replacement of the original calcium carbonate, and also because of the diagenetic process in general (Fig. 4.1). However, in some cases layers of the endo and exocuticle can be defined (Fig. 4.2), and even fine details such as glandular ducts and pore canals have been preserved (Vega *et al.* 1994). Recent studies have revealed cell-like structures on top of the epicuticle of *Costacopluma mexicana* (Fig 4.3), *Dakoticancer australis* (Fig. 4.4) and *Carcineretes planetarius* (Fig 4.5) specimens. The peculiar structure of a Paleocene Astacidae from Coahuila is also under study (Fig. 4.6). Cuticle structure of extinct crustacean families can be compared with extant families in attempts to define physiological and paleocological implications.

Acknowledgments

Authors wish to express their gratitude to CONACyT for the support from the grant titled "Paleobiogeografía y evolución de crustáceos y moluscos del Eoceno-Oligoceno de Baja California Sur y Chiapas, México".

References

Alencáster, G., 1977, Crustáceo (Malacostraca-Decapoda) del Cretácico Inferior del Estado de Guerrero, México, *Univ. Nal. Autón, México, Inst. Geol., Rev.*, 1:74-77.

FIGURE 4. 1, Cuticular structure of *Carcineretes planetarius* Vega, Feldmann, Ocampo and Pope. Note lack of detail on lamellae due to replacement. 2, Thin section of cuticle from *Costacopluma mexicana* Vega and Perrilliat. Note endocuticle (En), exocuticle (Ex) and epicuticle (Ep), as well as pore canals (Pc). 3, Cuticle of *C. mexicana* Vega and Perrilliat, showing cell like (CL) structures on epicuticle. 4, Cuticle of *Dakoticancer australis* Rathbun. Note cell like (CL) structures on epicuticle. 5, Cuticle section of *Carcineretes planetarius* Vega, Feldmann, Ocampo and Pope, showing cell-like structures (CL). 6, Cuticular section of an astacid crayfish. Scale bars = 100μ.

Bárcena, M., 1875, Descripción de un Crustáceo fósil del género *Sphaeroma (S. burkartii)* y reseña geológica del Valle de Ameca, Jalisco, *La Naturaleza*, 3:355-361.

Bishop, G. A., 1983, Fossil decapod crustaceans from the Late Cretaceous Coon Creek Tongue, Ripley Formation, Union County, Mississippi, *J. Crust. Biol.*, 3:417-430.

Bishop, G. A., 1986, Occurrence, preservation, and biogeography of the Cretaceous crabs of North America, in: *Crustacean Biogeography, Crustacean Issues,* Vol. 4 (R. H. Gore and K. L. Heck, eds.), A. A. Balkema, Rotterdam, The Netherlands, p. 111-142.

Bishop, G. A., 1988, Two crabs, *Xandaros sternbergi* (Rathbun, 1926) n. gen., and *Icriocarcinus xestos* n. gen., n. sp., from the Late Cretaceous of San Diego County, California, USA, and Baja California Norte, México, *Trans. San Diego Soc. Nat. Hist.,* 21:245-267.

Bishop, G. A., Feldmann, R. M., and Vega, F. J., 1998, The Dakoticancridae (Decapoda: Brachyura) from the Late Cretaceous of North America and Mexico, *Contr. Zool.*, 67:237-275.

Bouvier, E. L. 1899, Calappa zurcheri, Crabe noveau des Terrains Miocénes de Panama, *Bull. Mus. Hist. Nat. Paris*, 5:189-192.

Burkenroad, M. D., 1963, The Higher Taxonomy and Evolution of Decapoda (Crustacea), *Trans. San Diego Soc. Nat. His.*, 19:251-268.

Collins, J. S. H., and Morris, S. F., 1975, A new crab *Costacopluma concava* from the Upper Cretaceous of Nigeria, *Palaeontology*, 18:823-829.

Collins, J. S. H., and Rasmussen, H. W., 1992, Upper Cretaceous-Lower Tertiary decapod Crustaceans from West Greenland, *Grøn. Geolog. Uder., Bull.*, 162:1-46.

Collins, J. H. S., Higgs, R., and Corticula, B., 1994, A new crab, *Costacopluma bifida* (Crustacea, Decapoda) from the Paleocene of Venezuela, *Mizunami Fossil Mus. Bull.*, 21:29-34.

Dana, J. D., 1852, Crustacea. United States Exploring Expedition during the years 1838, 1839, 1840, 1841, 1842 Under the Command of Charles Wilkes, U. S. N., 13, C. Sherman, Philadelphia.

Feldmann, R. M., and Martins-Neto, R. G., 1995, *Costacopluma nordestiana* n. sp. (Decapoda: Retroplumidae) from the Maria Farinha Formation (Paleocene) of Brazil, *J. Paleont.,* 69:610-611.

Feldmann, R. M., Vega, F. J., Applegate, S. P., and Bishop, G. A., 1998, Early Cretaceous arthropods from the Tlayúa Formation at Tepexi de Rodríguez, Puebla, Mexico, *J. Paleont.,* 72:79-90.

Feldmann, R. M., Casadío, S., Chirino-Gálvez, L., and Aguirre-Urreta, M., 1995a, Fossil decapod crustaceans from the Jaguel and Roca Formations (Maastrichtian-Danian) of the Neuquén Basin, Argentina, *J. Paleont., Mem.* 43:1-22.

Feldmann, R. M., Vega, F. J., García-Barrera, P., Rico-Montiel, R., and Martínez-López, L., 1995b, A new species of *Meyeria* (Decapoda: Mecochiridae) from the San Juan Raya Formation (Aptian: Cretaceous), Puebla State, México, *J. Paleont.,* 69:402-406.

Feldmann, R. M., Vega, F. J., García-Barrera, P., Tucker, A., and Avendaño, J., 1996, The oldest record of *Lophoranina* (Decapoda: Raninidae) from the Late Cretaceous of Chiapas, southeastern Mexico, *J. Paleont.*, 70:296-303.

Feldmann, R. M., and Villamil, T., 2002, A new carcineretid crab (Upper Turonian, Cretaceous) of Colombia, *J. Paleont.*, 76:718-724.

Feldmann, R. M., Villamil, T., and Kauffman, E., 1999, Decapod and stomatopod crustaceans from mass mortality lagerstatten: Turonian (Cretaceous) of Colombia, *J. Paleont.*, 73:91-101.

Frayeé, R. H. B., 1996, Late Cretaceous swimming crabs: radiation, migration, competition, and extinction, *Acta Geol. Polonica*, 46:269-278.

Gaetani, M., Nicora, A., Premoli Silva, I., Fois, E., Garzanti, E., and Tintori, A., 1983, Upper Cretaceous and Paleocene in Zanskar Range (NW Himalaya), *Riv. Ital. Pal. Strat.,* 89:81-118.

Gill, T., 1894, A new bassalian type of crabs, *Am. Natur.,* 28:1043-1045.

Glaessner, M. F., 1969, Decapoda, in: *Treatise of Invertebrate Paleontology,* Part R, Arthropoda 4(2) (R. C. Moore, ed.), Geological Society of America and University of Kansas Press, Lawrence, p. R399-R533.

Haack, W., 1918, Ueber einen Isopoden aus dem Sepulit des weislichen, Osnegs *(Archaeoniscus Brodei,* Milne Edwards), *Jahr. König. Preus. Geolog. Land.* 39:73-102.

Haan, W. de, 1833-1850, Crustacea, in: *fauna Japonica sive descriptio animalium, quae in itinere per japoniam, jussu et auspiciis superiorum, qui summum in India batava Imperium tenent, sucepto, annis 1823-1830 collegit, notis, observationibus et adumbrationibus illustrativ* (P. F. de Siebold), A. Arnz, London, p. 109-164.

Hrebst, J. F. W., 1794, Versuch einer Naturgeschichte der Krebben und Krebse nebst einer systematischen beschreibung ihrer verscheidenen arten, vol. 2, 161 p.

Hessler, R., 1969, Peracarida, in: *Treatise of Invertebrate Paleontology,* Part R, Arthropoda 4(2) (R. C. Moore, ed.), Geological Society of America and University of Kansas Press, Lawrence, p. R360-R393.

Latreille, P. A., 1802-1803, *Histoire naturelle, général et particuliére, des crustacés et des Insectes,* Vol. 3. F. Dufart, Paris, 468 p.

Latreille, P. A., 1817, Les crustacés, les arachnides et les insectes, in: G. Cuvier, *Le regne animal distribue d'après son organisation, pour servir de base a l'histoire naturelle des animaux et d'introduction a l'anatomie comparée*, Ed. 1, Vol. 3, D'eterville, Paris, p. 1-653.

Latreille, P. A., 1831, *Cours d'Entomologie, ou de l'histoire naturelle des Crustacés, des arachnides, des Myriapodes et des insectes, etc.*, annales I, Atlas, Roret, Paris, France.

MacLeay, W. S., 1838, On the Brachyurous Decapod Crustacea brought from the Cape by Dr. Smith, in: *Illustrations of the Annulosa of South Africa; being a portion of the objects of Natural History chiefly collected during an expedition into the interior of South Africa, under the direction of Dr. Andrew Smith, in the years 1834, 1835, and 1836; fitted out by "The Cape of Good Hope Association for Exploring Central Africa*", London, p. 53-71.

Milne Edwards, H., 1840, *Histoire Naturelle des Crustacés, comprenant l'anatomie, la physiologie et la classification des ces animaux*, v. 3. Paris.

Morris, S. F., 1993, The fossil arthropods of Jamaica, in: *Biostratigraphy of Jamaica* (R. M. Wright and E. Robinson, eds.), Boulder, Colorado, *Geol. Soc. Am., Mem.*, 182:115-124.

Nyborg, T., Vega, F. J., and Filkorn, H., 2003, New Late Cretaceous and Early Tertiary decapod crustaceans from California, USA: implications for the origination of taxa in the eastern North Pacific. *Contr. Zool.*, 72:165-168.

Ortmann, A. E., 1892, Die Decapoden-Krebse des Strassburger Museums, V Teil, Die abteilungen Hippidae, Dromiidae, und Oxystomata, *Zool. Jahrb.*, 6:532-588.

Rafinesque, C. S., 1815, *Analyse de la nature, ou tableau de l'universe et des corps organiseés,* Palermo, Italia.

Rathbun, M. J., 1917, New species of South Dakota Cretaceous crabs, *Proc. U. S. Nat. Mus.*, 52:385-391.

Rathbun, M. J., 1926, The fossil stalk-eyed Crustacea of the north Pacific slope of North America, *U. S. Nat. Mus., Bull.* 138:1-155.

Rathbun, M. J., 1930, Fossil decapod crustaceans from Mexico, Proc. *U. S. Nat. Mus.,* 78:1-10.

Rathbun, M. J., 1935, Fossil Crustacea of the Atlantic and Gulf Coastal Plain, *Geol. Soc. Am. Spec. Pap.,* 2:1-160.

Rémy, J. M., 1959, Sur les Décapodes éocénes du Sénégal occidental. Etude géologique par A. Gorodiski. Etude Paléontologique par J. M. Rémy, *Bull. Soc. Géol. Fr.* 7th ser. 1:315-319.

Roberts, H. B., 1962, The Upper Cretaceous decapod crustaceans of New jersey and Delaware, in: *The Cretaceous fossils of New Jersey* (H. G. Richards, ed.), New Jersey Bureau of Geology and Topography, Bull. 61:163-191.

Saint-Laurent, M. de, 1989, La nouvelle superfamille des Retroplumoidea Gill, 1894 (Decapoda, Brachyura): systématique, affinities et evolution, in: *Résultats des Campagnes Musorstom* (J. Forest, ed.), *Mém. Mus. Nat. Hist. Natur.* (A), 5:103-179.

Samouelle, G., 1819, *The Entomologist's Useful Compendium, or an Introduction to the Knowledge of British Insects, etc.,* T. Boys, London.

Sars, G. O., 1893-1919, *An account of the Crustacea of Norway: Isopoda, Vol 2,* Alb. Cammermeyers Forlag and bergen Museum (Christiania and Copenhagen).

Savazzi, E., 1981, Functional morphology of the cuticular terraces in *Ranina (Lophoranina)* (brachyuran decapods; Eocene of NE Italy). *Neues Jahrb. Geol. Paläontol. Abh.,* 162:231-243.

Schweitzer, C., Feldmann, R. M., González-Barba, G., and Vega, F. J., 2002, New Crabs from the Eocene and Oligocene of Baja California Sur, Mexico, *J. Pal. Mem.,* 59:1-43.

Sharpton, L. E., Marín, J. L., Carney, S., Lee, G., Ryder, B. C., Schuraytz, P., Sikora, P. D., Spudis, A.

1996. A model of the Chicxulub impact basin based on evaluation of geophysical data, well logs, and drill core samples, in: *The Cretaceous-Tertiary Event and Other Catastrophes in Earth History* (G. Ryder, D. Fastovsky, and S. Gartner, eds.), Boulder, Colorado, *Geol. Soc. Am., Spec. Pap.*, 307:55-74.

Squires, R. L., and Demetrion, R. A., 1992, Paleontology of the Eocene Bateque Formation, Baja California Sur, Mexico, *Nat. Hist. Mus. Los Angeles County, Contr. Sci.*, 434:1-55.

Stenzel, H. B., 1944, A new Cretaceous crab *Graptocarcinus muiri* from Mexico, *J. Paleont.*, 18:550-552.

Stimpson, W., 1870-1881, Brachyura, in: *Preliminary report on the Crustacea dredged in the Gulf Stream in the Straits of Florida* (L. F. de Poutalès, ed.), Pt. 1, *Bull. Mus. Comp. Zool. Harvard Coll.*, 2:150-156.

Straelen, V. Van, 1925, Contribution à l'étude des crustacés décapodes de la période Jurassique, *Mem. Acad. royale Belgique, Class Sci., Part 4, Ser.* 2, 7:1-462.

Tucker, A., 1995, A systematic evaluation of fossil Raninidae from the Twin River Formation, Olympic Peninsula, Washington, and a reexamination of the Raninidae, Unpublished Ph. D. dissertation, Kent State University.

Vega, F. J., and Feldmann, R. M., 1991, Fossil crabs (Crustacea: Decapoda) from the Maastrichtian Difunta Group, northeastern Mexico, *Ann. Carnegie Mus.*, 60:163-177.

Vega, F. J., and Feldmann, R. M., 1992, Occurrence of *Costacopluma* (Decapoda: Brachyura: Retroplumidae) in the Maastrichtian of southern Mexico and its paleobiogegraphic implications, *Ann. Carnegie Mus.*, 61:133-152.

Vega, F. J., Feldmann, R. M., and Dávila-Alcocer, V., 1994, Cuticular structure in *Costacopluma mexicana* Vega and Perrilliat, from the Difunta Group (Maastrichtian) of northeastern Mexico, and its paleoenvironmental implications, *J. Paleont.*, 68:1074-1081.

Vega, F. J., and Perrilliat, M. C., 1989, Una especie nueva del género *Costacopluma* (Arthropoda: Decapoda) del Maastrichtiano de Nuevo León, México, *Univ. Nal. Autón. Méx., Inst. Geol. Rev.*, 8:84-87.

Vega, F. J., Feldmann, R. M., and Sour-Tovar, F., 1995, Fossil crabs (Crustacea: Decapoda) from the Late Cretaceous Cárdenas Formation, East-Central Mexico, *J. Paleont.*, 69:340-350.

Vega, F. J., Feldmann, R. M., and Villalobos-Hiriart, J. L., 1995, Additions to the crustacean (Decapoda) fauna from the Potrerillos Formation (Late Cretaceous) in northeastern Mexico, *Ann. Carnegie Mus.*, 64:39-49.

Vega, F. J., Feldmann, R. M., Ocampo, A. O., and Pope, K. O., 1997, A new species of Late Cretaceous crab (Brachyura: Carcineretidae) from Albion Island, Belize, *J. Paleont.*, 71:615-620.

Vega, F. J., Feldmann, R. M., Villalobos-Hiriart, J. L., and Gío-Argáez, R., 1999, A new decapod fauna from the Miocene Tuxpan Formation, Eastern Mexico, *J. Paleont.*, 73:407-413.

Vega, F. J., Feldmann, R. M., García-Barrera, P., Filkorn, H., Pimentel, F., and Avendaño, J., 2001a, Maastrichtian Crustacea (Brachyura: Decapoda) from the Ocozocuautla Formation in Chiapas, southeast Mexico, *J. Paleont.*, 75:319-329.

Vega, F. J., Cosma, T., Coutiño, M., Feldmann, R. M., Nyborg, T., Schweitzer, C., and Waugh, D., 2001b, New Middle Eocene crabs (Crustacea: Decapoda) from Chiapas, Mexico, J. Paleont., 75:929-946.

Vega, F. J., García-Barrera, P., Coutiño, M., Nyborg, T., Cifuentes-Ruiz, P., González-Rodríguez, K., Martens, A., Delgado, R., and Carbot, G., 2003, Early Cretaceous Arthropods from plattenkalk facies in Mexico, *Contr. Zool.*, 72:187-189.

Vía, L., 1969, Crustáceos Decápodos del Eoceno español (thése), *Pirineos*, 91-94:1-480.

White, A., 1847, *List of the specimens of Crustacea in the collection of the British Museum, British Museum*, London.

Withers, T. H., 1922, On a new brachyurous crustacean from the Upper Cretaceous of Jamaica, *Ann. Mag. Nat. Hist.*, 10:534-541.

Withers, T. H., 1926, Decapod crustaceans (Callianassa) from the Scotland Beds of Barbados, *Geological* Magazine, 63:104-108.

Chapter 6

Cretaceous Echinoids of Mexico

ITZIA NIETO-LÓPEZ and PEDRO GARCÍA-BARRERA

Museo de Paleontología, Departamento de Biología Evolutiva, Facultad de Ciencias, UNAM. Ciudad Universitaria 04510, Coyoacán, Mexico, D. F. nlie@minervaux2.fciencias.unam.mx

1. Abstract

This chapter offers an up to date list of Upper Cretaceous Mexican fossil echinoids. There are 149 known Cretaceous echinoids from 16 states of Mexico including: Baja California Sur, Chiapas, Chihuahua, Coahuila, Guerrero, Hidalgo, Jalisco, Michoacán, Nuevo León, Oaxaca, Puebla, San Luis Potosí, Sonora, Tamaulipas, Tlaxcala and Colima. Up until the first half of the XIX century reports on the fossil echinoids of Mexico were only vaguely mentioned within geological papers. After the XIX century date specific paleontological work on echinoids began to frequently appear. Recently, diverse paleontological studies are in progress, in the State of Chiapas which has generated a local list of nine species registered for the fist time in Mexico belonging to the following genera: *Orthopsis* sp., *Salenia* sp., *Goniopigus* sp., *Neotlingaster* sp., *Gompechinus* sp., *Lanieria* sp., *Petalobrissus* sp., *Gitolampas*, sp., and *Heteraster* sp. These species show great affinities with Caribbean, European and Asiatic faunas.

In Mexico there are two principal problems referring to the study of fossil echinoids: one is the lack of investigations focused on biogeographic, evolution and paleobiological aspects; and the second is the lack of a taxonomic revision of echinoid species reported in the literature.

F. Vega et al. (eds.), Studies on Mexican Paleontology, 101–114.

2. Introduction

Cretaceous echinoids of Mexico were first described in 1840 from Tehuacán, Puebla (Nyst and Galeotti, 1840). Since this initial report, 227 Cretaceous echinoid species have been described from Mexico. Paleontological studies of Cretaceous echinoids of Mexico are relatively rare when it is taken into account that Mexico was almost completely covered by a shallow sea during the Mesozoic era. Published papers on Cretaceous echinoid occurrences of Mexico are usually strictly systematic or are part of general fauna assemblage lists. Few reports have dealt with the paleobiological implications of Cretaceous echinoids of Mexico.

From 1840-1940 Cretaceous echinoids of Mexico were almost entirely reported by foreign workers. Therefore most collected material during this time was placed in museums outside of Mexico. In spite of this, there are diverse collections in Mexico. For example, the fossil echinoid collection at the Museo de Paleontología, Instituto de Geología (IGM), at the Universidad Nacional Autónoma de México (UNAM) has Cretaceous fossils from 16 Mexican states: Baja California Sur, Chiapas, Chihuahua, Coahuila, Guerrero, Hidalgo, Jalisco, Michoacán, Nuevo León, Oaxaca, Puebla, San Luis Potosí, Sonora, Tamaulipas, Tlaxcala and Colima.

In 1968 Dr. Blanca E. Buitrón produced a catalogue of known Cretaceous echinoids of Mexico. The main purpose of this paper is to update Dr. Buitrón's list. On doing this, the authors realized that there are two main problems with the study of Mexican Cretaceous echinoids: one is the scarcity of publications with a biogeographic, evolutionary or biological approach; and the second is the need to review the species reported as uncertain or *affinis*, in order to give them a correct taxonomic assignment.

3. History

Although the native human groups that lived in Mexico in pre-columbian times were undoubtedly familiar with fossils, it wasn't until the 19th century that the first formal report on Cretaceous echinoids of Mexico was published (Nyst and Galeotti, 1840). Nyst and Galeotti (1840) collected the Cretaceous echinoids *Cidarites propinquus, C. glandiferus* and *C. pustulosus* from the area of Tehuacán, Puebla. The Tehuacán locality was originally assigned to the Jurassic (Nyst and Galeotti, 1840). Later work modified the stratigraphy and taxonomy of these findings, such that *C. propinquus, C. glandiferus* were established as *Cidaris galeotti,* and *C. pustulosus* was redefined as *Tetragramma malbosii.* In addition the Tehuacán locality was reassigned to the Aptian. From 1840-1900 only three additional studies that included Cretaceous echinoids of Mexico were published: two from the northern part of Mexico (Schlütter, 1877; Cotteau, 1890); and one more from the south (Felix, 1891). By the end of the century only six Cretaceous echinoid species were known from Mexico.

The Instituto Geológico de Mexico was established in 1888, under the direction of Engineer José Guadalupe Aguilera. The Instituto Geológico de Mexico is the main repository for fossil material of Mexico including fossil echinoids.

During the first half of the 19th century, interest in the geology of Mexico brought about new discoveries of fossil echinoids. Most of the geologic work in Mexico was still conducted by foreign investigators during this time, however most work was published within Mexican journals and placed in Mexican museums.

Since about 1950 fossil echinoids of Mexico have increasingly been reported by Mexican paleontologists and published within Mexican journals mainly within the Revista of the Instituto de Geología, presently known as the *Revista Mexicana de Ciencias Geológicas.*

In 1975 the paleontologic collections of the IGM were moved to the Museo de Paleontología, Instituto de Geología (IGM), UNAM. IGM has the largest collection of Mexican fossil echinoids. 227 Cretaceous echinoid species, 95 of them type species, are contained within the collections of IGM.

4. Types of Publications

Discoveries of Cretaceous echinoids of Mexico have mainly come from geologic studies. In fact almost 50% of known Cretaceous echinoids of Mexico were from geologic studies of Mexico. Many of these types of studies only mention the occurrences of fossil echinoids providing no illustrations or photographs of the specimens. Later paleontologists were able to use these geologic reports to locate areas where these fossil echinoids occurred making collections and publishing on numerous new species and redescriptions of old fossil echinoid species possible.

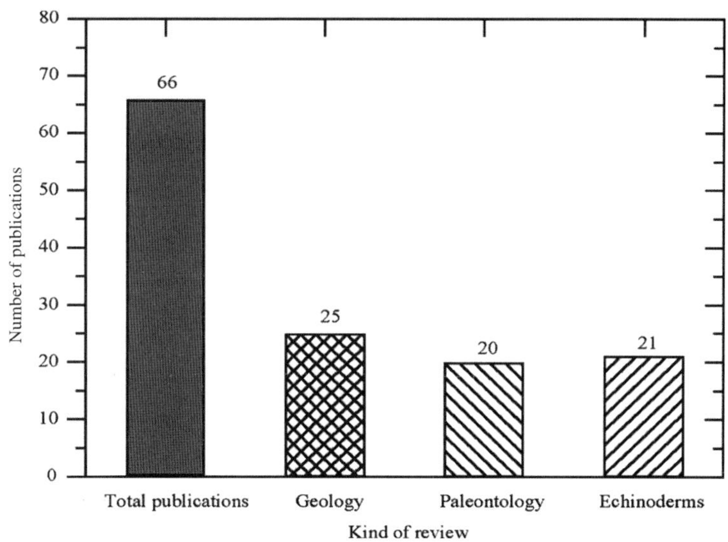

FIGURE 1. Geologic vs. Paleontologic publications.

Although in 1950 more than 80% of presently known fossil echinoids had been already described, and from 1968 on, no more than 30 new species have been described from Mexico, specific reports on echinoids have been increasingly more numerous, as locations with abundant echinoderms are being studied in detail (Fig. 1).

Before 1950 most reports on Cretaceous echinoids of Mexico had been published by foreign specialists, and only a few by national paleontologists. With the discovery of many new fossiliferous locations, and the integration of paleontologic studies at the Instituto de Geología, fossil echinoid research by national paleontologists has grown since the 1950's (Fig. 2). In spite of the recent interest in fossil echinoids by national paleontologists the only Mexican scientist who is dedicated formally and professionally to the study of echinoids is Dr. Blanca Estela Buitrón. It should also be mentioned that other investigators have worked on recent echinoids, such as Dr. Elena Caso (†), Dr. Alfredo Laguarda and Dr. Francisco Solís, of the Instituto de Ciencias del Mar y Limnología, UNAM.

The references used to construct Figures 1 and 2 are (Nyst and Galeotti, 1840; Desor, 1858; Meek, 1864; Virlet D'Aoust, 1865; Gabb, 1869; Loriol, 1875-1876; Schlüter, 1883-1892; Cotteau, 1890; Heilprin, 1890; Felix, 1891; Felix and Lenk, 1891; Aguilera and Ordóñez, 1893; Hill, 1893; Boyle, 1893; Aguilera *et al.* 1896; Stanton and Vaughan, 1896; Felix, 1899; Villada, 1905; Aguilera , 1906; Böse, 1906; Burckhardt and Villarelo, 1909; Böse, 1910, 1913; Adkins, 1918; Burckhardt, 1919; Böse, 1923; Israelsky, 1924; Burckhardt, 1925; Jordan and Hertlein, 1926; Böse and Cavins, 1927; Burckhardt, 1930; Müllerried, 1934; Lambert, 1935, 1936; Muir, 1936; Jones, 1938; Imlay, 1944; Kellum, 1944; King and Adkins, 1946; Maldonado,

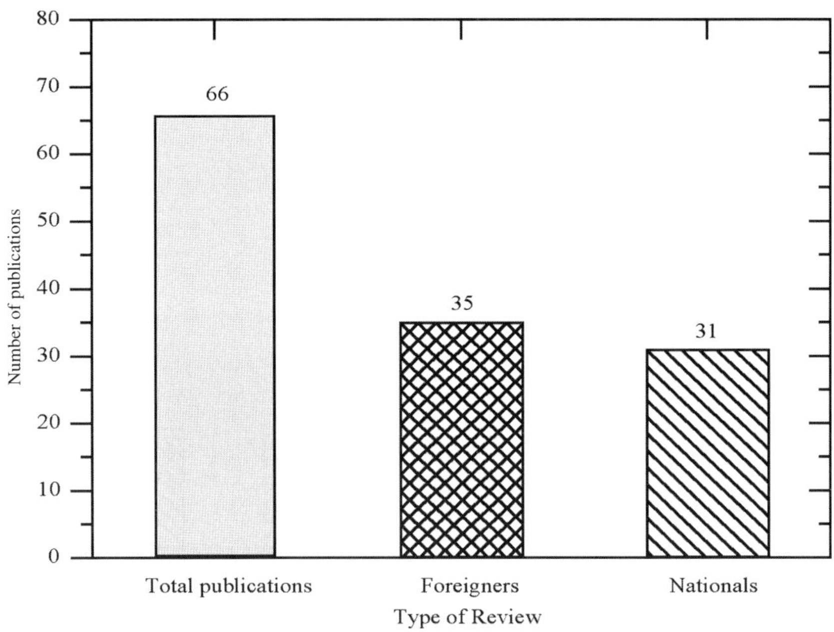

FIGURE 2. National vs. Foreign publications.

1950; Caso, 1951; Cooke, 1953; Maldonado, 1953; Allison, 1955; Cooke, 1955; Calderón, 1956; Erben, 1956; Kellum, 1956; Durham and Allison, 1960; Perkins, 1960; Alencáster and Buitrón, 1965; Buitrón, 1968; Myers, 1968; Buitrón, 1970, 1971, 1973a, 1973b, 1974a, 1974b, 1976, 1977; Almazán and Buitrón, 1984; García-Barrera, 1991; Sánchez, 1997.

5. Result of Investigations

Although further research is needed, there is a diverse and abundant representation of Cretaceous echinoids of Mexico. Buitrón (1968) was the first to compile published reports on Cretaceous echinoids of Mexico. Buitrón (1968) presented a list of species, geologic formations, ages and corresponding references. In the present report we update that list, eliminate synonyms and species considered as *affinis*, and include species reported in journals, congress memoirs and thesis/dissertation studies published after 1968 (Table 1). In addition synonyms and *affinis* species were deleted from Buitrón's (1968) list because these species had not been correctly reviewed for placement into a given species and possibly may represent redundant species names.

On Table 1 we show the Cretaceous echinoids of Mexico updated from Buitrón's (1968) list. First column represents the number of echinoid species present in Mexico. Second column represents the author(s) who described the species. Third column indicates in which state of the Mexican Republic the fossils were collected. Species that don't have a reference in the second column are part of the Colección Paleontológica Nacional, Instituto de Geología, UNAM, however these species have never been formally described.

TABLE 1. Cretaceous fossil echinoids reported from Mexico. *Museo de Paleontología, Instituto de Geología, UNAM. + Museo de Paleontología, Facultad de Ciencias, UNAM.

List of species	1st. Report	Locations
Ananchites sulcata Goldfuss	Aguilera and Ordoñez, 1893	Hidalgo
Astrodapsis bajasurensis Squires and Demetrion *		Baja California Sur
Cassidulus aequoreus Morton *	Sánchez, 1997	San Luis Potosí
Cidaris galeottii Desor	Meek, 1864	Puebla
Cidaris glandifera Goldfuss	Schlüter, 1883-1892	Puebla
"Cidaris" muellerriedi Lambert *	Müllerried, 1934	Puebla
Cidaris nobilis Agassiz	Felix, 1891	Oaxaca
Cidaris (?) postulosus Galeotti	Meek, 1864	North of Mexico
Cidaris submarginata Felix *	Felix, 1891	Puebla
Cidaris texanus Clark	Maldonado, 1953	Puebla
Cidarites glandiferus Goldfuss	Nyst and Galeotti, 1840	Puebla
Cidarites propinquus Münster	Nyst and Galeotti, 1840	Puebla
Cidarites postulosus Nyst and Galeotti	Nyst and Galeotti, 1840	Puebla
Clypeaster pileus Israelsky *		

TABLE 1. Continued.

List of species	1st. Report	Locations
Clypeaster rogersi (Morton) *		
Codiopsis stephensoni Cooke *	Sánchez, 1997	Nuevo León
Coenholectypus planatus (Roemer) *		
Coenholectypus transpecosensis (Cragin) *		
Conulus antillensis Cotteau	Lambert, 1935	Chiapas
Conulus chiapasensis Lambert *	Lambert, 1936	Chiapas
Conulus cookei Buitrón *		Chiapas
Conulus lamberti Buitrón *		
Conulus muellerriedi Lambert		
Conulus raulini (d'Orbigny) *		
Cyphosoma aquitanicum Cotteau		Chiapas
Cyphosoma texana Roemer	Felix, 1891	Puebla
Cyphosoma texanum Roemer	Aguilera and Ordoñez, 1893	N and E of Mexico
Diplopodia malbosi (Agassiz)	Gabb, 1869	Sonora
Desor	Cotteau, 1890	Puebla
Diplopodia (Pseudodiadema)	Heilprin, 1890	Puebla
malbosi	Böse, 1910	Sonora
Diplopodia texana (Roemer)	Stanton and Vaughan, 1896	Chihuahua
Diplopodia texanum (Roemer)	Burckhardt, 1930	Puebla
Diplopodia variolaris Brongniart		
Discoides cordobai Buitrón *		
Dumblea symmetrica Cragin		
Enallaster bravoensis Böse *	Cooke, 1955	Chihuahua
Enallaster (Washitaster)	Böse, 1906	Chihuahua
bravoensis Böse	Cooke, 1955	Chihuahua
Enallaster delgadoi Loriol		
Enallaster mexicanus Cotteau	Aguilera and Ordoñez, 1893	Chihuahua
Enallaster texanus (Roemer)	Cotteau, 1890	Chihuahua, Colima
	Stanton and Vaughan, 1896	Chihuahua, Baja California Sur and Coahuila
Encope grandis subsp. *inezana* Durham *		
Encope loretoensis Durham *		
Encope michoacanensis Durham *		
Encope perspective Agassiz *		
Encope shepherdi Durham *		
Encope tatetlaensis Böse *		
Epiaster aguilerae Böse		
Epiaster chiapasensis Lambert	Böse, 1910	
Epiaster dartoni Cooke	Lambert, 1935	Chihuahua
Epiaster elegans (Shumard)	Cooke, 1955	Chiapas
	Stanton and Vaughan, 1896	Coahuila
Epiaster texanus Roemer		Tamaulipas and
Epiaster whitei Clark	Aguilera and Ordoñez, 1893	Chihuahua
Galerites gigas Cotteau *	Villada, 1905	N and E of Mexico

TABLE 1. Continued.

List of species	1st. Report	Locations
Gonipygus bolaensis Jones	Sánchez, 1997	Puebla
Goniopygus zitteli telostocensis	Jones, 1938	Guerrero
Maldonado-Koerdell	Maldonado, 1953	Coahuila
Haimea bajasurensis		Puebla
Squires and Demetrion *		
Hardouinia potosiensis Lambert		
Hemiaster bexari Clark	Lambert, 1936	
	Jones, 1938	San Luis Potosí
Hemiaster calvini Clark *		Coahuila and
	Böse, 1906	San Luis Potosí
Hemiaster cholamensis Kew *		Chihuahua and
Hemiaster humphreyanus	Sánchez, 1997	Coahuila
Meek and Hayden *	Sánchez, 1997	San Luis Potosí
Hemiaste jackoni Maury		San Luis Potosí and
Hemiaster lacunosus Slocum *	Sánchez, 1997	Nuevo León
Hemiaster mirabilis	Sánchez, 1997	Guerrero
Peron and Gauthier *	Sánchez, 1997	San Luis Potosí
Hemiaster oldhami Noetling *		Guerrero
Hemiaster sphericus Lambert	Sánchez, 1997	
Hemiaster stella Morton *	Lambert, 1935	San Luis Potosí
Hemiaster texanus Roemer	Sánchez, 1997	Chiapas
Hemiaster ungula Morton *	Aguilera and Ordoñez, 1893	San Luis Potosí
	Sánchez, 1997	N and E of Mexico
Hemiaster welleri Clark *		San Luis Potosí and
Hemiaster wetherbyi Loriol *	Sánchez, 1997	Guerrero
Hemiaster whitei (Clark) *	Sánchez, 1997	San Luis Potosí
Hemicidaris (Pseudocidaris)	Kellum, 1956	San Luis Potosí
clunifera (Agassiz)	Maldonado, 1953	
Hemicidaris villadai		Puebla
Maldonado-Koerdell	Maldonado, 1953	
Heteraster aguilerai Buitrón *		Puebla
Heteraster alencasterae Buitrón		
Heteraster bravoensis Böse		Puebla
	Adkins, 1918	Michoacán
Heteraster mexicanus (Cotteau) *		Chihuahua and
	Adkins, 1918	Michoacán
Heteraster oblicuatus (Adkins) +		Chihuahua and
Heteraster texanus Roemer		Colima
Heteraster (Euaster) texanus Roemer	Aguilera, 1893	Chihuahua
Heteraster trahthi Kühn *	Lambert, 1936	Chihuahua
Heteraster wenonensis (Adkins) +	Sánchez, 1997	Chihuahua
Heterosalenia tlaxiacensis Buitrón *		San Luis Potosí
Holaster nanus Cragin		Michoacán
Holaster simplex Shumard	Jones, 1938	Tlaxcala
Holectypus castilloi Cotteau	Cooke, 1955	Coahuila
Holectypus (Coenholectypus)	Cotteau, 1890	Coahuila

TABLE 1. Continued.

List of species	1st. Report	Locations
castilloi Cotteau	Cooke, 1946	Jalisco
Holectypus limitis Böse		Coahuila and Jalisco
Holectypus parvus Jones	Böse, 1906	
Holectypus planatus Roemer	Jones, 1938	Chihuahua
Holectypus (Coenholectypus)	Stanton, 1896	Coahuila
planatus Roemer	Allison, 1955	Chihuahua
Holectypus transpecosensis Cragin		Baja California Sur
Isomicraster mexicanus Lambert	Cooke, 1946	
Lanieria lanieri (D'Orbigny)	Lambert, 1935	Chihuahua
Duncan		Chiapas
Leptarbacia arguta Clark	Cotteau, 1890	
Linthia mullerriedi Lambert		Chiapas and
Linthia paronai Checchi-Rispoli *	Böse and Cavins, 1927	San Luis Potosí
Linthia variabilis Slocum *	Lambert, 1935	Coahuila
Loriolia picteti (Desor)	Sánchez, 1997	Chiapas
Macraster aguilerae (Böse) *	Sánchez, 1997	Nuevo León
	Maldonado, 1953	Nuevo León
Macraster dartoni (Cooke) *	Böse, 1927	Puebla
Macraster elegans Romer		Chihuahua, Nuevo
Macraster texanus Roemer		León and Tamaulipas
Micraster americanus Stephenson *	Hill, 1891	
Micraster hagenowi Desor *	Adkins, 1918	Nuevo León
Orthopsis aguilerai	Sánchez, 1997	Chihuahua
Maldonado-Koerdell	Sánchez, 1997	San Luis Potosí
Orthopsis sanfilippoi	Maldonado, 1953	San Luis Potosí
Checchia-Rispoli *		Puebla
Pedinopsis meridanensis Cotteau *	Sánchez, 1997	
Petalobrissus burckhardti Lambert		Guerrero
Phyllobrissus potosiensis Lambert		
Phymosoma aquitanicum	Lambert, 1936	
(Cotteau)	Cooke, 1953	Chiapas
Phymosoma mexicanum Böse *	Maldonado, 1953	San Luis Potosí
		Puebla
Phymosoma mortenseni	Böse, 1906	
Checchia-Rispoli *		Chihuahua and
Phymosoma texanum (Roemer)	Sánchez, 1997	Michoacán
Phymosoma volanum (Cragin)		Guerrero
Proraster dalli Clark *	Adkins, 1932	
Pseudoananchytes	Jones, 1938	Puebla and Sonora
completa? (Cragin)	Sánchez, 1997	Coahuila
Pseudocidaris clunifera (Agassiz) *	Perkins, 1960	Nuevo León
Pseudocidaris galeotti Desor		Coahuila
Pseudocidaris lusitanica De Loriol *		
Pseudocidaris parryi Hall	Villada, 1905	Puebla
Pseudocidaris saussurei Loriol		Puebla
Pseudocidaris thurmanni (Agassiz) *	Burckhardt, 1930	

TABLE 1. Continued.

List of species	1st. Report	Locations
Pseudodiadema aguilerai	Loriol, 1875	Sonora
(Maldonado) *		Puebla
Pseudodiadema (Diplopodia)		
malbosi? Desor		Puebla and Oaxaca
Pseudodiadema saussurei Loriol	Aguilera and Ordoñez, 1893	
Pseudopyrina clarki (Böse)		Puebla
	Aguilera, 1906	
Pseudopyrina inaudita Böse	Adkins, 1928	Puebla
Pseudopyrina ochoterenai		Chiapas and
Müllerried	Lambert, 1935	Chihuahua
Pyrina clarki Böse		
Pyrina inaudita Böse	Müllerried, 1934	Puebla
Pyrina mexicana Lambert		
Pyrina parryi Hall	Böse, 1906	Chihuahua
	Böse, 1906	Chihuahua
Rachiosoma mortoni De Loriol *	Lambert, 1936	Chiapas
Salenia mexicana Schlüter *	Gabb, 1869	Sonora and
Salenia prestensis Desor		Chihuahua
	Sánchez, 1997	Guerrero
Salenia prestensis mexicana Schlüter	Schlüter, 1883-1892	Chihuahua
Salenia texana Credner *	Cotteau, 1890	Chihuahua and
Stereocidaris fusiformis (Cotteau) *		Puebla
Stomechinus semiplacenta (Agassiz) *	Maldonado, 1953	Chihuahua
Tetragramma bosei Jones		Colima
Tetragramma gloriae Buitrón *		
Tetragramma malbosii (Agassiz) *		
Tetragramma picteti (Desor) *	Jones, 1938	Coahuila and Sonora
Tetragramma streeruwitzi (Cragin)*	Buitrón, 1973 (b)	Michoacán
Tetragramma variolare (Brongniart) +	Lambert, 1926	Sonora
Toxaster elegans Schumard		
Washitaster bravoensis (Böse) *	Cooke, 1946	Chihuahua
Washitaster longisulcus		Michoacán
(Adkins and Winton) *	Aguilera and Ordoñez, 1893	N and E of Mexico

6. Types of Work

Research that addresses paleoecology or reconstruction of ancient environments of Cretaceous echinoids of Mexico are rare. Most of the publications on Cretaceous echinoids of Mexico have a systematic orientation, containing descriptions or redescriptions of species, or records of new locations for the same species. The only paper with an evolutionary approach to Cretaceous echinoids of Mexico is the one by Buitrón (1977), which includes the study of Mesozoic species from San Juan Raya, Puebla. Considering the fact that the fossil record of Cretaceous echinoids of Mexico is considerable, future work is needed to complete not only a taxonomic list of Cretaceous echinoids, but also papers dealing with the paleoecology and paleoenvironment of this group.

It should also be considered that most of the Cretaceous echinoids of Mexico have a narrow stratigraphic range and geographic distribution. Therefore Cretaceous echinoids of Mexico could be used as regional index fossils especially if they are considered together with other fossil groups such as foraminifera.

7. Main Locations

During the Cretaceous Mexico was covered by the western extension of the Tethys Sea. In the Lower Cretaceous the sea covered all of the southeast areas of modern day Mexico including the states of Oaxaca, Puebla, Guerrero and Chiapas, and a large portion of northern Mexico. During the Late Cretaceous, epicontinental seas reached their maximum extension. Therefore during the Cretaceous Mexico would have had a diverse and abundant echinoid history.

8. Present Work

Considering the fact that Mexico was covered by seas during the Cretaceous it is possible that fossil echinoids could be found within any of the extensive Cretaceous exposures of Mexico. Further prospecting of fossil echinoids within Cretaceous rocks is needed before a complete list of Cretaceous echinoids of Mexico can be achieved.

In the last two decades, with the creation of the Museo de Paleontología in Tuxtla Gutiérrez, Chiapas, the study of several fossil groups (including echinoids) has received renewed interest and support. Further field collecting and continued studies (systematic, stratigraphic, sedimentologic, taphonomic and palynologic) are currently underway.

One new Cretaceous locality in the state of Chiapas is currently under study. The ongoing project "Rudistas del Cretácico Superior de Chiapas y fauna Cretácico-Terciaria asociada", coordinated by Dr. Pedro García Barrera, Museo de Paleontología, Facultad de Ciencias, UNAM, has greatly increased the knowledge of Cretaceous echinoids of Mexico. There is only one previous report from this part of the country (Buitrón, 1974), in which two species are reported (*Conulus cookei* and C. *lamberti*).

Another study within the state of Chiapas, "Equinodermos Cretácicos de Chiapas", has, for the first time from this state, described nine species including: *Orthopsis* sp., *Salenia* sp., *Goniopigus* sp., *Neotlingaster* sp., *Gompechinus* sp., *Lanieria* sp., *Petalobrissus* sp., *Gitolampas*, sp., and *Heteraster* sp. These genera indicate a great affinity with Caribbean and Cuban faunas, and to a lesser extent, with faunas from Europe and Asia.

Cretaceous echinoids of Mexico have recently been used to confirm the age of the upper part of the Ocozocoautla Formation in Chiapas, formerly established using inoceramids and foraminifera. This demonstrates that fossil echinoids can, in association with other fossil groups, be used as index fossils.

Cretaceous echinoids of Mexico have also been recently used in paleoenvironment reconstructions, such as the project currently being worked on in Chiapas, where subdivision of the environment by the association of echinoid species, considering several taphonomic aspects, and analysis of pollen grains such as achritarcs and the pollen grain *Classopollis*, all together indicate that these sediments and fossils were deposited in a coastline marine environment with particular conditions.

References

Adkins, W. S., 1918, The Weno and Pawpaw formations of the Texas Comanchean, *Univ. Tex. Bull.*, 1856:1-146.

Adkins, W. S., The Mesozoic System in Texas, *Univ. Tex. Bull.*, 2838, 385p.

Aguilera, J. G., 1906, Excursión de Tehuacán a Zapotitlán San Juan Raya, *Congreso Geológico Internacional Xa. Ses. México,* Libreto guía 7:1-27.

Aguilera, J. G., and Ordóñez E., 1893, *Datos para la Geología de México*, Imprenta "Cosmos", Tacubaya, D. F., México.

Aguilera, J. G., Ordóñez, E., and Buelna, R., 1896. Bosquejo Geológico de México, *Bol. Inst. Geol. Méx.*, 4-6:1-268.

Alencáster, G., and Buitrón, B. E., 1965, Fauna del Jurásico Superior de la región de Petlalcingo, Estado de Puebla, *Univ. Nal. Autón. Méx., Inst. Geol., Paleont. Mex.*, part. II, 21:1-53.

Almazán, V. E., and Buitrón, B. E., 1984, Bioestratigrafía del Cretácico inferior de Baja California Norte, México. *III Congreso Latinoamericano de Paleontología,* Memorias, p. 23.

Allison, E. C., 1955, Middle Cretaceous Gastropoda from Punta China, Baja California, *J. Paleontol.*, 29:400-433.

Böse, E., 1906, Excursión au Cerro de Muleros près de Ciudad Juárez, Chihuahua, *Congreso Geológico Internacional. Xa. Sesión. México,* Libreto-guía 20, 24 p.

Böse, E., 1910, Monografía geológica y paleontológica del Cerro de Muleros cerca de Ciudad Juárez, Estado de Chihuahua. *Bol. Inst. Geol. Méx.*, 25:1-189.

Böse, E. 1913, Faunas del Cretácico Superior de Coahuila, y regiones limítrofes. *Bol. Inst. Geol. Méx.*, 30:1-56.

Böse, E., 1923, Algunas faunas cretácicas de Zacatecas, Durango y Guerrero. *Bol. Inst. Geol. Méx.*, 42:1-219, 19.

Böse, E., and Cavins, A., 1927, The Cretaceous and Tertiary of Southern Texas and Northern Mexico, *Univ. Tex. Bull.*, 2748:7-142.

Boyle, C. B., 1893, A catalogue and bibliography of North American Mesozoic Invertebrata, *U. S. Geol. Surv. Bull.*, 102:1-315.

Buitrón, S., B., 1968, Catálogo de equinoides fósiles de México, *Univ. Nal. Autón. Méx., Inst. Geol., Paleont. Mex.*, 26:1-50.

Buitrón, S. B., 1970, Equinoides del Cretácico Inferior de la región de San Juan Raya, Zapotitlán, Edo. de Puebla, *Univ. Nal. Autón. Méx., Inst. Geol., Paleont. Mex.,* 30:1-46.

Buitrón, S. B., 1971, Equinoides del Cretácico del Norte de México, *Univ. Nal. Autón. Méx., Inst. Geol., Paleont. Mex.,* 31:70.

Buitrón, S. B., 1973a, Equinoides del Cretácico Medio de Colima, *Univ. Nal. Autón. Méx., Inst. Geol., Paleont. Mex.*, 36:22-29.

Buitrón, S. B., 1973b, *Tetragramma gloriae* n. sp., equinoide de la Formación San Lucas (Hauteriviano-Aptiano) de la Región de Huetamo, Michoacán, *Univ. Nal. Autón. Méx., Inst. Geol., Paleont. Mex.,* 36:35-40.

Buitrón, S. B., 1974a, *Caenocidaris* cf. *C. cucumifera* (Agassiz) Equinoide del Jurásico Superior de Chihuahua. *Univ. Nal. Autón. Méx., Inst. Geol., Paleont. Mex.,* 40:44-48.

Buitrón, S. B., 1974b, Algunas especies de la familia Conulidae Lambert, 1911 (Echinoidea) del Cretácico Superior de Chiapas, *Univ. Nal. Autón. Méx., Inst. Geol., Paleont. Mex.,* 39:7-19.

Buitrón, S. B., 1976, Tres especies de equinoides (Echinodermata-Echinoidea) del Cretácico Superior de Atenango del Río, Guerrero, *Bol. Soc. Geol. Méx.,* 37:6-24.

Buitrón, S. B., 1977, Algunos aspectos evolutivos de los Equinoides de San Juan Raya, Puebla, *Rev. Soc. Mex. Hist. Nat.*, 35:277-285.

Burckhardt, C., 1919, Faunas Jurásicas del Symon (Zacatecas) y faunas Cretácicas de Zumpango del Río (Guerrero), *Bol. Inst. Geol. Méx.*, 33:136.

Burckhardt, C., 1925, Faunas del Aptiano de Nazas (Durango), *Bol. Inst. Geol. Méx.*, 45:75.

Burckhardt, C., 1930, Étude Synthétique sur le Mesozöique mexicaine. *Mém. Soc. Paléont. Suiss.,* 49-50:1-280.

Burckhardt, C., and Villarelo, I., 1909, Estudio geológico de los alrededores de una parte del Río Nazas. Parergones, *Bol. Inst. Geol. Méx.*, 3:115-135.

Calderón, G. A., 1956, Estratigrafía del Mesozoico y tectónica del Sur del Estado de Puebla, Presa de Valsequillo, Sifón de Huexotillanapa y problemas hidrológicos de Puebla, *Congreso Geológico Internacional XXa. Sesión, México*, Libreto-guía Excursión A-11, 94 p.

Caso, M. E., 1951, Los Equinoides fósiles del Cenozoico de México, *Bol. Asoc. Mex. Geól. Petrol.,* 3:57-96.

Cooke, C.W., 1946. Comanche echinoids, *J. Paleontol.,* 23:193-237

Cooke, C. W., 1953, American Upper Cretaceous Echinoidea. *U.S. Geol. Surv. Prof. Paper*, 254-A:1-44.

Cooke, C. W., 1955, Some Cretaceous Echinoids from the America, *U. S. Geol. Surv. Prof. Paper,* 264-E:87-112.

Cotteau, G. H., 1890, Note sur quelques Équinides du terrain Crétacé du Mexique, *Bull. Soc. Géol. Fr. Ser* 3, 18:292-299.

Desor, E., 1858, Sypnosis des Échinides fósiles, Paris, Wiesbaden.

Durham, J. W., and Allison, C., 1960, The geologic history of Baja California and its marine faunas, *Syst. Zool.,* 9:47-91.

Erben, H. K., 1956, El Jurásico Inferior de México y sus amonites, *Congreso Geológico Internacional, XXa. Sesión, México, Monografía,* 1-393.

Felix, J., 1891, Versteinerungen aus der mexicanischen Jura und Kreide Formation, *Pa-laeontographica*, 37:140-199.

Felix, J., 1899, Uebersicht über die geologischen Verhältnisse des Mexicanischen Staates Oaxaca. Beiträge zur Geologie und Palaontologje der Republik Mexico, Schweizerbart'sche Vert. Stuttgart, p. 1-194.

Felix, J., and Lenk. H., 1891, Uebersicht über die geologischen Verhältnisse des mexicanischen Staates Puebla. *Palaeontographica*, 37:117-139.

Gabb, W. M., 1869, Description of Cretaceous fossils from Mexico, *Paleontol. Calif.,* 2:257-276.

García-Barrera, P., 1991, Equinoides del Albiano Tardío de la Formación Mal Paso, de la Región de Chumbítaro, estados de Guerrero y Michoacán, México, *Rev. Soc. Mex. Paleontol.,* 4:23-41.

Heilprin, A., 1890, The Geology and Paleontology of the Cretaceous deposits of Mexico, *Proc. Acad. Nat. Sci. Phila.,* p. 445-469.

Hill, R.T., 1891. The Comanche series of the Texas-Arkansas region. *Geol. Soc. Am. Bull.*, 2:503-528.

Hill, R. T., 1893, The Cretaceous formations of Mexico and their relations to North American geographic development, *Am. J. Sci.,* 45:307-324.

Imlay, R. W., 1944, Correlation of the Cretaceous formations of the Greater Antilles, Central America and Mexico, *Geol. Soc. Am. Bull.*, 55: 1005-1046.

Israelesky, M. C., 1924, Notes on some Echinoids from the San Rafael and Tuxpan beds of the Tampico region, Mexico, *Proc. Calif. Acad. Sci.*, Ser. 4, 13:137-145.

Jones, T. S., 1938, Geology of Sierra de la Peña and Paleontology of the Indidura Formation, Coahuila, México, *Geol. Soc. Am. Bull.*, 49:69-149.

Jordan, E. K., and Hertlein, L. G., 1926, Contribution to the Geology and Paleontology of the Tertiary of Cedros Island and adjacent parts of lower California, *Proc. Calif. Acad. Sci., Ser.* 4, 15:409-464.

Kellum, L. B., 1944, Geologic history of northern Mexico and its bearing on petroleum exploration, *Amer. Assoc. Petrol. Geol. Bull.*, 28:301-325.

Kellum, L. B., 1956, Cretaceous invertebrates of the Aurora limestone, *Mich. Acad. Sci. Arts, Letters,* 41: 205-231.

King, R. E., and Adkins, W. S., 1946, Geology of a part of the Lower Conchos Valley, Chihuahua, Mexico, *Geol. Soc. Am. Bull.*, 57:275-294.

Lambert, J., 1935, Notes sur quelques Échinides fossiles. II.-Sur un *Echinolampas* du Désert Libyque. III.-Échinides du Mexique, *Bull. Soc. Geol. Fr.*, Ser. 5, 5:364-374.

Lambert, J., 1936, Quelques nouveaux Échinides fossiles du Crétacé du Mexique, *Bull. Soc. Geol. Fr.,* Ser. 5, 6:1-6.

Loriol, P., 1875-1876, Note sur quelques espèces nouvelles appartenant à la classe des Echinodermes, *Mém. Soc. Sci. Phys. Hist. Nat. Gèneve,* 24:659-673.

Maldonado, K. M., 1950, Otro equinoide y un braquiópodo del Cretácico Medio lnferior del Estado de Colima, México, *Ciencia*, 10:206-208.

Maldonado, K. M., 1953, Los equinoides regulares del Mesozoico de México, *An. Esc. Nal. Cienc. Biol.,* 7:15-44.

Meek, F. B., 1864, Check list of the invertebrate fossils of North America, Cretaceous and Jurassic, *Smith. Misc. Coll.*, 177: 1-40.

Muir, J. M., 1936, *Geology of the Tampico region, Mexico,* Tulsa, Oklahoma.

Müllerried, F. K, 1934, Estudios paleontológicos y estratigráficos en la región de Tehuacán, Puebla, *An. Inst. Biol.*, 1:1-69.

Myers, R. L., 1968, Bioestratigrafía de la Formación Cárdenas (Cretácico Superior), San Luis Potosí, *Univ. Nal. Autón. Méx., Inst. Geol., Paleont. Mex.,* 24:1-89.

Nyst., H., et Galeotti, H., 1840, Description de quelques fossiles du calcaire jurassique de Tehuacán, Mexique, *Bull. Acad. Sci. Brux.*, 28 p.

Perkins, B., 1960, Biostratigraphic studies in the Comanche (Cretaceous) series of Northern Mexico and Texas, *Geol. Soc. Am. Mem.*, 83:1-138.

Sánchez, M. A., 1997, Paleobiogeografía de equinoideos del Cretácico Superior (Maastrichtiano) de Nuevo León, San Luis Potosí y Guerrero, Unpublished Professional thesis, Universidad Nacional Autónoma de México.

Schlüter, C., 1883-1892, Die regulären Echiniden der norddeutschen Kreide. *Abh. Geol. kör. Speckarte Preuss. Thür. St.*, 4, 243 p.

Stanton, T. W., and Vaughan, T., 1896, Section of the Cretaceous at El Paso, Texas, *Am. J. Sci.*, Ser. 4, 1:21-26.

Villada, M. M., 1905, Una exploración a la Cuenca fosilífera de San Juan Raya, Estado de Puebla, *Confer. Mus. Nal. Hist. Nat. Mex.,* 44 p.

Virlet D'Aoust, T., 1865, Coup d'oeil général sur la topographie et la géologie de Mexique et de l'Amérique Centrale, *Bull. Soc. Géol. Fr.*, Ser 2, 23:14-50.

Chapter 7

Fossil Ophiuroids of Mexico

LEONORA MARTÍN-MEDRANO and PEDRO GARCÍA-BARRERA

Museo de Paleontología, Departamento de Biología Evolutiva, Facultad de Ciencias, UNAM. Ciudad Universitaria 04510, Coyoacán, México, D.F. pgb@hp.fciencias.unam.mx

1. Abstract

Here we present an up to date study on fossil ophiuroids from Mexico. The specimens come from the following States within the Mexican Republic: Baja California Sur, Chiapas, Chihuahua, Coahuila, Oaxaca and Puebla. The total number of specimens is 20 and one ophiuroid bed that bears many specimens. The ophiuroids have been collected in rocks from upper Paleozoic age (Pennsylvanian) to Pliocene. Almost all specimens are well preserved and with the aid of several techniques such as X rays and detailed images taken from SEM, they are being described. We have recognized two families and four genera all belonging to the order Ophiurida. A new species will soon be published in a research paper.

2. Introduction

Ophiuroids are one of six living classes that form the phylum Echinodermata (Hendler *et al.,* 1995), they comprise more than 250 genera and 2000 species making them the most abundant and diverse class of Echinodermata (Smith *et al.* 1995).

F. Vega et al. (eds.), Studies on Mexican Paleontology, 115–131.

Ophiuroids have a cosmopolitan distribution almost exclusively restricted today to marine environments. However a few species can survive in brackish water, for example *Ophiophragmus filograneus,* which occurs in estuaries with reduced salinities and not in the open sea (Talbot and Lawrence, 2002). Ophiuroids are found buried in sediment, inhabiting cracks or natural hollows. These echinoderms also can form symbiotic associations with corals and sponges. The lifespans of ophiuroids vary between 10 and 20 years; deep-water species may generally live longer than shallow-water ones (Donovan, 1991).

The body of ophiuroids is composed of a central disk, which can reach a diameter from 2 to 65 mm. Ophiuroids also have five mobile arms, simple or branched, which allow them to keep their normal corporal position, with the oral surface directed towards the substrate.

The success of this group is partly due to their internal skeleton (the most specialized among echinoderms), which is responsible for their versatile movements. The skeleton is composed of many calcareous ossicles connected by mesodermic tissue. In the center of the oral surface and surrounding the mouth there is a complex chewing system made up of a series of ossicles: the oral plate, oral and adoral shields, teeth and oral papilla (Figure 1). Every arm joint is composed of articulated segments including two kinds of skeletons, internal and superficial. Internal ossicles are called vertebrae, which are paired and may also be fused; and the external or superficial, composed of four plates which are named according to their position, as dorsal, ventral or lateral (Figure 1). Each lateral plate has spines with variable sizes, forms and numbers. Arms posses apical development, such that the form of plates in each arm segment is not the same. A common mistake is to give the name of vertebrae to the internal and superficial plates that form every arm joint, but according to Rasmussen (1950) the term vertebrae refers only to the internal plates in each segment of the arm. The arms used by ophiuroids to move, feed and protect themselves from predators, are very variable in their shape, in their ability to flex, in the character of their integuments and in the degree to which they are armed. It is possible to get some information about the habits of a species by analyzing its arm mobility (Litvinova, 1994).

Ophiuroids are carnivorous, suspension, filtration or detritus feeders (Barnes and Ruppert, 1996). This group of echinoderms do not have an anus nor intestine, instead they get rid of their waste materials by means of the bursal slit (Figure 1). The bursal slit is a specialized pouchlike structure located at the base of the arms and is also involved in reproduction and respiration. The nervous system of ophiuroids is not centralized, however their whole body has epithelial cells sensitive to touch (Barnes and Ruppert, 1996). In some species there is a photosensitive system at the tip of the arms made of calcite crystals soaked in soft tissue, which can detect the direction and intensity of light (Aizenberg *et al.*, 2001).

Ophiuroids are food for fishes, crustaceans, polichaetes, asteroids and other ophiuroids. Predators have intensified their activities since Jurassic times, when they diversified (Aronson, 1987). These echinoderms are capable of losing one of their arms or part of it if they feel threatened from predator or environmental stress; the lost portion of the body grows again through the development of new plates.

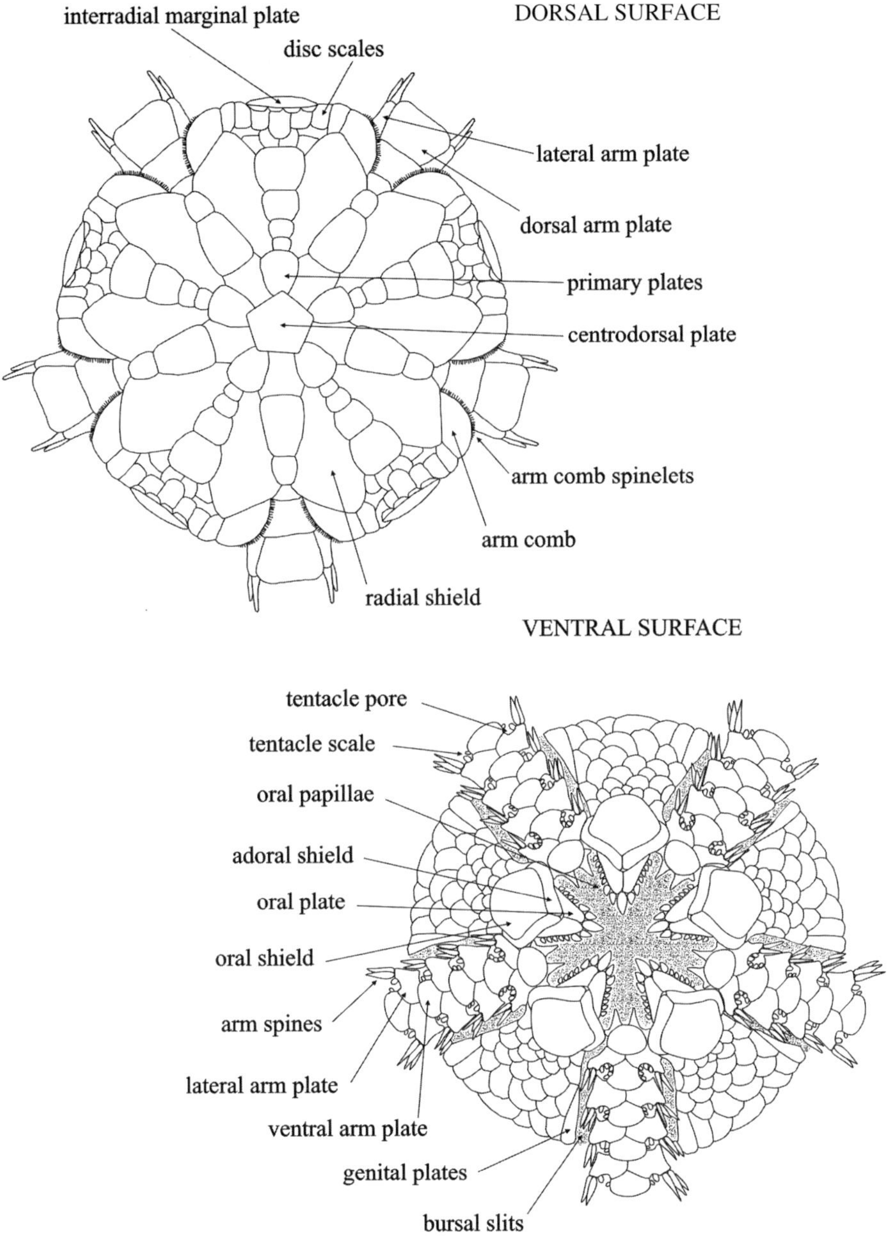

FIGURE 1. Illustrated glossary of morphological terms used in this study.

Although there is a lot of research concerning the higher taxonomy of ophiuroids (Matsumoto, 1917; Murakami, 1963; Fell, 1960; Spencer and Wright, 1966; Smith *et al.* 1995), there is still a need for a complete revision.

Spencer and Wright´s classification (1966) recognized four orders of fossil and recent ophiuroids: Stenurida, Oegophiurida, Phrynophiurida and Ophiurida. The first two orders appeared during early Paleozoic times (Spencer and Wright, 1966), and are the most primitive forms believed to have lived in shallow water, feeding on small organic suspended particles. With the exception of the family Ophiocanopidae, which belongs to order Oegophiurida, Stenurida and Oegophiurida became extinct in the Paleozoic (Bambach, 1985). Phrynophiurida and Ophiurida adopted different feeding habits. The members of the order Phrynophiurida developed branched arms, increasing the surface for food trapping. The members of the order Ophiurida had a greater mobility to their arms, which enabled them to become predators of many species, leading them to be the most abundant and diverse group of ophiuroids from the Mesozoic to present (Figure 2).

The first record of fossil ophiuroids dates back to the Ordovician, nearly 470 million years ago, with the genus *Stenaster*, which has been identified from isolated ossicles, arms, disc parts, and complete specimens collected from the United States, Canada, Ireland, Kasajstan and Scotland (Dean, 1999). Although ophiuroids are considered to be one of the most succesful groups among extant echinoderms, they

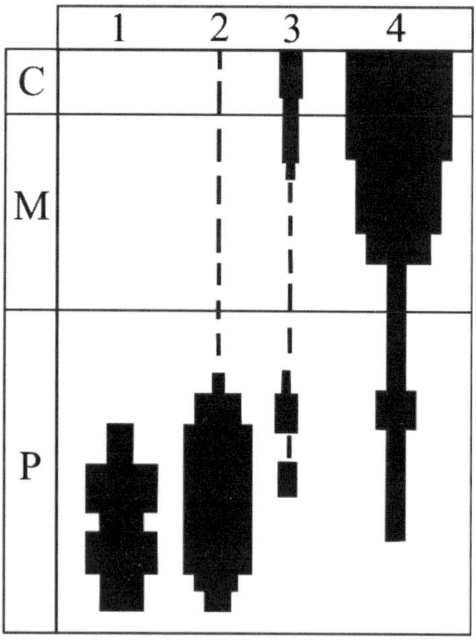

FIGURE 2. Geological ranges and diversity of the Orders within the Class Ophiuroidea. P, Paleozoic; M, Mesozoic; C, Cenozoic; 1, Order Stenurida; 2, Order Oegophiurida; 3, Order Phrynophiurida; 4, Order Ophiurida (Modified from Bambach, 1985).

almost went extinct at end of Permian. Only a small group of ophiuroids survived the Permian extinction event radiating to the numerous forms of ophiuroids found today.

Despite the fact that ophiuroid remains are quite common in Mesozoic and Cenozoic rocks, the fossil record of the ophiuroids is very scarce, when compared to Paleozoic deposits (Jagt, 1998). Ophiuroids are commonly preserved as disarticulated ossicles or fragments of discs and arms. Localities which yield more or less complete specimens are very few. However there are some extraordinary deposits around the world, called "brittlestars beds," from many geologic ages where ophiuroid specimen density is very high (Aronson and Sues, 1988). The specific identification of dissociated ossicles is often difficult, but not impossible. Ophiuroids seem to be evolutionary conservative, and it is common to recognize Mesozoic or Cenozoic taxa as representative of extant genera (Kutscher and Villier, 2003).

3. The Fossil Record of Mexican Ophiuroids

The fossil record of Mexican ophiuroids is relatively scarce when compared to other groups of echinoderms. The analysis of ophiuroids diversity and geologic distribution (Table 1, Figure 3) was recently published by Martín-Medrano (2003).

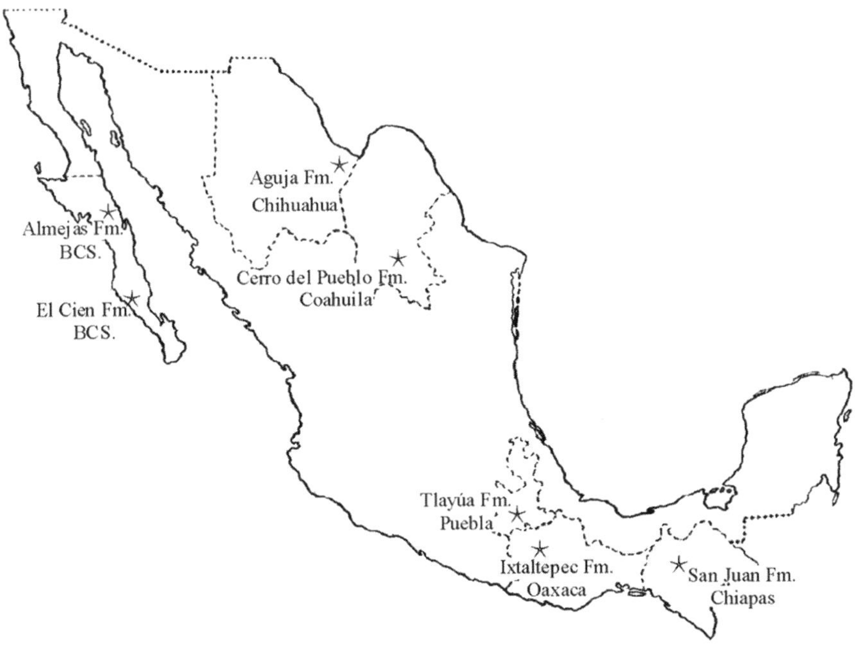

FIGURE 3. Geographical distribution of outcrops containing fossil ophiuroids in Mexico.

3.1. Ixtaltepec Fm., Oaxaca

Quiroz-Barroso and Sour-Tovar (1995) briefly described a Paleozoic ophiurinid from Ixtaltepec Formation (Pennsylvanian), Oaxaca. This is the oldest ophiuroid found in Mexico. The specimen is deposited at the Paleontology Museum, Facultad de Ciencias, U.N.A.M., and is represented by negative impressions of the dorsal and ventral surfaces.

The disc is pentagonal. Oral plates and adoral shields are very large and thin. The arms are large and thin without dorsal plates; the ventral arm plates are inconspicuous, while the tentacle pores are very wide (Quiroz-Barroso and Sour-Tovar, 1995).

This specimen represents the most austral finding in Mexico.

3.2. Tlayúa Fm., Puebla

The occurrence of Cretaceous ophiuroids at Tlayúa Quarry, Puebla was the first report of ophiuroids from Mexico (Applegate, 1987; Buitrón-Sánchez and Malpica-Cruz, 1987). The occurrences were also later confirmed by several additional workers (Applegate, 1996; Feldmann *et al.* 1998; Espinosa-Arrubarrena *et al.* 2000). However, no detailed descriptive work has been done with these specimens. In addition, Buitrón and Solís-Marín (1993) indicate the existence of a Cretaceous ophiuroid specimen, genus *Ophiura* collected in Puebla, but they do not give enough information to clarify the exact taxonomic and geographical placement of this specimen.

To date, thirteen different fossil ophiuroids from Tlayúa Quarry (Lower Cretaceous, Albian) have been collected and are under study by the first author of this work. Twelve of them are part of the Colección Nacional de Paleontología, Instituto de Geología, U.N.A.M. and the remaining specimen is deposited at the Paleontology Museum, Facultad de Ciencias, U.N.A.M. These are complete specimens, five of them with the dorsal surface exposed and eight with the ventral surface exposed.

The Tlayúa Quarry´s specimens show affinities with representatives of the *Ophiura* genus. The dorsal and ventral surfaces of the ophiuroid disc is covered by overlapping small scales of very different size and shape. The spinelets of the arm comb are very few and small. There are enlarged mouth slits with some small and thin teeth. The shape of the proximal dorsal arm plates is rhomboidal, while the distals are triangular. Ventral arm plates are pentagonal with 1 to 4 tentacle scales. The lateral arm plates are long with 3 to 5 spines (Figure 4).

In some specimens the arms are extended and oriented in different ways, in several others they are contracted under the central disk; these postures can help to infer the behavior of the animal before death. The excellent preservation of these specimens enables fine details of their anatomy to be easily seen (Figure 4). Environmental conditions that account for sediment deposition at Tlayúa Quarry have been interpreted in several different ways: Malpica-Cruz *et al.* (1989) and Pantoja-Alor (1992) suggest that the deposit was formed at a post-reef lagoon connected with the open sea;

FIGURE 4. Oral surface of ophiuroid specimen from Tlayúa Quarry in Puebla deposited at the Paleontology Museum, Facultad de Ciencias, U.N.A.M.

whereas, Applegate (1987) and Espinosa-Arrubarrena and Applegate (1996) present an up to date model, where they suggest an extremely shallow post-reef, stagnant, anaerobic and hypersaline lagoon, surrounded on one side by semi-arid land and on the other side by a deeper, well oxygenated lagoon with a high biodiversity; finally, Kashiyama *et al.* (2004) suggest a deposit formed within deep sea conditions (see Applegate *et al.* in this book).

3.3. Aguja Fm., Chihuahua

Martin-Medrano in 2003 reported the first record of an ophiuroid belonging to the family Amphiuridae from the Cretaceous of Chihuahua. In that work, she only briefly described it, however the complete description of that fossil is one of the subjects in her Masters Degree Thesis.

The complete specimen is represented by the ventral surface with two extended arms exposed and the distal parts of the other three arms folded near the central disk. The arms are thin and flexible. The ventral arm plates are pentagonal; the lateral arm plates are wider at the base with 2 to 3 large and thin spines. The mouth slits are very wide with small oral papilla. The oral shields are triangular; the bursal slits are large and convex (Figure 5). This specimen shows affinities with representatives of the genus *Amphiura*.

This specimen was collected at Aguja Formation, Chihuahua, in sediments considered to be transitional brackish, marine and fluvial environments (Lehman, 1997). However considering the marine nature of the ophiuroids, probably the deposit resembles a marine entrance in a coastal flood plain. The specimen is deposited at the Paleontology Museum, Facultad de Ciencias, U.N.A.M.

3.4. Cerro del Pueblo Fm., Coahuila

Only one specimen of ophiuroid has been reported from Cerro del Pueblo Formation, Coahuila (Martin-Medrano, 2003) (Figure 6). It is currently being studied by the same author. Due to permineralization the dorsal surface of this specimen is very well preserved. Unfortunately this specimen is incomplete, lacking one arm and the distal portions of the others. It shows affinities with representatives of the *Ophiomusium* genus. The dorsal surface of the disc is almost completely covered by the radial shields, primary and interradial marginal plates. Middle and distal dorsal arms plates are absent. Lateral arm plates are wing-shaped and cover almost completely the surface of the arm (Figure 6). The specimen is deposited at the Paleontology Museum, Facultad de Ciencias, U.N.A.M.

The Cerro del Pueblo Formation has been interpreted as a deltaic environment in a coastal flood plain, the deposit resembles a mixture of fresh, brackish and marine waters without any barriers to block the terrestrial vertebrate migrations (Kirkland *et al.* 2000).

3.5. San Juan Fm., Chiapas

A complete specimen of ophiuroid assigned to the family Ophioleucidae was briefly described by Martín-Medrano (2003), it belongs to the San Juan Formation, Chiapas, and has remarkably preserved extended arms and ventral surface (Figure

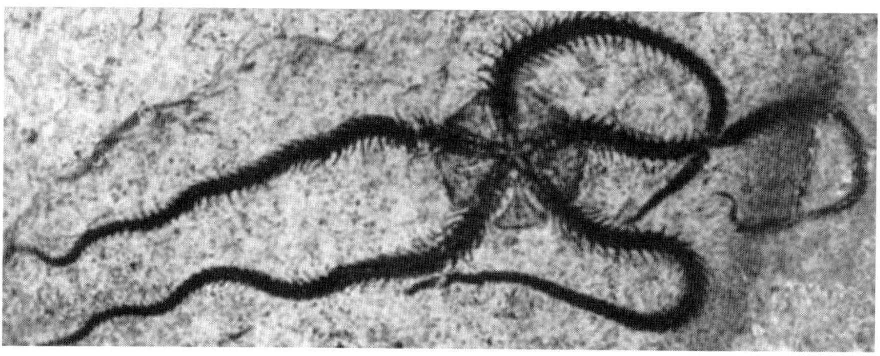

FIGURE 5. Oral surface of the amphiuroid found in Chihuahua (Martín-Medrano, 2003).

FIGURE 6. Dorsal surface of the ophiuroid specimen from Coahuila reported by Martin-Medrano, 2003.

7). Calcium carbonate preservation enables observation of tridimensional features of this specimen.

The shape of disc is pentagonal; bursal slits large. The mouth slit is very wide. The distal parts of oral plates are in contact. Ventral arm plates are rhomboidal. The lateral arm plates are very wide and cover almost completely the arm surface; two spines are present (Figure 7).

The San Juan Formation is believed to represent an open platform, shallow marine environment (Aguilar-Piña, 1993) with terrestrial influx. The specimen is deposited at the Paleontology Museum, Instituto de Historia Natural del Estado de Chiapas.

3.6. El Cien and Almejas Formations, Baja California Sur

The ophiuroids reported from El Cien Formation, in Baja California Sur by Martin-Medrano (2003) are represented by impressions of three specimens with exposed ventral surfaces and complete body disks; however the distal portions of the arms are absent (Figure 8). The specimens are deposited at the Paleontology Museum, Facultad de Ciencias, U.N.A.M. The detailed description of these specimens is being completed by the same author.

The interradial margins of the disc are convex. The oral shields cover almost all the interradial area of the disc; they are large and pyrifoms. Bursal slits large and convex. The arms are short with the widest part at the base and extremely acute at the tips. Ventral arm plates small and quadrangular; lateral arm plates curved with short

FIGURE 7. Ventral surface of the ophiuroid specimen assigned to the family Ophioleucidae found in Chiapas by Martin-Medrano (2003).

spines compressed to the body arm (Figure 8). These fossil ophiuroids show affinity with the representatives of the *Stegophiura* genus.

The Late Oligocene-Middle Miocene El Cien Formation has been interpreted as a transitional environment from terrestrial to deep sea facies (Applegate, 1986), but the deposit where the fossil ophiuroids have been found is characterized by a stable shallow platform.

Buitrón *et al.* (1994) reported the presence of Pliocene brittlestar beds within

FIGURE 8. Ventral surface of three ophiuroid specimens from El Cien Formation, Baja California Sur.

the Almejas Formation, Baja California Sur. The brittlestar layer is composed of sediments interpreted as a shallow marine environment deposited under normal salinity conditions (Buitrón *et al.* 1994). More than one hundred specimens were collected and assigned to the genus *Ophiura*. However there is no detailed description of these fossils in the literature, and the final taxonomic placement of these specimens remains uncertain.

The fossil record of Mexican ophiuroids is represented by four genera and two families all belonging to the order Ophiurida. These fossils cover a wide stratigraphic range from the Pennsylvanian to the Pliocene (Table 1, Figure 3). It is important to note that fossil ophiuroids of Mexico have been, up to now, reported by Mexican investigators who work and study at the Universidad Nacional Autónoma de México, where the majority of specimens are deposited.

4. Present Studies

Taxonomic determination of fossil ophiuroids is difficult because the dorsal and ventral complete surfaces are rarely preserved in the fossil record; even if fossil ophiuroids are well preserved it is still difficult to make an accurate taxonomic placement (Cornell *et al.* 1991). Accurate taxonomic assignment of Mexican fossil ophiuroids is especially difficult due to the small number of known specimens from Mexico. This is exemplified when there is only one specimen, of a given new taxonomic position, that has only the one surface, ventral or dorsal, exposed when both surfaces are needed for taxonomic assignment. This is true with specimens from: Cerro del Pueblo Formation, Coahuila; Aguja Formation, Chihuahua; San Juan Formation, Chiapas, and El Cien Formation, Baja California Sur. However good preservation of these fossils does allow the observation of some diagnostic structures of the disk and arms.

The fragility of fossil ophiuroids, caused by the degradation of organic matter, provokes a rapid disarticulation of the skeleton. Isolated ossicles should be represented more often in the fossil record than the relatively rare whole body (Hotchkiss *et al.* 1999a and b). There are no studies on mexican ophiuroids that report the finding of skeletal ossicles. In addition, there is no report in the literature that refers to the finding of marks or tracks in sediments that would indicate activity by ophiuroid echinoderms in Mexico (Bell, 2004).

At present the study of Mexican fossil ophiuroids is part of a Masters Degree project by Martín-Medrano at the Facultad de Ciencias, U.N.A.M. This project includes all known specimens of ophiuroids (Table 1), with the exception of those reported by Quiroz-Barroso and Sour-Tovar (1995) and Buitrón *et al.* (1994). With this Masters research it is hoped that more elements to refine the taxonomic determination of Mexican fossil ophiuroids can be accomplished. In addition this study will include isolated skeletal ossicles and in the future, when the number of specimens and the preservation of material permit it, some aspects of evolution, ethology, functional

TABLE 1. Fossil record of ophiuroids collected in Mexico. Abbreviations: FCMP, Museo de Paleontología, Facultad de Ciencias, UNAM; IGM, Colección Nacional de Paleontología, Instituto de Geología, UNAM; MPCH, Museo de Paleontología, Chiapas.

TAXA	FORMATION, STATE and AGE	COLLECTION	REFERENCE
Ophiura	Almejas Fm, BCS, Pliocene	?	Buitrón et al. 1994.
aff. *Stegophiura*	El Cien Fm, BCS, Late Oligocene	MPFC	Martin-Medrano (pers. obs.)
Fam. *Ophioleucidae*	San Juan Fm, Chiapas, Middle Eocene	MPCH	Martin-Medrano, 2003
aff. *Ophiomusium*	Cerro del Pueblo Fm, Coahuila, Late Cretaceous	MPFC	Martín-Medrano (pers. obs.)
aff. *Amphiura*	Aguja Fm, Chihuahua, Late Cretaceous	MPFC	Martin-Medrano (pers. Obs.)
aff. *Ophiura*	Tlayúa Fm, Puebla, Early Cretaceous (Albian)	IGM, MPFC	Martín-Medrano (pers. obs.)
Fam. Ophiurinidae	Ixtaltepec Fm, Oaxaca, Pennsylvanian	MPFC	Quiróz-Barroso and Sour-Tovar, 1995

anatomy, taphonomy, paleoecology and paleobiogeography.

Complete and well preserved fossil ophiuroid specimens can be related to extant families and genera. Therefore the Masters research will include a detailed comparative analysis of fossil and extant ophiuroids, including some families among the 10,000 specimens deposited at the Colección Nacional de Equinodermos de México of the Instituto de Ciencias del Mar y Limnología, U.N.A.M., and the recent ophiuroids Collection at the Natural History Museum of Los Angeles County.

Preperation of ophiuroid specimens is done by chemical and mechanical treatment, which enables the fine anatomy of the fossil ophiuroid to be viewed under the stereoscopic microscope, scanning electron microscope (Figure 9) and X-rays (Figure 10).

5. Other Fossil Echinoderms of Mexico

The degree and quality of preservation among the classes of echinoderms is based upon, in part, to the proportion of organic and inorganic matter that composes them. These differences cause a bias in the fossil record, favoring those forms with greater quantities of calcareous skeletal material and narrower sutures among plates.

This is reflected in the fossil record of echinoderms all over the world (Boardman *et al.*, 1987) including Mexican fossil echinoderms.

Echinoids are the group of echinoderms best represented in Mexico. In 1989

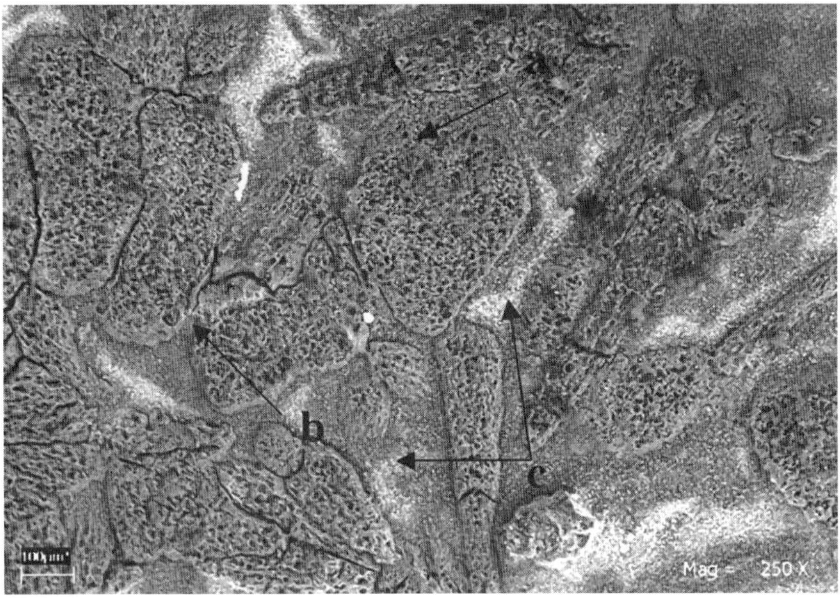

FIGURE 9. Image obtained with scanning electron microscope (250x). Detail of dorsal plates (a), lateral plates (b) and spines (c) of the arms of an ophiuroid fossil from Puebla.

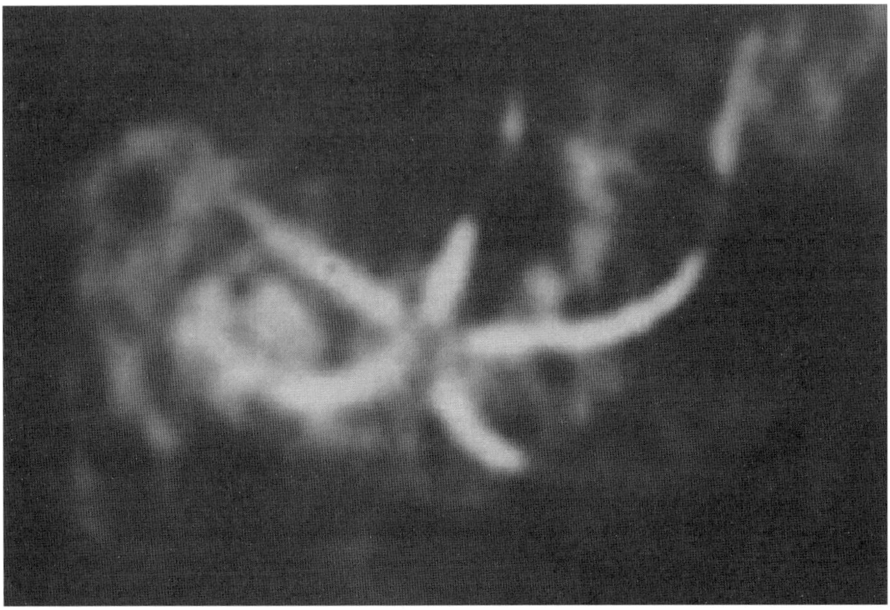

FIGURE 10. Image obtained with X-ray technique of the fossil ophiuroid from the San Juan Formation, Chiapas.

there were 33 Mesozoic and 34 Cenozoic type species described from Mexico (Perrilliat, 1989). In 1993 the numbers reached 186 Mesozoic and 117 Cenozoic species described from the states of Baja California, Sonora, Chihuahua, Coahuila and Nuevo León (Buitrón and Solís-Marín, 1993). Nieto-López and García-Barrera (in this volume) report 150 species of cretaceous echinoids from México. This diversity is closely related to the ossified calcareous material that composes the echinoids body morphology, which allows for the preservation of complete organisms in the fossil record.

The fossil record of Mexican crinoids is represented by only a few species. Up until 1989 there were only 1 Paleozoic and 3 Mesozoic type species described (Perrilliat, 1989). Four years later the number had grown to 11 Paleozoic and 4 Mesozoic species collected from the states of Puebla, Guerrero, Oaxaca, Tamaulipas, Hidalgo, Chiapas and Coahuila (Buitrón and Solís-Marín, 1993). The crinoid body consists of tightly sutured calcareous plates making them good candidates for fossil preservation.

In the class Holothuroidea there are seven complete Cretaceous species collected from Tlayúa Quarry, Puebla, Mexico (Applegate *et al.* 1996). Five of these species are closely related to the Family Psolidae, one of them belongs to the Family Phyllophoridae and the last species belongs to a new family (Buitrón *et al.,* 1995a). A new genus and species, *Ocuilinpilli faustinoi*, was also described in 1995, representing the second fossil record in the world for the Family Caudinidae (Buitrón *et al.*, 1995b). One year later, Applegate *et al.* (1996) reassigned the holothuroids from Tlayúa Quarry, indicating that four of the seven specimens belong, or are closely related, to the Family Psolidae; one specimen is part of the Family Sclerodactylidae; one specimen represents a new Family; and the last specimen was placed in an *incertae sedis* Family. Further research is needed to correctly place these seven species into exact taxonomic assignment. These findings constitute the first record of complete fossil holothuroids in the American continent, the first record of complete skeletal holothuroids of Cretaceous age and the first record of this class of echinoderms in Mexico. The seven specimens are deposited in the Colección Nacional de Paleontología, Instituto de Geología, U.N.A.M.

Although holothuroids are well represented in the fossil record of Mexico the preservation of these specimens was favored by the exceptional conditions of the sedimentary environment in which they were deposited (see Applegate *et al.* in this book). There are only six localities worldwide where complete holothuroids have been previously reported (Buitrón *et al.* 1995a). Holothuroids are usually only represented by isolated ossicles because most of the body is composed of organic material, which is rapidly degraded.

The fossil record of Mexican asteroids is limited to one Cretaceous species from Puebla, Mexico (Buitrón and Solís-Marín, 1993); however, there is no information concerning the location of where it was found or where it was accessioned. Cretaceous asteroids have been mentioned by several workers from Tlayúa Quarry,

Puebla, however none of them are formal descriptive publications (Applegate, 1987; Applegate, 1996; Feldmann *et al.* 1998; Espinosa *et al.* 2000; and Kashiyama *et al.* 2004); Applegate *et al.* (in this volume) mention the presence of at least two or three forms of asteroids in the Tlayúa Quarry, that have not been classified. The asteroids body is formed by numerous calcareous ossicles sutured by organic matter that breaks down after burial, which facilitates the disgregation of the ossicles. Therefore preservation of complete fossil asteroids is rare.

6. Conclusions

Within extant forms of ophiuroids, the number of fossil specimens is relatively scarce (only 25% of all known genera have been found as fossils, Boardman *et al.*, 1987). The rarity of fossil ophiuroids is due to the fragility and composition of the skeleton. Also, it must be noted, that the scarcity of reports of fossil echinoderms in Mexico is probably due to the lack of specialists in this group. However, fossil ophiuroids have a wide stratigraphic range in Mexico. The first author of this chapter catalogued six forms of fossil ophiuroids, from seven different formations in Chihuahua, Coahuila, Puebla, Oaxaca, Chiapas and Baja California Sur, with ages ranging from Pennsylvanian to Pliocene. At present, the same author is preparing some of those fossils to refine their taxonomic identity by using comparative analysis with recent organisms, and innovative techniques like scanning electron microscopy and x-rays; v. gr. the ones collected from Tlayúa, Aguja, Cerro del Pueblo, San Juan and El Cien Formations.

These fossil echinoderms contribute important information about sedimentary environments, behavior and community ecology. Due to the preservation bias in the fossil record, ophiuroids may have always been as abundant and diverse in the fossil record as they are today. However, for now, ophiuroids are considered a low diversity fossil group in Mexican paleontological records.

Systematic work in Mexico concerning this group is reaching maturity, and it is expected, with ongoing research, that the fossil record of Mexican ophiuroids will continue growing, taxonomically, and in other important aspects such as behavior, evolution, taphonomy, paleoecology and paleobiogeography, and from those studies, finally, this important group of echinoderms could be better understood in Mexico.

References

Aguilar-Piña, M., 1993, Bioestratigrafía general del Terciario (Paleógeno) de la Localidad El Jobo, Tuxtla Gutiérrez, Chiapas, Unpublished Professional thesis, Universidad Nacional Autónoma de México.

Aizenberg, J., Tkachenko, A., Weiner, S., Addadi, L., and Hendler, G., 2001, Calcitic microlenses as part of the photoreceptor system in brittlestars, *Nature*, 412:819-822.

Applegate, S. P., 1986, The El Cien Formation, Strata of Oligocene and Early Miocene age in Baja California Sur, *Univ. Nal. Autón. Méx., Inst. Geol., Rev.* 6:145-162.

Applegate, S. P., 1987, A preliminary study of the Tlayúa Quarry near Tepexi de Rodríguez, Puebla, *Rev. Soc. Mex. Paleont.,* 1:40-50.

Applegate, S. P., 1996, An Overview of the Cretaceous fishes of the Quarries near Tepexi de Rodríguez, Puebla, México, in: *Mesozoic Fishes-Systematics and Paleoecology* (Arratia, G., and Viohl, G., eds.), 529-538.

Applegate, S., Buitrón.Sánchez, B. E., and Solís-Marín, F. A., 1996, Seven new taxa of holothurians (Holothuroidea: Echinodermata) from the Tlayua Quarries near Tepexi de Rodríguez, Puebla, México Lower Cretaceous (Albian) age. *XIII Convención Geológica Nacional*, Libro de resúmenes, 51-52.

Aronson, R., 1987, Predation on fossil and Recent ophiuroids, *Paleobiol.* 13:187-192.

Aronson, R., and Sues, H., 1988, The fossil record of Brittlestars beds, in: *Echinoderm biology* (Burke, R., Mladenov, P., Lambert, P., and Parsley, R., eds.), A.A. Balkema Publishers, Netherlands, 147-148.

Bambach, R., 1985, Classes and Adaptive Variety: The Ecology of Diversification in Marine Faunas through the Phanerozoic, in: *Phanerozoic Diversity Patterns, Profiles in Macroevolution (*Valentine, J., ed.), Princeton University Press, Princeton, 191-253.

Barnes, R., and Ruppert, E., 1996, *Zoología de los Invertebrados,* McGraw-Hill Interamericana, México, 1114 p.

Bell, C. M., 2004, Asteroid and Ophiuroid trace fossils from the Lower Cretaceous of Chile, *Palaeontol.* 47:1-66.

Boardman, R., Cheetham, A., and Rowell, A., 1987, *Fossil Invertebrates.* Blackwell Science Editorial, England, 713 p.

Buitrón, B., and Solís-Marín, F. A., 1993, La Biodiversidad en los equinodermos Fósiles y Recientes de México, in: *Diversidad Biológica en México* (Gío-Argáez, R., y López-Ochoterena, E., eds.), *Univ. Nal. Autón. Méx., Inst. Cienc. Mar Limnol.,* 209-231.

Buitrón, B., Solís-Marín, F. A., Miranda, J. C., and Miranda, J. B., 1994, El Hallazgo de un Ofiuroidea Pliocenico de la Región de Vizcaíno, Baja California Sur, *XII Convención Geológica Nacional, Soc. Geol. Mex.,* Libro de Resúmenes, p. 26-27.

Buitrón-Sánchez, B. E., Applegate, S. P., and Solís-Marín, F. A., 1995a, El primer hallazgo de holoturias en México, procedentes de la Cantera Tlayúa (Albiano) de Tepexi de Rodríguez, Puebla, *V Congreso Nalcional de Paleontología*, Memoria, 5:3-4.

Buitrón-Sánchez, B. E., and Malpica-Cruz, R., 1987, Tepexi de Rodríguez, Puebla, Una localidad fosilífera famosa de México, *Primer Congreso Nacional de Paleontología,* Libreto guía de la excursión, p. 24.

Buitrón-Sanchez, B. E., Solís-Marín, F. A., and Applegate, S., 1995b, Una nueva especie de holoturia (Echinodermata: Holothuroidea) del Cretácico de Tepexi de Rodríguez, Puebla, *XIII Congreso Nacional de Zoología*, Programa y resúmenes, p. 16.

Cornell, W., Lemone, D., and Norland, W., 1991, Albian ophiuroids from Cerro de Cristo Rey, Dona Ana County, New Mexico, *J. Paleontol.,* 65:1009-1013.

Dean, J., 1999, What makes an ophiuroid? A morphological study of the problematic Ordovician stelleroid *Stenaster* and the paleobiology of the earliest asteroids and ophiuroids, *Zool. J. Linn. Soc.*, 126:225-250.

Donovan, S. K., 1991, The taphonomy of echinoderms: calcareous multi-element skeletons in the marine environment, in: *The Processes of Fossilization* (Donovan, S.K.,ed.), Belhaven Press, London, 241-269.

Espinosa-Arrubarrena, L., and Applegate, S. P., 1996, A paleoecological model of the vertebrate bearing beds in the Tlayúa quarries, near Tepexi de Rodríguez, Puebla, México, in: *Mesozoic Fishes – Systematics and Paleoecology* (Arratia, G., and Viohl, G., eds.), Verlag Dr. Friedrich Pfeil, Munchen, Germany, 539-550.

Espinosa-Arrubarrena, L., Montellano-Ballesteros, M., and Applegate, S., 2000, Paleontological highlight of the Mixteca Poblana in Central Mexico, in: *Guidebook of the field trips: Society of Vertebrate Paleontology* (Carranza-Castañeda, O., ed.), *Univ. Nal. Autón. Méx, Inst. Geol.,* 85-130.

Feldmann, R., Vega, F., Applegate, S., and Bishop, G., 1998, Early Cretaceous Arthropods from the Tlayúa Formation at Tepexi de Rodríguez, Puebla, México, *J. Paleontol.*, 72:79-90.

Fell, H. B., 1960, Synoptic keys to the genera of Ophiuroidea. *Victoria Univ. Wellington, Zool. Publ.,* 26:1-44.

Hendler, G., Miller, J., Pawson, D., and Kier, P., 1995, *Sea Stars, Sea Urchins, and Allies, Echinoderms of Florida and the Caribbean,* Smithsonian Institution Press, Washington, 390 p.

Hotchkiss, F., Prokop, R., and Petr, V., 1999a, Isolated skeletal ossicles of a new brittlestar of the Family Cheiropterasteridae Spencer, 1934 (Echinodermata: Ophiuroidea) in the Lower Devonian of Bohemia (Czech Republic), *J. Czech Geol. Soc.*, 44: 121-214.

Hotchkiss, F., Prokov, R., and Petr, V., 1999b, Isolated Vertebrae of Brittlestars of the Family Klasmuridae Spencer, 1925 (Echinodermata: Ophiuroidea) in the Devonian of Bohemia (Czech Republic), *J. Czech Geol. Soc.*, 44:329-333.

Jagt, J. W. M., 1998, Fossil explained 21: Post-Paleozoic ophiuroids, *Geology Today*, March-April:77-80.

Kashiyama, Y., Fastovsky, D. E., Rutheford, S., King, J., and Montellano, M., 2004, Genesis of a locality of exceptional fossil preservation: paleoenvironments of Tepexi de Rodríguez (mid-Cretaceous, Puebla, México), *Cret. Res.*, 25:153-177.

Kirkland, J. I., Hernández, R., Aguillón, M. C., Delgado de Jesús, C. R., Gómez, R., and Vallejo, I., 2000, The Late Cretaceous Difunta Group of the Parras Basin, Coahuila, México, and its vertebrate fauna, in: *Guide book of the field trips: Society of Vertebrate Paleontology* (Carranza-Castañeda, O., ed.), *Univ. Nal. Autón. Méx., Inst. Geol.*, 133-172.

Kutscher, M., and Villier, L., 2003, Ophiuroid remains from the Toarcian of Sainte-Verge (Deux-Sevres, France): paleobiological perspectives, *Geobios*, 36:179-194.

Lehman, T. M., 1997, Late Campanian dinosaur Biogeography in the Western Interior of North America, in: *DinoFest International* (Wolberg, D.L., Stump, E., and Rosenberg, G.D., eds.), *P. Acad. Nat. Sci. Phila.*, 223-240.

Litvinova, N. M., 1994, The life forms of ophiuroidea (based on the morphological structures of their arms), in: *Echinoderms Through time* (David, Guille, Féral and Roux, eds.), Balkema, Rotterdam.

Malpica-Cruz, V., Pantoja-Alor, J., and Galguera-Rosas, G., 1989, Microfacies de la Cantera Tlayúa, Puebla, México. *UNAM, Inst. Geol. lll Simposio Geología Regional, México, Memoria,* 48-50.

Martin-Medrano, L., 2003, El registro de ofiuroideos fósiles de México y la descripción de dos nuevas formas para el Cretácico y Terciario, Unpublished Professional thesis, Universidad Nacional Autónoma de México.

Matsumoto, H., 1917, A monograph of Japanese Ophiuroidea arranged according to a new classification, *J. Coll. Sci., Imp. Univ. Tokyo*, 38:1-408 .

Murakami, S., 1963, The dental and oral plates or the Ophiuroidea, *Royal Soc. New Zealand, Trans.*, 4:1-48.

Pantoja-Alor, J., 1992, Geología y paleoambientes de la Cantera Tlayúa, Tepexi de Rodríguez, Estado de Puebla, *Univ. Nal. Autón. Méx., Inst. Geol. Rev.*, 9:56-176.

Perrilliat, M. C., 1989. Echinodermata, *in: Fósiles Tipo Mexicanos* (Perrilliat, M.C., y Carreño, A.L., eds.), *Univ. Nal. Autón. Méx., Inst. Geol., p. 399-413*

Quiroz-Barroso, S., and Sour-Tovar, F., 1995, Nuevo Registro de Ofiuroideo (Ophiurinidae) para el Pensilvánico de América del Norte, proveniente de la Formación Ixtaltepec, Oaxaca, *Memoria del V Congreso Nacional de Paleontología, Resúmenes, Soc. Mex. Paleont.*, p. 31.

Rasmussen, H. W., 1950, Cretaceous Asteroidea and Ophiuroidea, with special reference to the species found in Denmark, *Danm. Geol. Unders., Ser* 2, 77:1-134 .

Smith, A. B., Paterson, G. L. J., and Lafay, B., 1995, Ophiuroid phylogeny and higher taxonomy: morphological, molecular and paleontological perspectives, *Zool. J. Linn. Soc.*, 114:213-243.

Spencer, W. K., and Wright, C. W., 1966, Asterozoans, in: *Treatise on Invertebrate Paleontology, Part U: Echinodermata 3* (Moore, R.C., ed.), *Geol. Soc. Am. and Lawrence, Kansas, Univ. Kansas Press,* U4-U107.

Talbot, T. D., and Lawrence, J. M., 2002, The effect of salinity on respiration, excretion, regeneration and production in Ophiophragmus filograneus (Echinodermata: Ophiuroidea), *J. Exp. Mar. Biol. Ecol.,* 275:1-14.

Chapter 8

Fossil record of Upper Paleozoic Marine Invertebrates from Mexico

SARA A. QUIROZ-BARROSO and FRANCISCO SOUR-TOVAR

Museo de Paleontología, Departamento de Biología Evolutiva, Facultad de Ciencias, UNAM. Ciudad Universitaria 04510, Coyoacán, México, D. F. saqb@hp.fciencias.unam.mx

1. Abstract

Upper Paleozoic fossil invertebrates of Mexico represent communities of marine environments associated with continental margins. A high percentage of very diverse brachiopods and mollusks, accompanied by other lesser diverse groups such as echinoderms, arthropods, bryozoans and corals occur throughout the Upper Paleozoic sequence of Mexico. Outcrops of Carboniferous and Permian rocks are especially abundant and Mexican fossil faunas from those systems exhibit an affinity to similar age fossil faunas from the rest of North America; this is especially true in the youngest faunas. This similarity may be explained by the configuration and geographic position of the Late Paleozoic crustal portions that form modern day Mexico. During the Late Paleozoic Pangea was forming and the ocean that existed between Gondwana and the northern continents was narrowing: during the entire Carboniferous and Early-Middle Permian, that ocean covered large regions of North America and extended over the majority of what constitutes present-day Mexico.

F. Vega et al. (eds.), Studies on Mexican Paleontology, 133–167.

2. Introduction

Throughout the territory of Mexico, outcrops of rocks from all periods of the Paleozoic are represented. Stratigraphic sequences of the Carboniferous and Permian are especially abundant (Figure 1). Carboniferous and Permian sequences contain mostly biotic communities that developed in marine environments associated with continental margins. The diversity of the biotic communities consists of representatives of most of the known invertebrate phyla. It is important to note that a large part of that diversity has still not been formally described. The research that has been published indicates that the most abundant and diverse groups for the Upper Paleozoic of Mexico are brachiopods and mollusks, followed by echinoderms and arthropods. Other groups, such as bryozoans and corals, are also common, but have not received much scientific study. Despite this, the information that has been generated regarding the invertebrates of the Upper Paleozoic of Mexico has established a similarity between the marine faunas from Mexico and those from the rest of North America. This similarity corresponds to the ideas that have been proposed with respect to the geographic configuration of the Earth and the changes that took place during the Carboniferous-Permian. The objective of this work is to present a synthesis of the existing knowledge concerning invertebrate fossils that have been reported for the Upper Paleozoic of Mexico.

3. Tectonic Framework

The tectonic evolution of Mexico is a peculiar and difficult problem. Although some aspects of the tectonic evolution of Mexico kinematic history are relatively well known, for example, the collision between North and South America during the Late Paleozoic, other aspects, such as the origin and date of the accretion of the terranes that constitute the region, are little understood. This disparity is due, in large measure, to the enormous extension and volume of Cenozoic volcanism in Mexico, which places limitations upon gaining access to the more ancient rocks and other structures necessary to establish the spatial-temporal evolution of the territory that Mexico encompasses (Ortega-Gutiérrez *et al.* 2000).

Existing information indicates that at least part of Mexico had its origins in various allochtonous tectonic terranes, floored by Proterozoic metamorphic basements. For example, the sedimentary rocks of northwestern Mexico form part of the Chihuahua, Caborca and Cortez terranes and are related to the evolution of the margins to the west and south of Laurentia. On the other hand, the stratigraphic sequences of the Early-Middle Paleozoic from central and southern Mexico are very different than those of North America, suggesting that these terranes evolved independently of Laurentia, at least until the Early Carboniferous (Sánchez-Zavala *et al.* 1999).

FIGURE 1. Upper Paleozoic localities of Mexico. Baja California: 1. Near San Felipe, Permian; 2. Northern Sierra Pinta, south of Laguna Salada, Carboniferous?; 3. Arroyo Zamora, near El Volcán, Zamora, Permian (Leonardian?); Chiapas: 4. Aguacate River, northwestern Chicomuselo, Santa Rosa Inferior Formation, Pennsylvanian; 5. Comalapa River-Cuilco River-Cashú area, southeastern Chicomuselo, Paso Hondo Formation, Middle Permian; 6. Monte Redondo-Guadalupe Victoria area, Grupera Formation, Lower Permian; Chihuahua: 7. Mina Plomosas-Placer de Guadalupe area, Monillas Formation (Upper Member), Upper Mississippian to Lower Pennsylvanian; Coahuila: 8. Las Delicias Valley, Las Sardinas Formation Lower Permian (Leonardian); Palo Quemado, La Difunta and La Colorada formations, Upper Permian (Wordian, Capitanian and Amarissian); Guerrero: 9. Olinalá, Los Arcos/Olinalá Formation, Middle Permian ?; Hidalgo: 10. Otlamalacatla, Tuzancoa Formation, Lower Permian; 11. Pemuxco-Mimiaguaco, Tuzancoa (not Guacamaya) Formation, Permian (Upper Wolfcampian-Lower Leonardian); 12. Calnali, Vicente Guerrero Formation, Mississippian; Del Monte Formation, Pennsylvanian; Oaxaca: 13. Santiago Ixtaltepec, Santiago Formation, Mississippian (Osagean); Ixtaltepec Formation, Lower-Middle Pennsylvanian; Puebla: 14. San Salvador Patlanoaya, Patlanoaya Formation, Lower Mississippian- Permian; Sonora: 15. Near Bisani, Mississippian; 16. Sierra del Tule, Caliza Escabrosa?, Mississippian (Kinderhookian-Osagean); Horquilla Formation, Pennsylvanian (Atokan-Virgilian); Earp Formation, Permian (Wolfcampian); 17. El Antimonio, Monos Formation, Permian; 18. North of Cananea, (Colina Limestone), Permian (Wolfcampian-Leonardian); Tamaulipas: 19. Peregrina Canyon, north of Ciudad Victoria, Vicente Guerrero Formation, Lower Mississippian.

4. Communities of the Upper Paleozoic

The Upper Paleozoic marine faunas of Mexico represent shallow water communities. It is also possible that turbidity currents to deeper zones of the marine environment could have transported some of them; such has been the interpretation given recently to the Pennsylvanian fauna from the Ixtaltepec Formation within the Mexican state of Oaxaca (Buitrón *et al.* 2000) and the Permian fauna from the Tuzancoa within the Mexican state of Hidalgo (Sour *et al.*, 2005b; Rosales-Lagarde *et al. in press*). Comparing the fossil associations in which these faunas are found with those that have been described for the Lower Paleozoic (Cooper *et al.* 1954; Robison and Pantoja, 1968; Sour-Tovar, 1990), it is notable that the patterns that can be observed in the faunas from Mexico correspond to those discovered in the rest of the world. The comparisons reflect the existence of important changes within the biological groups associated with epicontinental sea floors. The geological record of the Paleozoic of Mexico, although scarce and intermittent, also clearly reflects important changes in the biological groups associated with epicontinental sea floors.

The marine record of the Early Paleozoic from Mexico includes, as dominant groups, trilobites and brachiopods, associated with an abundant and diverse fauna in which archaeocyatids, mollusks and graptolites are common within associations comprised of bentonic epifaunal organisms. At the onset of the Late Paleozoic, life-forms changed notably with the presence of: 1) active predators such as nautiloids and ammonoids, common in Permian localities, or like *Helicoprion*, a shark reported from Permian localities in the north and south of Mexico (Bridges, 1964; Sour-Tovar *et al.* 2000); 2) deposit feeding organisms among which are found the nuculoid bivalves and; 3) suspension feeding organisms, especially corals, bryozoans, brachiopods, bivalves and sessile echinoderms. The suspension feeders exhibit remarkable diversification: as part of the epifauna, the crinoids, that possessed a stem and whose mobile arms would produce currents of water that permitted them to trap particles of food, were the dominant organisms above the sea bed; as part of the infauna, the bivalves presented a great diversity, and some of them, such as some Pholadomyoida, could have burrowed to greater depths into the substratum.

5. Groups of Marine Invertebrates Related to the Upper Paleozoic of Mexico

Tables 1-50 present a synthesis of Upper Paleozoic localities of Mexico and the taxa of invertebrates that have been reported from them. Other Carboniferous and Permian localities were not included in the tables due to the fact that: they are not fossiliferous; they are of continental origin; or only conodonts, fusilinids, or other organisms not considered in this study, were collected from them. The localities are listed according to the name of the Mexican State in which they are encountered (Figure 1). Also within the locality listings, the species are ordered systematically, following the arrangement established in the most recent volumes of the Treatise on Invertebrate Paleontology.

Baja California

Near San Felipe, Baja California, there is a conglomerate of Miocene or earlier Tertiary age with clasts containing fecal pellets, gastropods, bryozoans, crinoids blue-green algae and Permian fusulinids of early Leonardian age (Gastil *et al.*, 1973). Probable Carboniferous crinoids, solitary corals and brachiopods have been discovered in the Northern Sierra Pinta, just south of Laguna Salada (McEldowney, 1970).

TABLE 1. Crinoids from Zamora Formation, Permian (Leonardian?), Arroyo Zamora near El Volcán, Baja California. The crinoid columnals found in the debris flows of the Zamora Formation are the largest in diameter known anywhere and are thought to be late Paleozoic in age. The relatively inflexible columns that these columnals and pluricolumnals formed (up to 9 cm in diameter) are interpreted to have lived in relatively strong currents along the edge of the shelf or upper slope (Webster and Delattre, 1993).

Species	Order	Family
Crinoids:		
Apletoanteris bajaensis Webster and Delattre	(Uncertain)	(Uncertain)

Chiapas

TABLE 2. Pennsylvaniian fossils from Santa Rosa Inferior Formation, Aguacate River, northwestern Chicomuselo. Crinoids are characteristic of the Pennsylvanian from Texas and Kansas. In the same locality bivalves and not formally described corals, bryozoans, and ammonites, have also been found (Buitrón, 1977b).

Species	Order	Family
Bivalves:		
Aviculopecten sp.	Pterioida	Aviculopectinidae
Palaeolima sp.	Pterioida	Limidae
Crinoids:		
Cylindrocauliscus fiski? Moore and Jeffords	(Uncertain)	Cyclomischidae
Lamprosterigma mirificum Moore and Jeffords	(Uncertain)	Floricyclidae

TABLE 3. Gastropod from Paso Hondo Formation, Middle Permian, near Comalapa River, southeastern Chicomuselo, Chiapas (Buitrón, 1977a).

Species	Order	Family
Gastropods:		
Bellerophon (Bellerophon) crassus Meek and Worthen	Archaeogastropoda	Bellerophontidae

TABLE 4. Fossils from Grupera Formation, Lower Permian, Monte Redondo-Guadalupe Victoria area, Chiapas (Reyeros de Castillo, 1976).

Species	**Order**	**Family**
Corals:		
Lophophyllidium kayseri major (Huang)	Rugosa	Lophophyllidiidae
Lophophyllidium zaphrentoidea (Huang)	Rugosa	Lophophyllidiidae

TABLE 5. These fossils are part of a list presented in the work of Aguilera (1987). Middle Permian from Chicomuselo area, Chiapas. (Localites and formation are not certain).

Species	**Order**	**Family**
Bryozoans:		
Fenestella? sp.	Cryptostomata	Fenestellidae
Brachiopods:		
Productus semireticulatus Martin	Productida	Productidae
Gastropods:		
Pleurotomaria? sp.	Archaeogastropoda	Pleurotomariidae

TABLE 6. List of fossils presented by Bosë (1905) Middle Permian from Chicomuselo area, Chiapas. Undeterminated bivalves and crinoids have also been collected. (Localites and formation are not certain).

Species	**Order**	**Family**
Brachiopods:		
Productus semireticulatus Martin	Productida	Productidae
Productus sp.	Productida	Productidae
Dielasma sp.	Terebratullida	Dielasmatidae
Bryozoans		
Fenestella sp.	Cryptostomata	Fenestellidae
Gastropod:		
Pleurotomaria sp.	Archaeogastropoda	Pleurotomariidae

TABLE 7. Middle Permian Fossils from Chicomuselo area, Chiapas and probable adjacent portions of Guatemala. List from Aguilera (Müllerried *et al.* 1941).

Species	**Order**	**Family**
Corals:		
Zaphrentis sp.	Rugosa	Zaphrentidae
Brachiopods:		
Productus rogersi Norwood and Pratten	Productida	Productidae

TABLE 7. Continued.

Species	Order	Family
Productus semireticulatus Martin	Productida	Productidae
Productus scabriculus Martin	Productida	Productidae
Producus nodosus Newberry	Productida	Productidae
Productus occidentalis Newberry	Productida	Productidae
Productus mexicanus Shumard	Productida	Productidae
Martinia lineata Sowerby	Spiriferida	Ingelarellidae
Spirifer cameratus Morton	Spiriferida	Trigonotretidae
Spirigera (Athyris) subtilita Hall	Athyridida	Athyrydidae

TABLE 8. Fossils from Todos Santos, along Cuilco River, Chiapas. List from Müllerried, 1941.

Species	Order	Family
Sponge:		
Stromatidium cf. S. *typicale* Girty	Lithistida	(Uncertain)
Bryozoan:		
Fenestella cf. F. *capitanensis* Girty	Cryptostomata	Fenestellidae
Brachiopods:		
Dictyoclostus aff. *D. semireticulatus capitanensis* (Girty)	Productida	Productidae
Gastropods:		
Pleurotomaria? cf. *P.*? *carinifera* Girty	Archaeogastropoda	Pleurotomariidae
Pleurotomaria cf. *P. neglecta* Girty	Archaeogastropoda	Pleurotomariidae

TABLE 9. Ammonoids from Paso Hondo Formation, Permian (Leonardian), North-northeast of Cashú road to San José Montenegro, Chiapas (Müllerried, *et al.* 1941; Gutierrez-Gil, 1956; Thompson, 1956).

Species	Order	Family
Cephalopods:		
Perrinites hilli (Smith)	Ammonoidea	Perrinitidae
Peritrochia mullerriedi Miller and Furnish	Ammonoidea	Popanoceratidae

Chihuahua

TABLE 10. Fauna from Mina Plomosas-Placer de Guadalupe Area, Chihuahua, Monillas Formation, Upper Member, Upper Mississippian to Lower Pennsylvanian. Undescribed conulariids have also been found at this locality (Bridges, 1964).

Species	Order	Family
Brachiopods:		
Buxtonia sp.	Productida	Productidae
Chonetes sp.	Productida	Chonetidae
Overtonia sp.	Productidae	Linoproductidae
Linoproductus sp.	Productidae	Linoproductidae
Leiorhynchus sp.	Rhynchonellida	Leiorhynchidae
Crurithyris sp.	Spiriferida	Ambocoeliidae
Punctospirifer sp.	Spiriferinida	Punctospiriferidae

TABLE 11. Fauna from Pastor Limestone, Lower Pennsylvanian to Lower Permian (Upper Wolfcampian) age. Desmoinesian. Missourian, Virgilian, and Wolfcampian fusulinids have been identified (Bridges, 1964).

Species	Order	Family
Corals:		
Ivanovia? Sp.	Rugosa	Lonsdaleiidae
Chaetetes sp.	Tabulata	Chaetetidae
Brachiopods:		
Antiquatonia sp.	Productida	Productidae
Schuchertella or Derbyia sp.	Orthotetida	Schuchertellidae
Cleiothyridina sp.	Athyridida	Athyrididae
Huestedia sp.	Athyridida	Neoretziidae
"Spirifer" texanus	Spiriferida	Trigonotretidae

TABLE 12. Invertebrates from Plomosas Formation, age uncertain. Age could range from Wolfcampian into the Triassic. Permian shark *Helicoprion* has been found in the same locality (Bridges, 1964).

Species	Order	Family
Brachiopods:		
Leiorhynchoidea cf. L. *bisulcata*	Rhynchonellida	Petasmariidae
Composita sp.	Athyridida	Athyrididae
Gastropods:		
Babylonites cf. *B. acutus*	Archaeogastropoda	Omphalotrochidae
Discotropis? sp.	Archaeogastropoda	Omphalotrochidae
Baylea sp.	Archaeogastropoda	Raphistomatidae

TABLE 13. Sponges from Plomosas Formation (Lower Permian). The specimens are within a reefal limestone characterized by the presence of *Tubiphytes* (Bridges, 1964).

Species	**Order**	**Family**
Sponges:		
Actinocoelia cf. A. *malandrina*	Thalamida	?
Girtyocoelia dunbari	Thalamida	Celyphiidae
Guadalupia aff. G. *cylindrical*	Thalamida	?
Heliospongia vokesi	Lithisthida	Cylindrophymatidae

Coahuila

TABLE 14. Lower Permian (Leonardian) invertebrate fossils from Las Sardinas Formation, Las Delicias Valley, Coahuila Mexico (Cloud, 1944; Wardlaw *et al.*, 1979).

Species	**Order**	**Family**
Brachiopods:		
Elliotella sp.	Productida	Productellidae
Compressoproductus sp.	Productida	Monticuliferidae
Coscinophora sp.	Productida	Lyttoniidae
Derbyia sp.	Orthotetida	Derbyiidae
Leioryhnchoidea sp.	Rhynchonellida	Petasmariidae
Crurithyris sp.	Spiriferida	Ambocoeliidae
Neospirifer sp.	Spiriferida	Trigonotretidae
Neophricadothyris sp.	Spiriferida	Elythidae
Spiriferellina sp.	Spiriferinida	Crenispiriferidae
Cephalopods		
Perrinites vidriensis Böse	Ammonoidea	Perrinitidae
Medlicottia costellifera Miller and Furnish	Ammonoidea	Medlicottidae
Eumedlicottia whitneyi (Böse)	Ammonoidea	Medlicottidae

TABLE 15. Upper Permian (Wordian, Capitanian, Amerarassian) invertebrate fossils from Palo Quemado, La Difunta and La Colorada formations, Las Delicias Valley, Coahuila Mexico (Cloud, 1944; Wardlaw *et al.* 1979).

Species	**Order**	**Family**
Brachiopods:		
Orbiculoidea ovalis Cloud	Lingulida	Discinidae
Chonetes deliciasensis King	Productida	Chonetidae
Dyoros cf. D. extensus Cooper and Grant	Productida	Rugosochonetidae
Avonia costata King	Productida	Productellidae
Avonia subquadrata Cloud	Productida	Productellidae
Echinauris sp.	Productida	Productellidae

TABLE 15. Continued.

Species	Order	Family
Marginifera popei minor King	Productida	Productellidae
Paucispinifera sp.	Productida	Productellidae
Liosotella cf. *L. popei* (Shumard)	Productida	Productellidae
Liosotella sp.	Productida	Productellidae
Yakovlevia demunitivus (Cloud)	Productida	Productellidae
Dictyoclostus deminutivus Cloud	Productida	Productidae
Kochiproductus?	Productida	Productidae
Waagenoconcha sp.	Productida	Echinoconchidae
Cancrinella rugosa Cloud	Productida	Linoproductidae
Ctenalosia sp.	Productida	Aulostegidae
Echinosteges sp.	Productida	Aulostegidae
Prorichthofenia likharewi King	Productida	Richthofeniidae
Prorichthofenia permiana (Shumard)	Productida	Richthofeniidae
Prorichthofenia sp.	Productida	Richthofeniidae
Meekella enigma Cloud	Orthotetida	Meekellidae
Streptorhynchus cf. *S. gregarium* Girty	Orthotetida	Schuchertellidae
Enteletes sp.	Orthida	Enteletidae
Leiorhynchoidea schucherti Cloud	Rhynchonellida	Petasmariidae
Leioryhnchoidea laveis Cloud	Rhynchonellida	Petasmariidae
Pugnoides sp.	Rhynchonellida	Petasmariidae
Paranorella imperiales Cloud	Rhynchonellida	Paranorellidae
Stenoscisma kingi Cloud	Rhynchonellida	Stenocismatidae
Rhynchopora taylori Girty	Rhynchonellida	Rhynchoporidae
Petasmatherus sp.	Rhynchonellida	Petasmatheridae
Composita emarginata affinis Girty	Athyridida	Athyrididae
Hustedia cf. H. meekana (Shumard)	Athyridida	Neoretziidae
Hustedia mexicana (Haack)	Athyridida	Neoretziidae
Crurithyris atenuata Cloud	Spiriferida	Ambocoeliidae
Attenuatella atenuata (Cloud)	Spiriferida	Ambocoeliidae
Martinia sp.	Spiriferida	Ingelarellidae
"*Spiriferinida*" *laxa* Girty	Spiriferinida	Spiriferinidae?
Heterelasma shumardium Girty	Terebratulida	Dielasmatidae
Dielasmina guadalupensis Girty	Terebratulida	Dielasmatidae
Dielasma sp.	Terebratulida	Dielasmatidae
Cephalopods:		
Bitaunioceras coahuilense Miller	Nautiloidea	Orthoceratidae
Stearoceras sp.	Nautiloidea	Grypoceratidae
Titanoceras sp.	Nautiloidea	Grypoceratidae
Liroceras? sp.	Nautiloidea	Liroceratidae
Bactrites mexicanus Miller	Ammonoidea	Bactridae
Agathiceras girtyi Böse	Ammonoidea	Agathiceratidae
Agathiceras suessi Gemmellaro	Ammonoidea	Agathiceratidae

TABLE 15. Continued.

Species	Order	Family
Perrinites hilli (Smith)	Ammonoidea	Perrinitidae
Popanoceras bowmani (Böse)	Ammonoidea	Popanoceratidae
Stacheoceras toumanskyae Miller and Furnish	Ammonoidea	Popanoceratidae
Stacheoceras gemmellaroi Miller	Ammonoidea	Popanoceratidae
Waagenoceras guadalupense Girty	Ammonoidea	Cyclolobidae
Waagenoceras dieneri Böse	Ammonoidea	Cyclolobidae
Waagenoceras karpinskyi Miller	Ammonoidea	Cyclolobidae
Waagenoceras girtyi Miller and Furnish	Ammonoidea	Cyclolobidae
Timorites schucherti Miller and Furnish	Ammonoidea Ammonoidea	Cyclolobidae Gastrioceratidae
Pseudogastrioceras roadense (Böse)	Ammonoidea	Neoicoceratidae
Pseudogastrioceras haacki Miller	Ammonoidea	Neoicoceratidae
Altudoceras sp.	Ammonoidea	Neoicoceratidae
Strigogoniatites kingi Miller	Ammonoidea	Neoicoceratidae
Adrianites dunbari Miller and Furnish	Ammonoidea	Adrianitidae
Adrianites newelli Miller and Furnish	Ammonoidea	Adrianitidae
Adrianites plummeri Miller	Ammonoidea	Adrianitidae
Doryceras difuntense (Miller)	Ammonoidea	Adrianitidae
Pseudagathiceras spinosum Miller	Ammonoidea	Adrianitidae
Epithalassoceras ruzencevi Miller and Furnish	Ammonoidea	Thalassoceratidae
Medlicottia burckhardti Böse	Ammonoidea	Medlicottidae
Medlicottia whitneyi Böse	Ammonoidea	Medlicottiidae
Medlicottia girtyi Miller and Furnish	Ammonoidea	Medlicottidae
Medlicottia sp.	Ammonoidea	Medlicottiidae
Propinacoceras beyrichi Gemmmellaro	Ammonoidea	Medlicottiidae
Propinacoceras sp.	Ammonoidea	Medlicottiidae
Episageceras nodosum Wanner	Ammonoidea	Medlicottiidae
Episageceras sp.	Ammonoidea	Medlicottiidae
Paraceltites ornatus	Ammonoidea	Paraceltitidae
Miller and Furnish *Paraceltites altudensis* (Böse)	Ammonoidea	Paraceltitidae
Paraceltites hoeferi Gemmmellaro	Ammonoidea	Paraceltitidae
Paraceltites rectangularis Miller	Ammonoidea	Paraceltitidae
Kingoceras kingi Miller	Ammonoidea	Paraceltitidae
Xenodiscites waageni	Ammonoidea	Xenodiscidae
Miller and Furnish *Cibolites mojsisovicsi* Miller	Ammonoidea	Xenodiscidae

TABLE 15. Continued.

Species	**Order**	**Family**
Cibolites waageni (Miller and Furnish)	Ammonoidea	Xenodiscidae
Syrdenites sp.	Ammonoidea	?
Eoaraxoceras ruzhencevi Spinoza, Furnish and Glenister	Ammonoidea	?
Stracheoceras gemmellaroi Miller	Ammonoidea	?
Epibalassoceras ruzhencevi Miller and Furnish	Ammonoidea	?
Mexicoceras guadalupense (Girty)	Ammonoidea	?
Neogeoceras gyrtyi (Miller and Furnish)		

TABLE 16. Crinoid from Permian rocks of Valle de Las Delicias, Coahuila (Strimple, 1971).

Species	**Order**	**Family**
Crinoids:		
Parspaniocrinus beinerti Strimple	Cladida	Spaniocrinidae

Guerrero

TABLE 17. Brachiopods and crinoids (Esquivel-Macías, 1996, 2000; López-López, 1997), and trilobites (Garibay-Romero *et al.* 1998) from Los Arcos/Olinalá Formation, Permian of Olinalá, Guerrero.

Species	**Order**	**Family**
Brachiopods:		
Paramarginifera? sp.	Productida	Marginiferidae
Dictyoclostus depressus Cooper	Productida	Productidae
Linoproductus sp.	Productida	Linoproductidae
Cancrinella rugosa Cloud	Productida	Linoproductidae
Paranorella imperialis Cloud	Rhynchonellida	Camarotoechiidae
Wellerella lemasi minor Cooper	Rhynchonellidae	Wellerellidae
Martinia sp.	Spiriferida	Martinidae
Dielasma aff. *D. spatulatum* Girty	Terebratulida	Dielasmatidae
Crinoids:		
Pentaridica pentagonalis Moore and Jeffords	(Uncertain)	Pentacauliscidae
Trilobites:		
Anisopyge perannulata (Shumard)	Ptychopariida	Phillipsidae

TABLE 18. Brachiopods and cephalopods from Los Arcos/Olinalá Formation, Unit 2, Middle Permian, Olinalá, Guerrero (González-Arreola *et al.* 1994).

Species	**Order**	**Family**
Brachiopods:		
Orbiculoidea sp.	Lingulida	Discinidae
Krotovia sp.	Productida	Productellidae
Cancrinella sp.	Productida	Linoproductidae
Leiorhynchoidea schucherti Cloud	Rhynchonellida	Petasmariidae
Tomiopsis sp.	Spiriferida	Ingelarellidae
Cephalopods:		
Bitaunioceras coahuilensis Miller	Orthocerida	Orthoceratidae
Stacheoceras toumanskyae Miller and Furnish	Ammonoidea	Popanoceratidae
Pseudogastrioceras roadense (Böse)	Ammonoidea	Neoicoceratidae

TABLE 19. Conulariids, brachiopods, gastropods, bivalves and cephalopods from Los Arcos/ Olinalá Formation, Unit 4. Middle Permian, Olinalá, Guerrero (González-Arreola *et al.* 1994).

Species	**Order**	**Family**
Conulariids:		
Paraconularia sp.	Conulariida	Conulariida
Brachiopods:		
Orbiculoidea ovalis Cloud	Lingulida	Discinidae
Thamnosia depressa (Cooper)	Productida	Productidae
Cancrinella rugosa Cloud	Productida	Linoproductidae
Gastropods:		
Babylonites carinatus Yochelson	Archaeogastropoda	Omphalotrochidae
Bivalves:		
Posidoniella sp.	Pterioida	Myalinidae
Cephalopods:		
Stacheoceras toumanskyae Miller and Furnish	Ammonoidea	Popanoceratidae
Waagenoceras dieneri Böse	Ammonoidea	Cyclolobidae
Paraceltites elegans Girty	Ammonoidea	Paraceltitidae

TABLE 20. Cephalopod and crinoid from Los Arcos/Olinalá Formation, Unit 5. Middle Permian, Olinalá, Guerrero (González-Arreola *et al.* 1994).

Species	**Order**	**Family**
Cephalopods:		
Paraceltites elegans Girty	Ammonoidea	Paraceltitidae

TABLE 20. Continued.

Species	Order	Family
Crinoids:		
Preptopremnum rugosum Moore and Jeffords	(Uncertain)	Leptocarphiidae

TABLE 21. Brachiopods, gastropods and cephalopods from Los Arcos/Olinalá Formation, Unit 7. Middle Permian, Olinalá, Guerrero (according to González-Arreola *et al.* 1994).

Species	Order	Family
Brachiopods:		
Thamnosia depressa (Cooper)	Productida	Productidae
Wellerella sp.	Rhynchonellida	Wellerellidae
Composita sp.	Athyridida	Athyrididae
Hustedia sp.	Athyridida	Neoretziidae
Spiriferellina cristata (Von Schlotheim)	Spiriferinida	Crenispiriferidae
Gastropods:		
Babylonites carinatus Yochelson	Archaeogastropoda	Omphalotrochidae
Cephalopods:		
Metacoceras sp.	Nautilida	Tainoceratidae
Stacheoceras toumanskyae Miller and Furnish	Ammonoidea	Popanoceratidae

Hidalgo

TABLE 22. Brachiopods and trilobites from Tuzancoa Formation, Lower Permian, Otlamalacatla, Hidalgo. In the same locality fusulinids, rugose corals, bryozoans, bivalves and crinoids have also been collected (Sour-Tovar *et al.* 2005b).

Species	Order	Family
Brachiopods:		
Krotovia sp.	Productida	Productellidae
Dasysaria sp.	Productida	Productidae
Derbyoides cf. dunbari Cooper and Grant	Orthotetida	Orthotetidae
Neospirifer cf. N. amphigyus Cooper and Grant	Spiriferida	Trigonotretidae
Neospirifer sp.	Spiriferida	Trigonotretidae
Spiriferellina tricosa Cooper and Grant	Spiriferida	Crenispiriferidae
Holosia? sp.	Rhynchonellidae	Wellerellidae
Trilobites:		
Anisopyge whitei Pabian and Fagerstrom	Ptychopariida	Phillipsidae

TABLE 23. Crinoids from Tuzancoa Formation, Permian (upper Wolfcampian-lower Leonardian), Pemuxco- Mimiaguaco, Hidalgo (Arellano-Gil *et al.* 1998). These beds were considered as part of Guacamaya Formation, that crops out in Tamaulipas State.

Species	**Order**	**Family**
Crinoids:		
Cyclocaudex plenus Moore and Jeffords	(Uncertain)	Cyclomischiade
Cyclocaudex jucundus Moore and Jeffords	(Uncertain)	Cyclomischidae
Mooreanteris perforatus Moore and Jeffords	(Uncertain)	Cyclomischiidae
Cyclocrista cheneyi Moore and Jeffords	(Uncertain)	Leptocarphiidae
Preptopremnum rugosum Moore and Jeffords	(Uncertain)	Leptocarphiidae

TABLE 24. Brachiopods from Vicente Guerrero Formation, Mississippian, Calnali, Hidalgo (Moreno-Cano and Patiño-Ruiz, 1981).

Species	**Order**	**Family**
Brachiopods:		
Overtonia sp.	Productida	Productellidae
Allorhynchus cf. *A. heteropsis*	Rhynchonellida	Allorhynchidae
Septosyringothyris cf. *S.* de*mantei*	Spiriferinida	Syringothrididae

TABLE 25. Brachiopods (Moreno-Cano and Patiño-Ruiz, 1981), and crinoids (Buitrón *et al.* 1987) from Del Monte Formation, Pennsylvanian, Calnali, Hidalgo.

Species	**Order**	**Family**
Brachiopods:		
Neospirifer sp. cf. *fascimger*	Spiriferida	Trigonotretidae
Crinoid:		
Cylindrocauliscus fiski Moore and Jeffords	(Uncertain)	Cyclomischidae
Baryschyr anosus Moore and Jeffords	(Uncertain)	Cyclomischidae
Cyclocaudex jucundus Moore and Jeffords	(Uncertain)	Cyclomischidae
Cyclocaudex insaturatus Moore and Jeffords	(Uncertain)	Cyclomischidae
Cyclocaudex cf. C. costatus Moore and Jeffords	(Uncertain)	Cyclomischidae
Mooreanteris waylandensis Miller	(Uncertain)	Cyclomischidae
Plummeranteris sansaba Moore and Jeffords	(Uncertain)	Floricyclidae

TABLE 25. Continued.

Species	Order	Family
Heterostelechus keithi Miller	(Uncertain)	Leptocarphiidae

Oaxaca

TABLE 26. Brachiopods (Navarro-Santillán, *et al.* 2002) and rostroconchs (Quiroz-Barroso, *et al.* 2000) from Santiago Formation, Lower Mississippian (Osagean), Nochixtlán region, Oaxaca. These organisms formed part of a community with abundant sponges, tabulate corals, bryozoans, mollusks and crinoids.

Species	Order	Family
Brachiopods:		
Actinochonchus lamellossus (Léveillé)	Athyridida	Athyrididae
Torynifer pseudolineatus (Hall)	Spiriferida	Elythidae
cf. *Syringothyris* sp.	Spiriferinida	Syringothyrididae
Punctospirifer gigas Sour-Tovar and Navarro	Spiriferinida	Punctospiriferidae
Rostroconchs:		
Pseudomulceodens cancellatus (Hyde)	Conocardioida	Conocardiidae

TABLE 27. Conulariids, brachiopods and gastropods from Ixtaltepec Formation, Lower-Middle Pennsylvanian, Nochixtlán region, Oaxaca. Conulariids and gastropods from Buitrón *et al.* (2000); Brachiopods from Sour-Tovar (1994); Sour-Tovar and Quiroz-Barroso (1989); and Sour-Tovar and Martínez-Chacón (2004).

Species	Order	Family
Conulariids:		
Paraconularia sp.	Conulariida	Conulariidae
Brachiopods:		
Orbiculoidea missouriensis (Shumard)	Lingulida	Discinidae
Tornquistia sp.	Productida	Anopliidae
Neochonetes (*N.*) *granulifer* (Owen)	Productida	Rugosochonetidae
Neochonetes (*N.*) *mixteco* (Sour-Tovar and Martínez-Chacón)	Productida	Rugosochonetidae
Reticulatia huecoensis (King)	Productida	Productidae
Echinaria semipunctata (Shepard)	Productida	Echinoconchidae
Linoproductus planiventralis var. *ixtaltepensis* (Sour and Quiroz)	Productida	Linoproductidae
Linoproductus echinatus Hoare	Productida	Linoproductidae

TABLE 27. Continued.

Species	**Order**	**Family**
Orthotetes mixteca Sour and Quiroz	Orthotetida	Orthotetidae
Derbyia crassa (Meek and Hayden)	Orthotetida	Derbyiidae
Gastropods:		
Anomphalus sp.	Archaeogastropoda	Anomphalidae

TABLE 28. Bivalves (Quiroz-Barroso and Perrilliat, 1997, 1998); trilobites (Morón-Ríos and Perrilliat, 1988); and crinoids (Buitrón *et al.* 2000) from Ixtaltepec Formation, Lower-Middle Pennsylvanian, Santiago Ixtaltepec, Oaxaca. One specimen of undescribed ophiuroid has been found at this locality (Quiroz-Barroso and Sour-Tovar, 1995).

Species	**Order**	**Family**
Bivalves:		
Nuculopsis sp.	Nuculoida	Nuculidae
Palaeoneilo sinuosus Quiroz and Perrilliat	Nuculoida	Malletiidae
Paleyoldia sp.	Nuculoida	Nuculanidae
Phestia bellistriata (Stevens)	Nuculoida	Nuculanidae
Phestia sp.	Nuculoida	Nuculanidae
Parallelodon tenuistriatus Meek and Worthen	Arcoida	Parallelodontidae
Myalinella meeki (Dunbar)	Pterioida	Myalinidae
Septimyalina sp.	Pterioida	Myalinidae
Leptodesma (Leptodesma) sp.	Pterioida	Pterineidae
Pterinopectinella sp.	Pterioida	Pterinopectinidae
Aviculopecten cf. *A. basilicus* Newell	Pterioida	Aviculopectinidae
Aviculopecten sp. A	Pterioida	Aviculopectinidae
Aviculopecten sp. B	Pterioida	Aviculopectinidae
Aviculopecten sp. C	Pterioida	Aviculopectinidae
Aviculopecten sp. D	Pterioida	Aviculopectinidae
Limipecten cf. *L. texanus* Girty	Pterioida	Aviculopectinidae
Streblopteria sp.	Pterioida	Aviculopectinidae
Euchondria cf. *E. levicula* Newell	Pterioida	Aviculopectinidae
Euchondria sp.	Pterioida	Aviculopectinidae
Schizodus cf. *S. ulrichi* Worthen	Trigonioida	Myophoriidae
Permophorus spinulosa (Morningstar)	Veneroida	Permophoridae
Permophorus cf. P. oblongus (Meek)	Veneroida	Permophoridae
Astartella concentrica (Conrad)	Veneroida	Astartidae

TABLE 28. Continued.

Species	Order	Family
Edmondia ovata Meek and Worthen	Pholadomyoida	Edmondiidae
Ectogrammysia prolata Hoare	Pholadomyoida	Grammysiidae
Trilobites:		
Griffithides ixtaltepecensis Morón and Perrilliat	Ptychopariida	Phillipsidae
Crinoids:		
Cylindrocauliscus fiski Moore and Jeffords	(Uncertain)	Cyclomischidae
Cyclocaudex insaturatus Moore and Jeffords	(Uncertain)	Cyclomischidae

TABLE 29. Invertebrates from the Lower Mississippian of Patlanoaya Formation, San Salvador Patlanoaya, Puebla (Villaseñor-Martínez *et al.* 1987).

Species	Order	Family
Corals:		
Neozaphrentis sp.	Rugosa	Hapsiphyllidae
Brachiopods:		
Orbiculoidea sp.	Lingulida	Discinidae
Barroisella sp.	Lingulida	Lingulidae
Chonetes sp.	Productida	Productellidae
Rugauris sp.	Productida	Productellidae
Rhytiopora cf. *blairi* (Miller)	Productida	Productellidae
Sinuatella sinuata (de Koninck)	Productida	Paulostegidae
Actinoconchus squamosus (Phillips)	Athyridida	Athyrididae
Actinoconchus planosulcatus (Phillips)	Athyridida	Athyrididae
Martinothyris lineatus (Sowerby)	Spiriferida?	?
Spirifer gregeri (Weller)	Spiriferida	Spiriferidae
Prospira sp.	Spiriferida	Spiriferidae
Unispirifer sp.	Spiriferida	Spiriferidae
Asyrinxia lata McCoy	Spiriferinida	Syringothrydidae
Punctospirifer sp.	Spiriferinida	Punctospiriferidae
Crinoids:		
Lomalegnum hormidium Moore and Jeffords	(Uncertain)	Cyclomischidae
Cyclostelechus turritus Moore and Jeffords	(Uncertain)	Exaesiodiscidae
Pentagonomischus sp.	(Uncertain)	

TABLE 30. Invertebrates from the Middle Pennsylvanian of Patlanoaya Formation, San Salvador Patlanoaya, Puebla (Villaseñor-Martínez *et al.* 1987).

Species	**Order**	**Family**
Corals:		
Lophophyllidium sp.	Rugosa	Lophophyllidiidae
Brachiopods:		
Productus sp.	Productida	Productidae
Koslowskia sp.	Productida	Productidae
Antiquatonia sp.	Productida	Productidae
Reticulatia sp.	Productida	Productidae
Linoproductus sp.	Productida	Linoproductidae

TABLE 31. Invertebrates from the Upper Pennsylvanian of Patlanoaya Formation, San Salvador Patlanoaya, Puebla (Villaseñor-Martínez *et al.* 1987).

Species	**Order**	**Family**
Brachiopods:		
Productus sp.	Productida	Productidae
Chaoiella gruenewaldti (Krotov)	Productida	Productidae
Echinaria sp.	Productida	Echinochonchidae
Spiriferella sp.	Spiriferida	Spiriferidae
Crenispirifer sp.	Spiriferinida	Crenispiriferidae
Gastropods:		
Meekospira sp.	Caenogastropoda	Meekospiridae
Crinoids:		
Cyclocaudex typicus Moore and Jeffords	(Uncertain)	Cyclomischidae
Megacanthopora fallacis Moore	(Uncertain)	

TABLE 32. Invertebrates from the Permian (Wolfcampian) of Patlanoaya Formation, San Salvador Patlanoaya, Puebla (Villaseñor-Martínez *et al.* 1987).

Species	**Order**	**Family**
Brachiopods:		
Wellerella sp.	Rhynchonellida	Wellerellidae
Bivalves:		
Palaeomutela (Palaeomutela) sp.	Unionoida	Palaeomutelidae
Schizodus sp.	Trigonioida	Myophoriidae
Cephalopods:		
Bactrites sp.	Bactritida	Bactridae
Properrinites sp.	Ammonoidea	Perrinitidae

TABLE 33. Invertebrates from the Permian (Leonardian) of Patlanoaya Formation, San Salvador Patlanoaya, Puebla (Villaseñor-Martínez *et al.* 1987).

Species	**Order**	**Family**
Gastropods:		
Keeneia sp.	Archaeogastropoda	Sinuopeidae
Omphalonema sp.	Archaeogastropoda	Holopeidae
Bivalves:		
Myalina apachesi Marcou	Pterioida	Myalinidae
Annuliconcha sp.	Pterioida	Aviculopectinidae
Permophorus sp.	Veneroida	Permophoridae
Cephalopods:		
Bactrites sp.	Bactritida	Bactridae
Perrinites hilli (Smith)	Ammonoidea	Perrinitidae
Medlicottia sp.	Ammonoidea	Medlicottiidae

TABLE 34. Brachiopods and crinoids from Patlanoaya Formation, Permian, San Salvador Patlanoaya, Puebla (Esquivel-Macias, 2000).

Species	**Order**	**Family**
Brachiopods:		
Dictyoclostus sp.	Strophomenida	Dictyoclostidae
Crinoids:		
Isocrinus sp.	Isocrinida	Isocrinidae
Cyclomischus alternatus Moore and Jeffords	(Uncertain)	Cyclomischidae
Rhysocamax cf. *R. cristata* Moore and Jeffords	(Uncertain)	Millericrinidae
Lomalegnum hormidium Moore and Jeffords	(Uncertain)	Cyclomischidae
Stiberostaurus aestimatus Moore and Jeffords	(Uncertain)	Cyclomischidae
Blothronagma cinctutum Moore and Jeffords	(Uncertain)	Cyclomischidae
Cyclocaudex typicus Moore and Jeffords	(Uncertain)	Cyclomischidae
Cyclocaudex jucundus Moore and Jeffords	(Uncertain)	Cyclomischidae
Mooreanteris perforatus (Moore and Jeffords)	(Uncertain)	Cyclomischidae
Mooreanteris. cf. *M. waylandensis* Miller	(Uncertain)	Cyclomischidae
Dierocalipter cf. *D. doter* Moore and Jeffords	(Uncertain)	Flucticharacidae
Ampholenium? cf. *A. apolegma* Moore and Jeffords	(Uncertain)	Exaesiodiscidae

TABLE 34. Continued.

Species	Order	Family
Floricyclus cf. *F.angustimargo* Moore and Jeffords	(Uncertain)	Floricyclidae
Cyclorista cheneyi Moore and Jeffords	(Uncertain)	Leptocarphiidae
Preptopremnum? cf. *P rugosum* Moore and Jeffords	(Uncertain)	Leptocarphiidae
Pterotocrinus cf. *P. bifurcatus* Wachsmuth and Springer	(Uncertain)	Hexacrinitacea

TABLE 35. Mississippian-Pennsylvanian crinoids from San Salvador Patlanoaya, Puebla Unit I. Mississippian (Osagean) (Velasco de León and Buitrón, 1992).

Species	Order	Family
Crinoids:		
Stiberostaurus aestimatus Moore and Jeffords	(Uncertain)	Cyclomischidae
Ampholenium apolegma Moore and Jeffords	(Uncertain)	Exaesiodiscidae

TABLE 36. Crinoids from San Salvador Patlanoaya, Puebla Unit III. Upper Pennsylvanian (Virgilian) (Velasco de León and Buitrón, 1992).

Species	Order	Family
Crinoids:		
Cyclocaudex jucundus Moore and Jeffords	(Uncertain)	Cyclomischidae

Sonora

TABLE 37. Invertebrates from Northwestern Sonora, Mississippian outcrops near Bisani (Easton *et al.* 1958).

Species	Order	Family
Corals:		
Rotiphyllum occidentale Easton	Rugosa	Metriophyllidae
Rotiphyllum vesiculosum Easton	Rugosa	Metriophyllidae
Trochophyllum (Trochophyllum) sp.	Rugosa	Laccophyllidae
Trochophyllum (Barrandeophyllum) sp.	Rugosa	Laccophyllidae
Cystelasma invaginatum Easton	Rugosa	Polycoeliidae
Cyathaxonia cordillerensis Easton	Rugosa	Cyathaxoniidae

TABLE 37. Continued.

Species	**Order**	**Family**
Triplophyllites (Homalophyllites) circularis Easton	Rugosa	Hapsiphyllidae
Triplophyllites (Homalophyllites) sp.	Rugosa	Hapsiphyllidae
Triplophyllites sp.	Rugosa	Hapsiphyllidae
Neozaphrentis tenella (Miller), emend. Easton	Rugosa	Hapsiphyllidae
Koninckophyllum sp.	Rugosa	Aulophyllidae
Caninia corniculum (Miller) emend. Easton	Rugosa	Cyathopsidae
Caninia sp.	Rugosa	Cyathopsidae
Caninophyllum sonorense Easton	Rugosa	Cyathopsidae
Lithostrotionella confluens Easton	Rugosa	Lonsdaleiidae
Pleurodictyum subramosum Easton	Tabulata	Favositidae
Syringopora tubifera Easton	Tabulata	Auloporidae
Syringopora sp.	Tabulata	Auloporidae
Brachiopods:		
Leptaena Dalman, 1828	Strophomenida	Rafinesquinidae
Leptaena cooperi Sanders	Strophomenida	Rafinesquinidae
Plicochonetes geniculatus (White)	Productida	Rugosochonetidae
Reticulatia cooperensis (Swallow)	Productida	Productidae
Schellwienella umbonata Sanders	Orthotetida	Pulsiidae
Schuchertella sp.	Orthotetida	Schuchertellidae
Rhipidomella cf. *R. missourensis* (Swallow)	Orthida	Rhipidomellidae
Perditocardinia cf. *P. dubia* (Hall)	Orthida	Rhipidomellidae
Schizophoria sulcata Sanders	Orthida	Schizophoriidae
Camarotoechia sp.	Rhynchonellida	Camarotoechiidae
Dorsisinus louisianensis (Weller)	Rhynchonellida	Nomina dubia
"Cleiothyridina" glenparkensis Weller	Athyridida	Athyrididae
"*Cleiothyridina*" sp.	Athyridida	Athyrididae
Composita obesa Sanders	Athyridida	Athyrididae
Hustedia circularis (Miller)	Athyridida	Neoretziidae
Cyrtospirifer? latior (Swallow)	Spiriferida	Cyrtospiriferidae
Crurithyris levicula (Rowley)	Spiriferida	Ambocoeliidae
Tylothyris? sp.	Spiriferida	Mucrospiriferida
Cyrtina burlingtonensis Rowley	Spiriferinida	Cyrtinidae
Punctospirifer sulcifer Sanders	Spiriferinida	Punctospiriferidae
Punctospirifer globosa Sanders	Spiriferinida	Punctospiriferidae
Girtyella sp.	Terebratulida	Dielasmatidae
Girtyella? sp.	Terebratulida	Dielasmatidae
Dielasmoides sp.	Terebratulida	Dielasmatidae
Dielasmoides?	Terebratulida	Dielasmatidae
Beecheria sp.	Terebratulida	Heterelasminidae

TABLE 37. Continued.

Species	**Order**	**Family**
Bivalves:		
Parallelodon sulcatus Weller	Arcoida	Parallelodontidae
Rostroconchs:		
Conocardium sp.	Conocardioida	Conocardiidae
Gastropods:		
Bellerophon sp.	Archaeogastropoda	Bellerophontidae
Straparollus (Euomphalus) sp. A	Archaeogastropoda	Euomphalidae
Straparollus (Euomphalus) sp. B	Archaeogastropoda	Euomphalidae
Phanerotinus cf. P. *paradoxus* Winchell	Archaeogastropoda	Euomphalidae
Platyschisma sp.	Archaeogastropoda	Sinuopeidae
Rhineoderma cf. *R. nystii* de Koninck	Archaeogastropoda	Sinuopeidae
Baylea aff. *B. yvanii* (Leveillé)	Archaeogastropoda	Raphistomatidae
Baylea sp.	Archaeogastropoda	Raphistomatidae
Borestus sp.	Archaeogastropoda	Phymatopleuridae
Platyceras (Platyceras) sp. A	Archaeogastropoda	Platyceratidae
Platyceras (Orthonychia) sp. A	Archaeogastropoda	Platyceratidae
Cephalopods:		
Triboloceras digonum (Meek and Worthen)	Nautilida	Trigonoceratidae?

TABLE 38. Mississippian (Kinderhookian-Osagean) invertebrates from Sierra del Tule, Sonora, Escabrosa Limestone. (González-León, 1986).

Species	**Order**	**Family**
Corals:		
Trochophyllum (Trochophyllum) sp.	Rugosa	Laccophyliidae
Cyathaxonia cordillerensis Easton	Rugosa	Cyathaxoniidae
Triplophyllites (Homalophyllites) circularis Easton	Rugosa	Hapsiphyllidae
Neozaphrenites sp.	Rugosa	Hapsiphyllidae
Sychnoelasma? sp.	Rugosa	Hapsiphyllidae
Stelchophyllum microstylum	Rugosa	Lithostrotionidae
Caninophyllum sp.	Rugosa	Cyathopsidae
Syringopora sp.	Tabulata	Auloporidae
Brachiopods:		
Orthotetes sp.	Orthotetida	Orthotetidae
Schuchertella cf. *S. potosiensis* (C. Bianson)	Orthotetida	Schuchertellidae
Schizophoria sulcata Sanders	Orthida	Schizophoriidae
Cleiothyridina sp.	Athyridida	Athyrididae
Composita sp.	Athyridida	Athyrididae

TABLE 38. Continued.

Species	**Order**	**Family**
Hustedia sp.	Athyridida	Neoretziidae
Cyrtospirifer cf. *C. latior* (Swallon)	Spiriferida	Cyrtospiriferidae
Crurythyris sp.	Spiriferida	Ambocoeliidae
Spirifer tenuicostatus Hall	Spiriferida	Trigonotretidae
Spirifer louisianensis Rowley	Spiriferida	Trigonotretidae
Punctospirifer sp.	Spiriferinida	Punctospiriferidae
Dielasma chouteaunenis Weller	Terebratulida	Dielasmatidae

TABLE 39. Pennsylvanian (Atokan, Desmoinesian, Missourian, and Virgilian) invertebrates from Horquilla Formation. Fusulinids and *Tubiphytes* sp. have been collected from the same localities (González-León, 1986).

Species	**Order**	**Family**
Corals:		
Chaetetes sp.	Tabulata	Chaetetidae
Bryozoans:		
Cyclotrypa sp.	Cyclostomata	Fistuliporidae
Prismopora sp.	Cyclostomata	Hexagonellidae
Brachiopods:		
Hustedia sp.	Athyridida	Neoretziidae
Crurythyris sp.	Spiriferida	Ambocoeliidae
Punctospirifer sp.	Spiriferinida	Punctospiriferidae

TABLE 40. Permian (Wolfcampian) invertebrates from Earp Formation (González-León, 1986).

Species	**Order**	**Family**
Brachiopods:		
Dictyoclostus sp.	Productida	Productidae
Wellerella multiplicata	Rhynchonellidae	Wellerellidae
Composita sp	Athyridida	Athyrididae
Gastropods:		
Bellerophon sp.	Archaeogastropoda	Bellerophontidae
Schizostoma sp.	Archaeogastropoda	Euomphalidae
Shansiella planicosta (Girty)	Archaeogastropoda	Portlockiellidae

TABLE 41. Invertebrates from Los Monos Formation, Permian of El Antimonio, Sonora (Cooper *et al.* 1965). Fusulinid *Parafusulina antimoniensis*, indeterminated sponges and corals have also been reported for the same formation.

Species	Order	Family
Brachiopods:		
Orbiculoidea sp.	Lingulida	Lingulidae
Chonetes foshagi Cooper	Productida	Chonetidae?
Chonetes gibberulus Cooper	Productida	Chonetidae?
Chonetes monosensis Cooper	Productida	Chonetidae?
Marginifera sp.	Productida	Productellidae
Liosotella angustata Cooper	Productida	Productellidae
Liosotella rugosa Cooper	Productida	Productellidae
Liosotella subrugosa Cooper	Productida	Productellidae
Liosotella magnirugosa Cooper	Productida	Productellidae
Muirwoodia sp.	Productida	Productellidae
Dictyoclostus depressus Cooper	Productida	Productidae
Waagenoconcha montpelierensis (Girty)	Productida	Echinoconchidae
Anidanthus alatus Cooper	Productida	Linoproductidae
Cancrinella phosphatica Cooper	Productida	Linoproductidae
Heteralosia mexicana Cooper	Productida	Strophalosiidae
Derbyia arellanoi Cooper	Orthotetida	Derbyiidae
Derbyia elongata Cooper	Orthotetida	Derbyiidae
Streptorhynchus sp.	Orthotetida	Schuchertellidae
Leiorhynchoidea cloudi Cooper	Rhynchonellidae	Petasmariidae
Stenoscisma sp.	Rhynchonellidae	Stenoscismatidae
Rhynchopora taylori Girty	Rhynchonellidae	Rhynchoporidae
Rhynchopora taylori rotunda Cooper	Rhynchonellidae	Rhynchoporidae
Rhynchopora bicostata Cooper	Rhynchonellidae	Rhynchoporidae
Wellerella hemiplicata Cooper	Rhynchonellidae	Wellerellidae
Wellerella lemasi Cooper	Rhynchonellidae	Wellerellidae
Wellerella lemasi minor Cooper	Rhynchonellidae	Wellerellidae
Wellerella rotunda Cooper	Rhynchonellidae	Wellerellidae
Wellerella sp.	Rhynchonellidae	Wellerellidae
Uncinunellina? pulchra Cooper	Rhynchonellidae	Wellerellidae
Composita grandis Cooper	Athyridida	Athyrididae
Hustedia meekana (Shumard)	Athyridida	Neoretziidae
Hustedia meekana plicatella Cooper	Athyridida	Neoretziidae
Hustedia elongata Cooper	Athyridida	Neoretziidae
Pseudomartinia martinezi Cooper	Spiriferida	Martiniidae
Neospirifer sp.	Spiriferida	Trigonotretidae
Spiriferella? scobinoidea Cooper	Spiriferida	Spiriferllidae
Spiriferella sp.	Spiriferida	Spiriferllidae
Punctospirifer convexus Cooper	Spiriferinida	Punctospiriferidae
Spiriferellina laxa (Girty)	Spiriferinida	Crenispiriferidae
Spiriferellina sonorensis Cooper	Spiriferinida	Crenispiriferidae

TABLE 41. Continued.

Species	Order	Family
Glossothyropsis magna Cooper	Terebratulida	Mutationellinae
Dielasma floresi Cooper	Terebratulida	Dielasmatidae
Dielasma cf. *D. prolongatum* Girty	Terebratulida	Dielasmatidae
Dielasma cf. *D. spatulatum* Girty	Terebratulida	Dielasmatidae
Heterelasma contrerasi Cooper	Terebratulida	Heterelasminidae
Heterelasma sp.	Terebratulida	Heterelasminidae
Scaphopods:		
Plagioglypta canna (White)		Dentaliidae
Gastropods:		
Euphemites subpapillosus (White)	Archaeogastropoda	Sinuitidae
Warthia sp.	Archaeogastropoda	Sinuitidae
Straparollus? sp.	Archaeogastropoda	Euomphalidae
Omphalotrochus? sp.	Archaeogastropoda	Omphalotrochidae
Pleurotomaria? sp. A	Archaeogastropoda	Pleurotomariidae
Pleurotomaria? sp. B	Archaeogastropoda	Pleurotomariidae
Pleurotomaria? sp. C	Archaeogastropoda	Pleurotomariidae
Orthonychia sp. A	Archaeogastropoda	Platyceratidae
Bivalves:		
Nucula sp.	Nuculoida	Nuculidae
Nuculana obesa White	Nuculoida	Nuculanidae
Myalina sp.	Pterioida	Myalinidae
Aviculopecten aff. *A. montpelierensis* Girty	Pterioida	Aviculopectinidae
Cyrtorostra sp.	Pterioida	Oxytomidae
Schizodus parvulus Cooper	Trigonioida	Myophoriidae
Pleurophorus sonorensis Cooper	Veneroida	Permophoridae
Cephalopods:		
Waagenoceras dieneri Böse	Ammonoidea	Cyclolobidae

TABLE 42. Invertebrates from lowermost faunal horizon of Colina Limestone, Permian (late Wolfcampian), north of Cananea, northern Sonora, (Blodgett, *et al.* 2002). Unstudied bryozoans and calcareous green alga, probably *Mizzia* are also present.

Species	Order	Family
Brachiopods:		
Meekella sp.	Orthotetida	Meekellidae
Gastropods:		
Straparollus sp.	Archaeogastropoda	Euomphalidae
cf. *Apachella*? *arizonensis* Winters	Archaeogastropoda	Eotomariidae
Worthenia sp.	Archaeogastropoda	Lophospiridae
Amaurotoma sp. (=*Yunnania* of some authors)	Archaeogastropoda	Trochonematidae

TABLE 42. Continued.

Species	**Order**	**Family**
Glyptospira sonorensis Blodgett, Moore and Gray	Archaeogastropoda	Microdomatidae
Anomphalus sp.	Archaeogastropoda	Anomphalidae
Trachydomia sp.	Archaeogastropoda	Neritopsidae
Bivalves:		
Edmondia sp.	Pholadomyoida	Edmondiidae

TABLE 43. Gastropod from Upper faunal interval (173.7 m above the base of the section) of Colina Limestone, Permian (Leonardian). North of Cananea, northern Sonora, (Blodgett *et al.* 2002).

Species	**Order**	**Family**
Gastropods:		
Euomphalus kaibabensis Chronic	Archaeogastropoda	Euomphalidae

TABLE 44. Permian mollusks from Colina Limestone (183.8-184.4 m above the base of the section), North of Cananea, northern Sonora, (Blodgett *et al.* 2002).

Species	**Order**	**Family**
Gastropods:		
Bellerophon sp.	Archaeogastropoda	Bellerophontidae
Euomphalus kaibabensis? Chronic	Archaeogastropoda	Euomphalidae
Cibecuia? sp.	Archaeogastropoda	Murchisoniidae
Goniasma sp.	Archaeogastropoda	Murchisoniidae
Orthonema sp.	Caenogastropoda	Turritellidae
Bivalves:		
Palaeonucula sp.	Nuculoida	Nuculidae
Aviculopinna sp.	Mytiloida	Pinnidae
Schizodus sp.	Trigonioida	Myophoriidae
Astartella subquadrata Girty	Veneroida	Astartidae

TABLE 45. Permian mollusks from Colina Limestone (191.1 m above the base of the section), North of Cananea, northern Sonora, (Blodgett, *et al.* 2002).

Species	**Order**	**Family**
Gastropods:		
Euphemites sp.	Archaeogastropoda	Sinuitidae
Bellerophon sp.	Archaeogastropoda	Bellerophontidae
Euomphalus sp.	Archaeogastropoda	Euomphalidae
Euomphalus kaibabensis Chronic	Archaeogastropoda	Euomphalidae

TABLE 45. Continued.

Species	Order	Family
Straparollus sp.	Archaeogastropoda	Euomphalidae
Naticopsis sp.	Archaeogastropoda	Neritopsidae
Goniasma terebra (White)	Archaeogastropoda	Murchisoniidae
Bivalves:		
Palaeonucula sp.	Nuculoida	Nuculidae
Schizodus sp.	Trigonioida	Myophoridae
Astartella subquadrata Girty	Veneroida	Astartidae
Sanguinolites? sp.	Pholadomyoida	Grammysiidae
Scaphopods:		
Plagioglypta sp.		Dentaliidae

TABLE 46. Permian gastropods from Colina Limestone (192.6 m above the base of the section), North of Cananea, northern Sonora, (Blodgett *et al.* 2002).

Species	Order	Family
Gastropods:		
Warthia sp.	Archaeogastropoda	Sinuitidae
Euomphalus kaibabensis Chronic	Archaeogastropoda	Euomphalidae
Oncochilus insolitus Chronic	Archaeogastropoda	Neritidae

TABLE 47. Permian mollusks from Colina Limestone (195.1 m above the base of the section), North of Cananea, northern Sonora, (Blodgett *et al.* 2002).

Species	Order	Family
Gastropods:		
Warthia sp.	Archaeogastropoda	Archaeogastropoda
Bellerophon sp.	Archaeogastropoda	Archaeogastropoda
Euomphalus kaibabensis Chronic	Archaeogastropoda	Archaeogastropoda
Worthenia cf. *W. crenulata* Batten	Archaeogastropoda	Archaeogastropoda
Glyptospira sp.	Archaeogastropoda	Archaeogastropoda
Glyptospira terebra (White)	Archaeogastropoda	Archaeogastropoda
Cibecuia cedarensis Winters	Archaeogastropoda	Archaeogastropoda
Bivalves:		
Sanguinolites? sp.	Pholadomyoida	Pholadomyoida
Aviculopinna sp.	Mytiloida	Mytiloida

TABLE 48. Permian invertebrates from Colina Limestone (214.9-216.5 m above the base of the section), North of Cananea, northern Sonora (Blodgett *et al.* 2002).

Species	Order	Family
Brachiopods:		
Penicularis cf. *P. subcostata* King	Productida	Productidae

TABLE 48. Continued.

Species	**Order**	**Family**
Pontisia aff. *P. stehli stehli* Cooper and Grant	Rhynchonellida	Pontisiidae
Dielasma sp.	Terebratulidae	Dielasmatidae
Gastropods:		
Euomphalus kaibabensis Chronic	Archaeogastropoda	Euomphalidae
Apachella sp.	Archaeogastropoda	Eotomariidae
Glyptospira sp.	Archaeogastropoda	Microdomatidae
Kinishbia sp.	Caenogastropoda	Procerithiidae

TABLE 49. Invertebrates from east of the measured section from Colina Limestone, North of Cananea, northern Sonora (Blodgett *et al.* 2002, Locality JS-96-45). Undetermined smooth, small brachiopods, pelmatozoan ossicles, and trilobite free cheeks and pygidia also are present at this locality.

Species	**Order**	**Family**
Brachiopods:		
Meekella sp.	Orthotetida	Meekellidae
Pontisia n. sp. aff. *P. stehli stehli* Cooper and Grant	Rhynchonellida	Pontisiidae
Gastropods:		
Bellerophon sp.	Archaeogastropoda	Bellerophontidae
Euomphalus kaibabensis Chronic	Archaeogastropoda	Euomphalidae
Baylea sp. 1	Archaeogastropoda	Raphistomatidae
Baylea sp. 2	Archaeogastropoda	Raphistomatidae
Worthenia? or *Platyworthenia*? sp.	Archaeogastropoda	Lophospiridae
Glyptospira sp.	Archaeogastropoda	Microdomatidae
Kinishbia nodosa Winters	Caenogastropoda	Procerithiidae
Bivalves:		
Bakevellia sp.	Pterioida	Bakevelliidae

Tamaulipas

TABLE 50. Lower Mississippian (Osagean) brachiopods from Vicente Guerrero Formation, Peregrina Canyon, North of Ciudad Victoria, Tamaulipas (Sour-Tovar, 1999; Sour-Tovar and Martínez-Chacón, 2004; Sour Tovar *et al.* 2005a).

Species	**Order**	**Family**
Brachiopods:		
Barroisella sp.	Lingulida	Lingulidae
Orbiculoidea sp.	Lingulida	Discinidae
Yagonia collinsoni Carter	Productida	Anopliidae
Derbyia? sp.	Orthotetidina	Derbyiidae

TABLE 50. Continued.

Species	Order	Family
Camarotoechia sp.	Rhynchonellida	Camarotoechiidae
Rotaia subtrigona (Meek and Worthen)	Rhynchonellida	Tetracameridae
Lamellosathyris lamellosa (Léveillé)	Athyridida	Athyrididae
Cleiothyridina tenuilineata (Rowley)	Athyridida	Athyrididae
Camarophorella sp.	Athyridida	Meristidae
Alispirifer tamaulipensis Sour-Tovar, Alvarez , and Martínez-Chacón	Spiriferida	Strophopleuridae
Tylothyris? sp.	Spiriferida	Mucrospiriferida
Torynifer pseudolineatus Hall	Spiriferida	Elythidae
Syringothyris cf. *S. typa* Winchell	Spiriferinida	Syringothyrididae
Syringothyris? sp.	Spiriferinida	Syringothyrididae
Punctospirifer sp.	Spiriferinida	Punctospiriferidae
Beecheria chouteauensis (Weller)	Terebratulida	Heterelasminidae

6. Conclusions

A cursory analysis of Upper Paleozoic invertebrate groups of Mexico may be summarized as follows:

1. Sponges. – The references for this group are restricted to a sponge of the order Lithistida from the Upper Paleozoic of Chiapas (Müllerried *et al.* 1941) and sponges belonging to the orders Lithistida and Thalamida from Permian rocks of Plomosas Formation, Chihuahua (Bridges, 1964).

2. Conulariids. – The presence of conulariids has been reported from the Upper Member of the Late Mississippian-Early Pennsylvanian Monillas Formation, in the area of Minas Plomosas-Placer de Guadalupe, Chihuahua (Bridges, 1964). The genus *Paraconularia* has been reported for the Middle Permian Los Arcos/Olinalá Formation in Guerrero (González-Arreola *et al.* 1994) and for the Pennsylvanian Ixtaltepec Formation in Oaxaca (Buitrón *et al.* 2000).

3. Corals. – During the Late Paleozoic the reefs were formed by a variety of organisms, such as sponges, bryozoans, calcareous algae and some classes of brachiopods. Rugose and tabulate corals were ahermatypical organisms, although they lived associated to the reefs. The most thorough study of rugose and tabulate corals of Mexico was from Mississippian rocks of Sonora (Easton, *et al.* 1958; González-León, 1986).

4. Bryozoans. – Prominent in this group are the Cyclostomata bryozoans of the Pennsylvanian Horquilla Formation in Sonora (González-León, 1986) and the undetermined Cryptostomata bryozoans from the Pennsylvanian Ixtaltepec Formation in Oaxaca (Ortíz-Lozano, 1996). Of the latter, the fenestellids were very diverse and abundant and dominate in the marine communities, however their abundance may be overestimated due to the fact that many are commonly preserved as small, broken fragments.

5. Brachiopods. – Brachiopods are well represented throughout the Carboniferous-Permian localities of Mexico, as they are also in Carboniferous-Permian rocks of the rest of North America. At the generic level great quantities of cosmopolitan brachiopod forms are recognized, but it is notable that the genera that have a restricted geographic distribution are typically North American. Amongst the biotic communities in which brachiopods are found, they represent the greatest relative abundance, but, in some cases, brachiopod diversity is not as extensive as that of bivalves. The best record of Mexican fossil brachiopods are from the: Mississippian of Tamaulipas, Oaxaca and Sonora (Sour-Tovar, 1999; Navarro-Santillán *et al.* 2002; Sour-Tovar and Martínez-Chacón, 2004; Sour-Tovar *et al.* 2005a; Easton, 1958; González-León, 1986); Pennsylvanian of Oaxaca (Sour-Tovar, 1994; Sour-Tovar and Quiroz-Barroso, 1989; Sour-Tovar and Martínez-Chacón, 2004); and Permian of Sonora and Coahuila (Cooper *et al.* 1965; Cloud, 1944; Wardlaw *et al.* 1979). The brachiopod faunas of the Carboniferous of Tamaulipas, Puebla, Guerrero and Oaxaca are similar to Carboniferous outcrops in the entire region of the Mid-Continent, while the Permian brachiopod faunas from Mexico are limited in their similarity to the faunas of the southeastern United States, in particular to those found in the States of Texas and New Mexico.

6. Scaphopods. – The only two examples of scaphopods referred to in the literature pertain to the genus *Plagioglypta* from the Permian of El Antimonio and Cananea, Sonora (Cooper *et al.* 1965; Blodgett, *et al.* 2002).

7. Gastropods. – Gastropods are well represented however their preservation usually does not allow identification to species level. The greatest diversity of gastropods has been reported from Permian rocks of Sonora (Blodgett *et al.* 2002; González-León, 1986; Knight *in* Cooper *et al.* 1965).

8. Rostroconchs. – Rostroconchs are an extinct class of mollusks represented in Mexico by *Pseudomulceodens cancellatus*, which has permitted confirmation of Osagean age (Early Mississippian) for the Santiago Formation located within the Nochixtlan region of Oaxaca (Quiroz-Barroso *et al.*, 2000). In previous years, *Conocardium sp.* was referred to as a bivalve from Mississippian rocks of northeastern Sonora (Easton *et al.* 1958).

9. Bivalves. – The greatest level of diversity of bivalves reported to date is from the Early-Middle Pennsylvanian Ixtaltepec Formation, with 26 species from the families Nuculidae, Malletiidae, Nuculanidae, Parallelodontidae, Myalinidae, Pterineidae, Pterinopectinidae, Aviculopectinidae, Myophoriidae, Permophoridae, Astartidae, Edmondiidae and Grammysiidae. These species include epifaunal and infaunal elements. Similar to the brachiopods of the Carboniferous from Ixtaltepec, bivalves demonstrate relationships to faunas of the Mid-Continent (Quiroz-Barroso and Perrilliat, 1997, 1998).

10. Cephalopods. – Rocks that contain Upper Paleozoic cephalopods are located in various states of Mexico. Permian cephalopod species predominate in both number and geographic distribution. This is exemplified in the systematic study of the Permian cephalopods from Delicias, Coahuila, which demonstrated the diversity attained by the group during the Upper Paleozoic (Miller, 1944). Permian cephalopods have also been reported from the Patlanoaya Formation of Puebla (Villaseñor-Martinez *et al.* 1987) and from Mississippian rocks of Sonora (Easton *et al.* 1958).

11. Trilobites. – Among the communities of invertebrates of the Upper Paleozoic of Mexico the trilobites are scarce and their record is limited to the report of: *Griffithides ixtaltepecensis* for the Pennsylvanian Ixtaltepec Formation in Oaxaca (Morón-Ríos and Perrilliat, 1988); *Anisopyge perannulata* from the Middle Permian Los Arcos/ Olinalá Formation of Guerrero (Garibay-Romero *et al.* 1998); and *Anisopyge whitei* from the Lower Permian Tuzancoa Formation of Hidalgo (Sour-Tovar *et al.* 2005b). These species, with exception of *Anisopyge perannulata* of which only one example has been found, are present in relative abundance, forming part of communities of terrigenous platform environments. The small diversity of Carboniferous-Permian trilobites is a trait that is shared with the majority of the contemporary localities of the world and is the principal product of the decline of this group throughout the Upper-Middle Paleozoic.

12. Crinoids. – Upper Paleozoic crinoids of Mexico are abundant, but with the exception of the crinoid described by Strimple (1971) from the Permian of Coahuila, the crinoid fossils consist principally of fragments and isolated plates of stems. The majority of the crinoid studies in Mexico have followed the classification proposed by Moore and Jeffords in 1968 mainly for describing parts disassociated from their columns (stems) (Buitrón, 1977b; Villaseñor-Martínez *et al.* 1987; Buitrón *et al.* 1987; Velasco de León and Buitrón, 1991, 1992; Esquivel-Macías, 1996; Esquivel-Macías *et al.* 2000). Although a vast quantity of isolated plates may have corresponded to one single organism, the abundance and the variety of crinoid forms indicate that the group was an important and diverse constituent of the marine communities of the Upper Paleozoic.

13. Ophiuroids. – The only reference to the Ophiuroid group for the Paleozoic of Mexico corresponds to an example of the family Ophiurinidae, collected from the argillaceous strata of the Ixtaltepec Formation (Quiroz-Barroso and Sour-Tovar, 1995). It is possible that it represents a new genus and, at the same time, constitutes the most austral record of Pennsylvanian ophiuroids known to date.

Acknowledgments

This research was supported by PAPIIT project number IN228803, DGAPA, UNAM.

References

Aguilera, J. G., 1897, Sinopsis de geología mexicana, *Bol. Inst. Geol. Méx.*, 4-6:187-250.

Arellano-Gil, J., Vachard, D., Yussim, S., and Flores de Dios-González, A., 1998, Aspectos estratigráficos, estructurales y paleogeográficos del Pérmico Inferior al Jurásico Inferior en Pemuxco, Estado de Hidalgo, México, *Rev. Mex. Cienc. Geol.*, 15:9-13.

Blodgett, R. B., Moore, T. E., and Gray, F., 2002, Stratigraphy and paleontology of Lower Permian rocks north of Cananea, northern Sonora, Mexico, *J. S. Am. Earth Sci.,* 15:481-495.

Böse, E., 1905, Reseña acerca de la geología de Chiapas y Tabasco, *Bol. Inst. Geol. Méx.*, 20:5-100.

Bridges, L. W., 1964, Geology of mina Plomosas-Placer de Guadalupe area, Chihuahua, Mexico, *W. Tex. Geol. Soc.,* 50-64:50-54.

Buitrón, B. E., 1977a, *Bellerophon (Bellerophon) crassus* Meek and Worthen (Mollusca, Gastropoda) en el Pérmico de Chiapas, *Univ. Nal. Autón. Méx., Inst. Geol., Rev.,* 1:69-73.

Buitrón, B. E., 1977b, Invertebrados (Crinoidea y Bivalvia) del Pensilvánico de Chiapas, *Univ. Nal. Autón. Méx., Inst. Geol., Rev.*, 1:144-150.

Buitrón, B. E., Flores de Dios-González, A., and Vachard, D., 2000, Nuevos hallazgos de invertebrados paleozoicos (Scyphozoa-Conulata, Achaeogastropoda-Trochina y Echinodermata-Crinoidea) en la región de Nochixtlan-Ixtaltepec, Oaxaca. *2ª Reunión Nacional de Ciencias de la Tierra, Bol. Inf. Un. Geof. Mex.,* 20, p. 298.

Buitrón, B. E., Patiño, J., and Moreno, A., 1987, Crinoides del Paleozoico Tardío (Pensilvánico) de Calnali, Hidalgo, *Rev. Soc. Mex. Paleont.*, 1:125-136.

Cloud, P. E., 1944, Permian brachiopods, pt. 3 of Geology and Paleontology of the Permian area northwest of Las Delicias, southwestern Coahuila, *Geol. Soc. Am. Spec. Pap.,* 52:1-49.

Cooper, G. A., Arellano, A. R. V., Johnson, J. H., Okulitch, V. J., Stoyanow, A., and Lochman, Ch., 1954, Geología y paleontología de la Región de Caborca, norponiente de Sonora, Parte 1, Paleontología y estratigrafía del Cámbrico de Caborca, *Univ. Nal.Autón. Méx., Inst. Geol., Bol.,* 58:1-259.

Cooper, G. A., Dunbar, C. O., Duncan, H, Miller, A. K., and Knight, J. B., 1965, Fauna pérmica de El Antimonio, oeste de Sonora, *Univ. Nal.Autón. Méx., Inst. Geol., Bol.,* 58:1-122.

Easton, W. H., Sanders, J. E., Knight, J. B., and Miller, A. K., 1958, Mississippian Fauna in Northwestern Sonora Mexico, *Smith. Misc. Coll.*, 119:1-87.

Esquivel-Macías, C., 1996, Invertebrados del Paleozoico Tardío de las regiones de Olinalá, Guerrero y Patlanoaya, Puebla, Unpublished M.S. thesis, Universidad Nacional Autónoma de México, Facultad de Ciencias.

Esquivel-Macías, C., 2000, Braquiópodos y Crinoides del Paleozoico tardío de las Formaciones Olinalá, Gro; Patlanoaya y Cuxtepeque, Puebla, Unpublished Ph.D. thesis, Universidad Nacional Autónoma de México, Facultad de Ciencias.

Esquivel-Macías, C., Ausich, W. I., Buitrón-Sanchez, B. E., and Flores de Dios, A., 2000, Pennsylvanian and Mississippian pluricolumnal assemblages (class Crinoidea) from southern México and a new occurrence of a column with a tetralobate lumen, *J. Paleontol.*, 74:1187-1190.

Garibay-Romero, L. M., Guerrero-Suástegui M., and Torres de León, R., 1998, Trilobite pérmico de la

Formación Olinalá (Los Arcos), Estado de Guerrero, *VI Congreso Nacional de Paleontología, Soc. Mex. Paleont.*, Resúmenes, p. 23.
Gastil, R. G., Lemone, D. V., and Stewart , W. J., 1973, Permian Fusulinids from near San Felipe, Baja California, *Am. Assoc. Petr. Geol., Bull.*, 57:746-747.
González-Arreola, C., Villaseñor-Martinez, A. B., and Corona-Esquivel, R., 1994, Permian fauna of the Los Arcos Formation, Municipality of Olinalá, State of Guerrero, Mexico. *Rev. Mex. Cienc. Geol.* 11:214-221.
González-León, C., 1986, Estratigrafía del Paleozoico de la Sierra del Tule, noreste de Sonora, *Univ. Nal. Autón. Méx., Inst. Geol., Rev.,* 6:117-135.
Gutiérrez-Gil, R., 1956, Bosquejo Geológico del Estado de Chiapas, in: *Geología del Mesozoico y estartigrafía pérmica del Estado de Chiapas* (Maldonado-Koerdell, M., ed.), *XX Congreso Geológico Internacional, Mexico*, Excursión C-15, p. 9-32.
López-López, M. B., 1997, Braquiópodos del Paleozoico Tardío de la región de Olinalá, Estado de Guerrero, Unpublished Professional thesis, Universidad Nacional Autónoma de México.
McEldowney, R. C., 1970, An occurrence of Paleozoic fossils in Baja California, Mexico, *Cordilleran Section 66th Annual Meeting, Geological Society of America,* Abstracts with Programs, p. 117.
Miller, A. K., 1944, Geology and paleontology of the permian area northwest of Las Delicias, southestern Coahuila, Mexico, part. 4: Permian Cephalopods, *Geol. Soc. Am. Spec. Pap.*, 52:71-127.
Moore, R. C., and Jeffords, R. M. 1968, Classification and nomenclature of fossil crinoids based on studies of dissociated parts of their columns, in: *Echinodermata* (Moore, R. C., ed.), The University of Kansas Paleontological Contributions. Univ. Kansas Paleont. Inst., 46:1-86.
Moreno-Cano, . A., and Patiño-Ruiz, J., 1981, Estudio del Paleozoico en la Región de Calnali, Hgo. (en la Sierra madre Oriental), Unpublished Professional thesis, Instituto Politécnico Nacional.
Morón-Ríos, A., and Perrilliat, M. C., 1988, Una especie nueva del género *Griffithides* Portlock (Arthropoda, Trilobita) del Paleozoico Superior de Oaxaca, *Univ. Nal. Autón. Méx., Inst. Geol., Rev.,* 7:67-70.
Müllerried, F. K. G., Miller, A. K., and Furnish, W. M., 1941, The Middle Permian of Chiapas, southernmost Mexico and its fauna, *Am. J. Sci.*, 239:397-406.
Navarro-Santillán, D., Sour-Tovar, F., and Centeno-García, E., 2002, Lower Mississippian (Osagean) brachiopods from the Santiago Formation, Oaxaca, Mexico: stratigraphic and tectonic implications, *J. S. Am. Earth Sci.,* 15:327-336.
Ortega- Gutiérrez, F., Sedlock, R. L., and Speed, R. C., 2000, Evolución tectónica de México durante el Fanerozoico, in: *Biodiversidad, taxonomía y biogeografía de artrópodos de México: hacia una síntesis de su conocimiento, volúmen II* (Llorente-Bousquets, J, González-Soriano, E., and Papavero, N., eds.), 676 p.
Ortiz-Lozano, J. A., 1996, Consideraciones Paleoambientales y Morfológicas de las Colonias de Briozoarios Pensilvánicios de la Formación Ixtaltepec en la Región de Nochixtlán, Oaxaca, Unpublished Professional thesis, Universidad Nacional Autónoma de México.
Quiroz-Barroso, S. A., and Perrilliat, M. C., 1997, Pennsylvanian nuculoids (Bivalvia) from the Ixtaltepec Formation, Oaxaca, Mexico, *J. Paleontol.*, 71:400-407.
Quiroz-Barroso, S. A., and Perrilliat, M. C., 1998, Pennsylvanian Bivalves from the Ixtaltepec Formation, Mexico, *J. Paleontol.*, 72:1011-1024.
Quiroz-Barroso, S. A., Pojeta Jr., J., Sour-Tovar, F., and Morales-Soto, S., 2000, *Pseudomulceodens*: a Mississippian Rostroconch from Mexico, *J. Paleontol*, 74:1184-1186.
Quiroz-Barroso, S. A., and Sour-Tovar, F., 1995, Nuevo registro de ofiuroideo (Ophiurinidae) para el Pensilvánico de América del Norte, proveniente de la Formación Ixtaltepec, Oaxaca., *V Congreso Nacional de Paleontología, Soc. Mex. Paleont.*, Resúmenes, p. 31.
Reyeros de Castillo, M. M., 1976, Corales del Pérmico Inferior del Estado de Chiapas, México, Univ. Nal. Autón. Méx., Inst. Geol., Paleontol. Mex., 41, 18 p.
Robison, R. A., and Pantoja-Alor, J., 1968, Paleozoic sedimentary rocks in Oaxaca, Mexico, *Science*, 17: 1033-1035.
Rosales-Lagarde, L., Centeno-García, E., Dostal, J., Sour-Tovar. F., Ochoa-Camarillo H., and Quiroz-

Barroso, S., In press, The Tuzancoa Formation of Hidalgo state: evidence of an Early Permian Submarine Continental Arc in Eastern Mexico, *Rev. Mex. Cienc. Geol.*
Sánchez-Zavala, J. L., Centeno-García, E., and Ortega-Gutierrez, F., 1999, Review of paleozoic stratigraphy of Mexico and its role in the Gondwana-Laurentia connections, in: *Laurentia-Gondwana connections before Pangea* (Ramos, V. A., and Duncan, K. J., eds), *Geol. Soc. Am. Spec. Pap.,* 336:211-226.
Sour-Tovar, F., 1990, Comunidades Cámbrico-Ordovícicas del área de Santiago Ixtaltepec, Oaxaca. (Formación Tiñú): Implicaciones Paleoambientales y Paleogeográficas, *Rev. Soc. Mex. Paleont.,* 3:7-24.
Sour-Tovar, F., 1994, Braquiópodos pensilvánicos del área de Santiago Ixtaltepec, Municipio de Nochixtlán, Oaxaca, Unpublished M.S. thesis, Universidad Nacional Autónoma de México.
Sour-Tovar, F., 1999, Braquiópodos del Misisípico Inferior (Osageano) de la Formación Vicente Guerrero en el Anticlinorio Huizachal-Peregrina, norte de Ciudad Victoria, Tamaulipas, México, Unpublished Ph.D. thesis, Universidad Nacional Autónoma de México.
Sour-Tovar, F. and Martínez-Chacón, M. L., 2004, Braquiópodos chonetoideos del Carbonífero de México, *Rev. Esp. Paleont.*, 19:125-138.
Sour-Tovar, F., and Quiroz-Barroso, S. A., 1989, Braquiopodos pensilvánicos (Strophomenida) de la Formación Ixtaltepec, Santiago Ixtaltepec, Oaxaca., *Rev. Soc. Mex. Paleont.*, 2:5-11.
Sour-Tovar, F., Alvarez, M. F., and Martínez-Chacón, M. L., 2005a, Lower Mississippian (Osagean) spire bearing brachiopods from Cañon de la Peregrina, north of Ciudad Victoria, Tamaulipas, Northeastern Mexico, *J. Paleontol.* 79: 475-491.
Sour-Tovar, F., Quiroz-Barroso, S. A., and Applegate, S. P., 2000, Presence of *Helicoprion* (Chondrichthyes, Elasmobranchii) in the Permian Patlanoaya Formation, Puebla, Mexico, *J. Paleontol.*, 74:363-366.
Sour-Tovar, F., Pérez-Huerta, A., Quiroz-Barroso, S. A., and Centeno-García, E., 2005b, Braquiópodos y trilobites del Pérmico del noroeste del estado de Hidalgo, México, *Rev. Mex. Cienc. Geol.,* 22:24-35.
Strimple, H. L., 1971, A Permian Crinoid from Coahuila, Mexico, *J. Paleontol.*, 45:1040-1042.
Thompson, M. L., 1956, Rocas paleozoicas del Sur de México, in: *Geología del mesozoico y estartigrafía Pérmica del Estado de Chiapas* (Maldonado-Koerdell, M., ed.), *XX Congreso Geológico Internacional, México,* Excursión C-15, 82 p.
Velasco de León, P., and Buitrón, B. E., 1991, Algunos crinoides del Pensilvánico de la Región de San Salvador Patlanoaya, estado de Puebla, México, *III Congr. Nal. Paleont., Soc. Mex. Paleont.*, Resúmenes, p. 138.
Velasco de León, P., and Buitrón, B. E., 1992, Algunos crinoides (Echinodermata-Crinoidea) del Misisípico-Pensilvánico de San Salvador Patlanoaya, Estado de Puebla, *Rev. Soc. Mex. Paleont.*, 5: 71- 81.
Villaseñor-Martínez, A. B., Martínez-Cotés, A., and Contreras-Montero, B., 1987, Bioestratigrafía del Paleozoico Superior de San Salvador Patlanoaya, Puebla, Méx., *Rev. Soc. Mex. Paleont.*, 1:396-417.
Wardlaw, B. R., Furnish, W. M., and Nestell, M. K., 1979, Geology and Paleontology of the Permian beds near Las Delicias, Coahuila, Mexico, *Geol. Soc. Am. Bull.,* 90:111-116.
Webster, G. D., and Delattre, M. P., 1993, Permian crinoid columnals from the Zamora Formation near El Volcan, Baja California Norte, México, in: *The Prebatholithic Stratigraphy of Peninsular California* (Gastil, R. G. and Miller, R. H., eds.), *Geol. Soc. Am. Spec. Pap.*, 279:91-95.

Chapter 9

Mesozoic Osteichthyans of Mexico

JESÚS ALVARADO-ORTEGA[1], KATIA A. GONZÁLEZ-RODRÍGUEZ[2], ALBERTO BLANCO-PIÑÓN[3], LUIS ESPINOSA-ARRUBARRENA[1], and ERNESTO OVALLES-DAMIÁN[4]

[1] • Instituto de Geología, Universidad Nacional Autónoma de México, Ciudad Universitaria, 04510, D. F., México.
[2] • Centro de Investigaciones Biológicas, Universidad Autónoma del Estado de Hidalgo, Km. 4.5 carretera Pachuca-Tulancingo, 42184 Pachuca, Hidalgo, México.
[3] • Centro de Investigaciones en Ciencias de la Tierra, Universidad Autónoma del Estado de Hidalgo, Km. 4.5 Carretera Pachuca-Tulancingo, 42184 Pachuca, Hidalgo, México.
[4] • Museo Eliseo Palacios Aguilera, Calz. de los Hombres Ilustres s/n Parque Madero, Colonia Centro, CP 29000, Tuxtla Gutiérrez, Chiapas, México.

1. Abstract

This paper provides an updated review of the Mesozoic fossil Osteichthyan record of Mexico. This catalogue includes multiple entries of papers, theses, and diverse scientific meeting reports regarding Cretaceous Osteichthyan fishes of Mexico that

F. Vega et al. (eds.), Studies on Mexican Paleontology, 169–207.

reflect the present state of knowledge. In addition, it provides a review of the Mexican paleoichthyology history, including the most recent fossil Osteichthyan discoveries. Within this framework some systematic and paleobiogeographical implications generated from the enrichment of the Mexican Mesozoic Osteichthyan record are briefly discussed providing future perspectives to this young field of Mexican science.

2. Introduction

Since 1840, when the first report of a Mexican fossil fish appeared, the advancement of Paleoichthyology has faced many difficulties in Mexico. For a long time, only foreign paleontologists lead the investigations on Mexican fossil fishes. Nevertheless, after a slow and progressive collaboration with Mexican scientists, today reach on systematic, taphonomical, paleobiogeographical, and paleoenviromental aspects of Mexican fossil fishes is underway.

The history of Mexican paleoichthyology shifted drastically in 1981, after the discovery of the Tlayúa quarry, considered to be the First Mexican Lagersttäte by some national and international paleontologists due to its abundant, diverse, and extraordinary well-preserved fossils, for which fossil fishes are numerous. After the scientific discovery of the Tlayúa quarry locality by Shelton P. Applegate and Luis Espinosa-Arrubarrena, paleontologists and students have been incorporated into projects focused on the preservation and study of the Tlayúa fossils. Recently, young scientists are studying fish remains from new and old localities, increasing the Mexican fossil fish record and the number of specimens housed in collections of Mexican Museums and Universities. Additionally, scientifically minded people and Paleontological Amateur Associations have started to support the research of Mexican scientists and often report the discovery of new fossil fish localities. Today, a small group that involves at least two generations of students and mature paleontologists constitutes the most important circle of Mexican Paleoichthyologists known to date.

In this chapter, we provide an updated assessment on the Mexican fossil record of the Mesozoic Osteichthyans or bony fishes. Future investigations related with this group of vertebrates, must include the large number of discoveries of fossil fishes from Mexican territories. This review reconstructs the history and attempts to give some insight into the future of the Mexican Paleoichthyology.

3. History of the Mexican Paleoichthyology

Mexican pre-Hispanic civilizations had some contact with fossil fishes; but unfortunately, the beliefs related to this contact is lost. Evidence is represented by fish remains found on the limestone blocks used during the construction of Palenque, an important Mayan archaeological site in Chiapas State (Sapper, 1899; Mülleried, 1951), and a *Carcharodon megalodon* tooth found in La Venta, an Olmecas archeological site in Tabasco (Applegate, pers. obs., 2000; Gío-Argáez and Rodríguez-Arévalo, 2003).

Scientific research on Mexican fossil fishes started in the XIX century (Figure 1), when Herman von Meyer (1840) described the shark *Carcharodon mexicanus* (a possible *C. megalodon* synonymous), based on a single tooth from an uncertain locality in Michoacán. The first Mexican fossil Osteichthyan known, *Prymnetes longiventer* Cope, 1871, was erected based on a single specimen from an uncertain locality near Tuxtla Gutiérrez, Chiapas. Later, Felix (1891) reported the occurrence of *Thrissops*? remains and described two Early Cretaceous actinopterygians species, *Otomitla speciosa* and *Belenostomus ornatus,* from Cerro de la Virgen, Oaxaca.

At the end of the XIX century, the creation of several Mexican Institutions dedicated to Natural Science Research encouraged the participation of national scientists in palentological studies. Aguilera (1896, 1906, 1907), the first Mexican geologist-paleontologist that identified fossil fishes from Mexico, reported the occurence of Cretaceous fishes in his geological papers, including shark and *Ceratodus* remains from Los Peyotes, into the Parras Basin, Coahuila; and the doubtfully occurrence of *Diplomystus* sp. and *Syllaemus latifrons* from unknown localities. Villada (1897) in his "Catálogo de la Colección de Fósiles del Museo Nacional" included the record of a possible Cretaceous and still undescribed teleost, collected near Tula, Hidalgo.

The Mexican Revolution, started in 1910, which interrupted all paleontological research in the country. However, during this time foreign paleontologists published the first reports on Cenozoic shark teeth from Veracruz (Dickerson and Kew, 1917) and Baja California (Wittich, 1913). Meanwhile, Böse (1913) continued his studies on Mesozoic sharks from the northern part of Mexico. These studies provide the bases for future additional reserach (Jordan and Hertlein, 1926; Hertlein and Jordan, 1927; Hanna and Hertlein, 1927; and Böse and Cavins, 1928). In addition, during this period, Brown (1912) reported the occurrence of the first Mexican Cenozoic freshwater fossil fishes from Jalisco.

Once the Mexican Revolution was over, a slow but progressive consolidation of Mexican paleontologists was possible. During the following years, F. K. Müllerried contributed three papers on fossil fishes, including his identification of *Pycnodus* sp. and teleosts on limestone remains from the Archeological ruins of Palenque, Chiapas, previously noted by Sapper in 1899 (Müllerried, 1951). He also reported the discovery of Cretaceous shark teeth from Sierra de San Miguel, Hidalgo State (Müllerried, 1939) and described *Helicoprion mexicanus* from Las Delicias, Coahuila State (Müllerried, 1945), which still represents the only Paleozoic fish known from Mexico; two additional reports of *Helicoprion* sp. are from Chihuahua (Bridges and Deford, 1962) and Puebla (Sour-Tovar *et al.* 2000).

The incoming of Manuel Maldonado-Koerdell, the "last Humboltian naturalist of modern Mexico" (de Cserna, 1972), brought an additional impulse to the consolidation of the Mexican Paleontology. Manuel Maldonado Koerdell included 8 bony fishes and 25 sharks in his Catalogue of Mexican Fossil Fishes (Maldonado-Koerdell, 1948a and b; 1949). Some years later, Dunkle and Maldonado-Koerdell (1953) described the

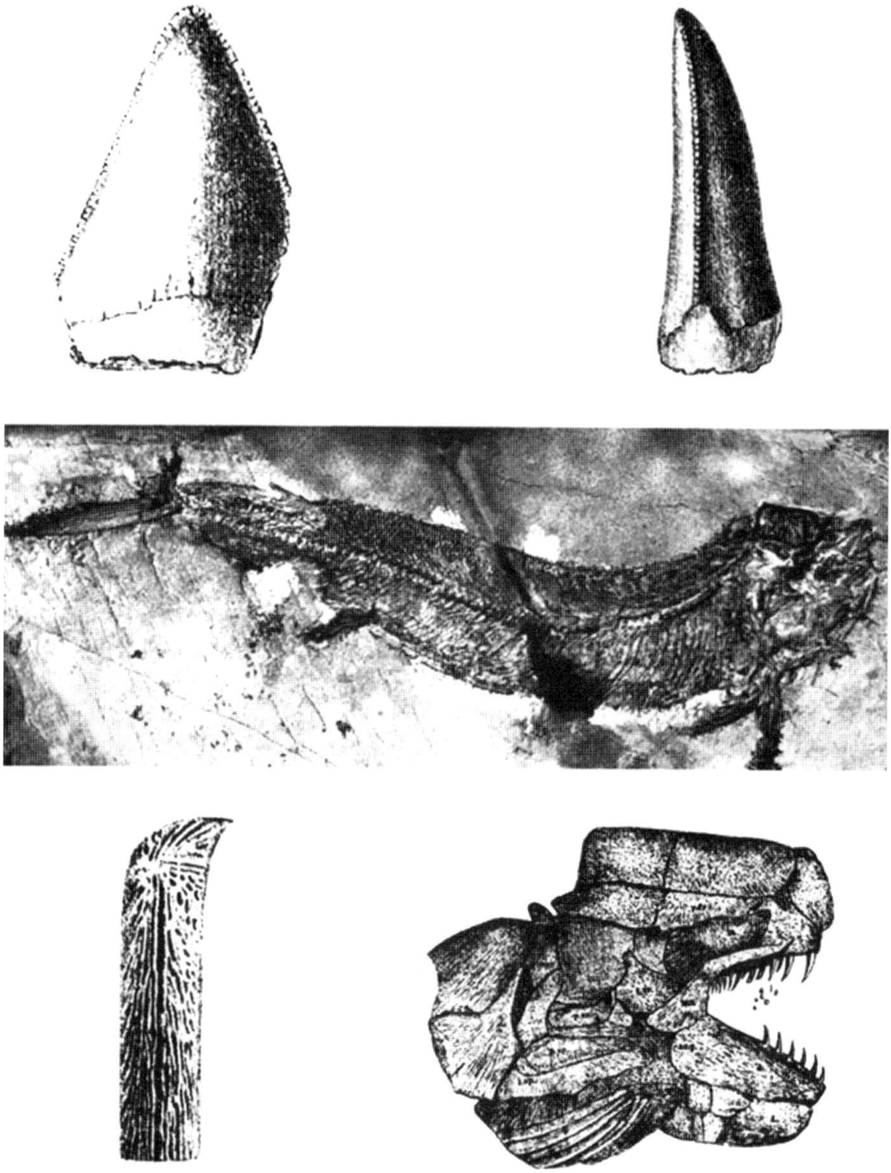

FIGURE 1. Remarkable Mexican fossil fishes described during the XIX century. A) *Carcharodon mexicanus* von Meyer, 1840, the first Mexican fossil fish described [frontal (left) and lateral (right) views as published by von Meyer (1867)]. B) *Prymnetes longiventer* Cope, 1871, the first Mexican fossil bony fish described (after Bardack, 1965, pl. I, fig. B). C) *Otomitla speciosa* Felix, 1891. D) Frontal bone of *Belenostomus ornatus* Felix, 1891, (C and D after Applegate, 1989, fig. 164, e and d respectively).

first Mexican Jurassic bony fish, *Gyrodus macrophthalmus*, and an Early Cretaceous teleosts, *Leptolepis tamanensis,* both discovered near Tamán, San Luis Potosí. Finally, Maldonado-Koerdell (1956) described the Upper Turonian ictiofauna from Xilitla, also in San Luis Potosí State.

Both World Wars halted the work of foreign paleontologists on Mexican fossil fishes, except for the report of Eocene-Oligocene teeth of *Carcharodon megalodon* from Tabasco (Leriche, 1938). After these conflicts, a new generation of foreign investigators assumed the studies on Mexican fossil sharks (Kuckow, 1957, 1959a and b, 1961; Bridges and Deford, 1962; Hertlein, 1966; Minch *et al.* 970; Ashby, 1987; Kruckow and Thies, 1990). Nevertheless, since that time the participation of Mexican institutions and paleontologists gradually became predominant, mainly in the area of fossil sharks and rays (Applegate, 1978, 1986, 1993; Applegate and Espinosa-Arrubarrena, 1981; Espinosa-Arrubarrena, 1979, 1987; Espinosa-Arrubarrena and Applegate, 1991; Durham, *et al.* 1981; Aranda-Manteca, 1994, 2002; Aguillón-Martínez *et al.*, 1998; González-Barba, 1995, 2000 a and b; González-Barba, *et al.* 1998; 2002; Blanco-Piñón, 1998; Ferrusquía-Villafranca *et al.* 1999; and Sour-Tovar *et a*l. 2000; Kirkland and Aguillón-Martínez, 1995; Kirkland et al., 2000; González-Barba *et al.* 2004; Alvarado-Ortega *et al.* 2004; Aranda-Manteca, 2004).

Studies on the Cenozoic freshwater osteichthyan fishes from the rich Plio-Pleistocene deposits of the Chapala Lake Basin, Jalisco, formally began in 1958 (Downs, 1958; Miller, 1965; Smith *et al.* 1975; Smith, 1980, 1981; Cavender and Miller, 1982). Eventually, these investigations included other areas, reaching the downtown of Mexico City (Bradbury, 1971) and a locality in Sonora (Shaw, 1982), which even today is substudied. Since the 1960´s Mexican paleontologists have been the sole researchers related to the freshwater fossil fishes of Mexico. Block-Iturriaga (1963) described "*Algasea paleomexicana*" (informal unpublished name) based on Pleistocene freshwater fish remains collected by Maldonado-Koerdell in Tlapacoya, Estado de México. Hernández-Junquera (1977) recognized a skull belonging to *Ictalurus* in his description of the Pleistocene fossils from Laguna de la Media Luna, San Luís Potosí State. Alvarez and Arriola (1972) erected the goodeid *Tapatia occidentalis* from Santa Rosalía, Jalisco State. Alvarez (1966, 1974) and Alvarez and Moncayo (1976) continued the studies on fossil fishes from Jalisco and Estado de México. Espinosa-Pérez and collaborators (1991) identified a badly-preserved cyprinodontid fish from Cenozoic deposits at the Los Ahuehuetes locality, near Tepexi de Rodríguez, Puebla State. Guzmán and collaborators (1998) performed a stratigraphical analysis of the type locality of *Tapatia.* Becerra-Martínez (2003) described the Pliocene goodeid fishes (based only on fins and tails) from Atotonilco El Grande Formation, Hidalgo State. Alvarado-Ortega and Carranza-Castañeda (2002a-c) identified Pliocene *Ictiobus* sp. and *Ictalurus* sp. from freshwater localities in Hidalgo State, previously reported for the latter author (in Carranza-Castañeda *et al.* 1982; Carranza-Castañeda and Miller, 2000, Castillo-Cerón *et al.* 1996). Recently, Castañeda-Posadas (2004) also reported the presence of fossils belonging to these two latter genera from Miocene outcrops near Panotla, Tlaxcala.

Investigations on Mexican Cenozoic marine bony fishes began with the identification of several species based on otholits from Veracruz (Weiler, 1959). Scarce additional works include the report of some fish remains from Baja California (Minch et al. 1970; Aranda-Manteca, 1994) and the report of the blue marlin *Makaira nigricans* (Fierstine, *et al.* 2001). These minimum studies only show a small part of the large array of the bony fish remains collected since the 70's along the Baja California Peninsula.

In the mid 50´s Dunkle predicted the presence of a Cretaceous lithographic limestone locality, existing somewhere in Mexico. During many years there was practically no progress in the discovery and study of Mesozoic bony fishes in Mexico, except for a vague report of uncertain fish remains from El Gallo Formation, Baja California (Lillegraven, 1972). Nevertheless, in 1981 Dunkle´s prediction came true with the scientific discovery of the Tlayúa Quarry by S. P. Applegate, L. Espinosa-Arrubarrena, and P. López-Neri.

Applegate, herein considered the "Father of the Mexican Paleoichthyology" due his participation in the formation of new generations of Mexican paleoichthyologists (Figure 2), arrived to Mexico during the 70's interested on the fossil sharks localities in the Baja California. Applegate and his first Mexican student, L. Espinosa-Arrubarrena, created one of the largest collections of fossil fishes from this State including shark and teleostean teeth, now housed in the Colección Nacional de Paleontología (IGM). Since the scientific discovery of the Tlayúa Quarry, both paleontologists have been in charge of projects dedicated to the collect, preservation, and study of the fossils from this locality (Applegate and Espinosa-Arrubarrena, 1981, 1982; Espinosa-Arrubarrena and Applegate, 1988, 1992, 1996; among others).

The incorporation of young students from the Facultad de Ciencias, UNAM, to these projects has yielded unpublished theses and scientific papers related to Cretaceous Osteichthyans (González-Rodríguez, 1989, 1996, 2004; González-Rodriguez and Reynoso, 2004; González-Rodríguez *et al.* 2004; and Alvarado-Ortega, 1998a and b, 2004). L. Grande and D. Bardack, both from the Field Museum of Chicago, were involved in the Tlayúa project being co-responsible in a National Science Foundation grant from 1983 to 1986. Nowadays, L. Grande is coauthor of the amiid *Pachyamia mexicana* Grande and Bemis, 1998, and his numerous personal observations are included into papers that deal with the Tlayúa fishes (i.e. Chang and Grande, 1997; Brito, 1997; Maisey, 2000; Grande, 1999).

In the mid-80´s, Barrios-Rivera (1985) reviewed the literature on Mexican fossil vertebrates; in this work, he recorded 22 Mesozoic Osteichthyan fossil fish taxa from 12 localities. This catalogue shows a significant progress from those 8 bony fishes reported from the 1948-1956 Maldonado-Koerdell´s Fossil Fishes of Mexico. Furthermore, González-Rodríguez (1990) evaluated more extensively these records, including the collections housed in the Instituto de Geología and Facultad de Ciencias, UNAM, she found 20 Cretaceous fossil fish localities in the States of Durango, Coahuila, Tamaulipas, Hidalgo, Chihuahua, Nuevo León, San Luis Potosí,

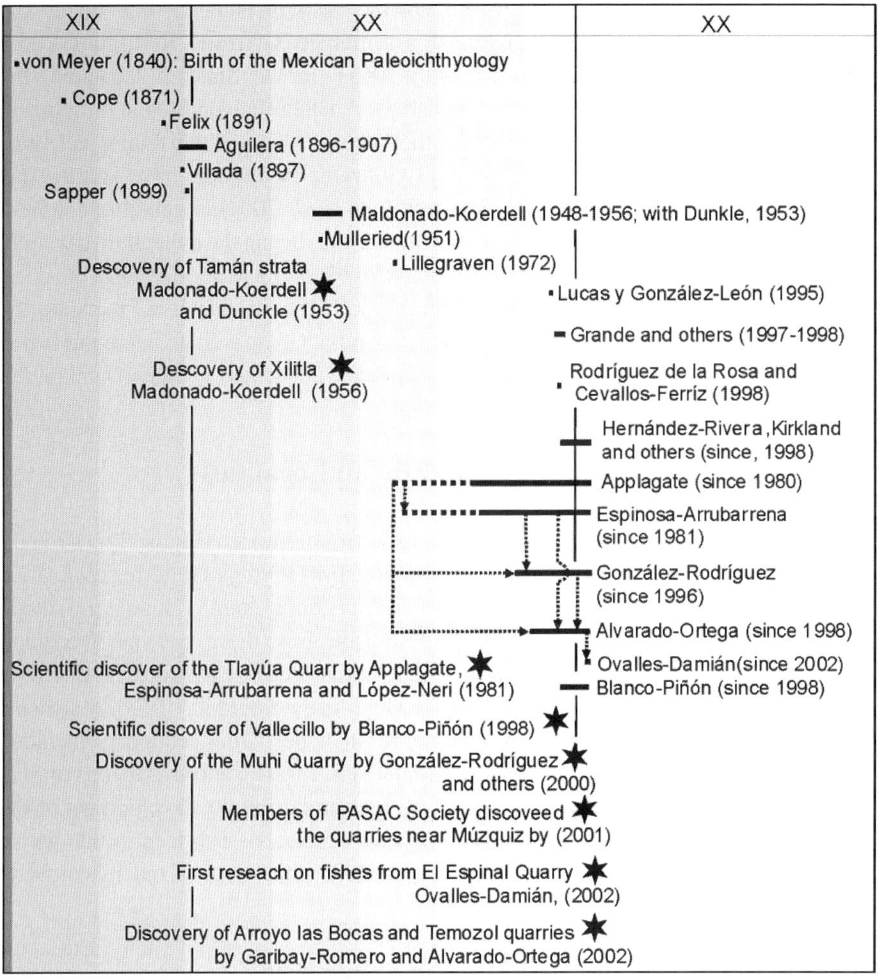

FIGURE 2. History of the Mexican fossil osteichthyan researches since the birth of the Mexican Paleoichthyology in XIX century. Fine arrowed lines show the teacher-student relationships. Parentheses show the period of active time of each author into the paleoichthyology field.

Michoacán, Oaxaca, Veracruz, Chiapas, and Puebla.

In 1998, Blanco-Piñón (1998) started to work in Vallecillo, a Late Cretaceous (Turonian) fossil locality near Monterrey City, Nuevo León State, reporting the existence of well-preserved bony fishes. In this same year, Paleontology students of Chiapas State began the field works throughout the State, discovering a good number of Mesozoic Osteichthyans (Avendaño-Gil *et al.* 1998, 1999).

Carranza-Castañeda and Applegate (1994) described an isolated uncertain pycnodont vomer from Hidalgo State. More recently, González-Rodríguez and collaborators (2000) reported the Muhi Quarry, also in Hidalgo, and began the taxonomical studies of its Cenomanian fish assemblage (González-Rodríguez, 2002; González-Rodríguez and Applegate, 2000, 2001). In the State of Coahuila, the members of PASAC (abbreviation in Spanish for Amateur Paleontologist of Sabinas) discovered some fossil fish localities near the town of Múzquiz, Coahuila, and since 2003, Alvarado-Ortega, Blanco-Piñón, and González-Rodríguez, have studied this new Late Cretaceous fish locality (Blanco-Piñón *et al.* 2004). Regarding southern Mexico, Garibay-Romero and Alvarado-Ortega (2002) began the collection and study of fishes from two Late Cretaceous localities, Arroyo Las Bocas and Temozol quarries, near Taxco, Guerrero State. In addition, during 2002 the first results of the study on the fishes from El Espinal Quarry, near Ocozocoautla, Chiapas State, were published (González-Rodríguez *et al.* 2002; Ovalles-Damián, 2004; and Ovalles-Damián and Alvarado-Ortega, 2002).

4. Main Mesozoic Mexican Osteichthyan Localities

Distribution of the Mesozoic fossil fish localities in Mexico (Figure 3) is strongly influenced by geological events (see López-Ramos, 1981; Ortega-Gutiérrez *et al.* 1992, 2000, among others). Although some marine, transitional, and continental Triassic outcrops are present in Mexico, reports of fossil fishes from these rocks are inexistent. Nowadays, the Kimmeridgian section of the Tamán beds near Moctezuma River, Tamán, San Luis Potosí State (Dunkle and Maldonado-Koerdell, 1953), represents the only Mexican Jurassic locality bearing fossil fishes. This section is included within the complex sedimentary system associated to Jurassic and Cretaceous marine platforms, islands, reefs, and basins that were present during the development of the Gulf of Mexico. Interestingly, the only Mexican Cretaceous fish locality known so far that is considered as a fresh water deposit is El Espinal Quarry, in Chiapas State (González-Rodríguez *et al.* 2002).

4.1. Tlayúa Quarry: the First Mexican Largestätte

The Tlayúa Quarry is a 30 to 34 m. thick outcrop named as the Middle Member of the Tlayúa Formation by Pantoja-Alor (1992). It is a sequence of thin laminated yellow-brown limestone strata with red-orange clay enriched for hematitic minerals forming the bedding planes. The Tlayúa fossil assemblage includes algae, bentonic foraminifers, sponges, gorgonians, annelids, gastropods, bivalves, ammonoids, belemnoids, crustaceans, arachnids, insects, echinoids, asteroids, holoturids, crinoids, ophiuroids, fishes, reptiles, poorly preserved wood, pollen, permineralized wood and plant prints (see Alvarado-Ortega, 2004; Applegate 1988, 1992, 1996; Applegate *et al.* 2000; Feldmann *et al.* 1998; González-Rodríguez *et al.* 2004; Grande and Bemis, 1998; among others).

Based on its ammonoids and belemnoids Middle to Late Albian age has been assigned to the Middle Member of the Tlayúa Formation (Seibertz and Buitrón, 1987; Buitrón and Malpica-Cruz, 1987, and Seibertz and Spaeth, 2002). Recently, Kashiyama *et al.* (2004), suggested a possible Aptian age for these strata based on the occurrence of several inner-shelf, benthonic foraminifers. Nevertheless, a paleomag netostratigraphical analysis performed by Benammi and collaborators (2003, 2004) confirms the Albian age assigned to this locality.

About 70% of the macrofossils collected in the Tlayúa Quarry belong to diverse Osteichthyan fish taxa (Table 1), including a sarcopterygian (Espinosa-Arrubarrena *et al.*, 1996) and several actinopterygian groups. Actually, a large part of these fishes are identified at different supra-specific levels (Applegate, 1996; Applegate *et al*, 2000) and being potential new species. Contrary to Pantoja-Alor (1992) and Maisey (2000) we remark the absence of shark remains in this locality.

The fossil fish species formally described from this locality involve the ionoscopiform *Teoichthys kallistos* Applegate, 1988; the pycnodontiform *Tepexichthys aranguthyorum* Applegate, 1992; the amiid *Pachyamia mexicana* Grande and Bemis, 1998; the macrosemiids *Macrosemiocotzus americanus* González-Rodríguez, Applegate and Espinosa-Arrubarrena, 2004, and *Notagogus novomundi* González-Rodríguez and Reynoso, 2004; as well as the Ichthyodectiform *Unamichthys espinosai* Alvarado-Ortega, 2004. Besides, the descriptions of several new fish taxa from Tlayúa are in preparation, including a *Lepidotes* sp., two macrosemiids, an ionoscopiform, three ichthyodectiforms, an ellimmichthyiform, and at least seven pycnodontiforms.

Diverse paleoecological models have been suggested to explain the extraordinary preservation of the vertebrate fossils of the Tlayúa Quarry. In one of these models, the Tlayúa deposit was possible in an Open Marine Basin with restricted circulation, with storm-dominated sedimentation and strong terrestrial influence (Kashiyama *et al.* 2004). In contrast, other authors suggest that the deposit of the Tlayúa limestone strata occurred in a transitional environment; according to Applegate and collaborators (in this volumen), it was into a restricted basin near to the coast, which is comparable to the lagoonal model proposed for the fossil bearing limestone beds at Solnhofen, Germany. Contrary to these models, Malpica-Cruz *et al.* (1989) and Pantoja Alor (1992) suggest a more open back-reef lagoon, nevertheless, recently Guerrero-Arenas (2004) [see also Guerrero-Arenas *et al.* (2002)] corroborated the restricted-basin condition for the Tlayúa deposit based on a Carbon and Oxygen isotopic analysis showing that freshwater was contained into Tlayúa basin during at least some depositional events.

4.2. Vallecillo

In 1998, inhabitants of Vallecillo, a small village located 120 Km north of Monterrey City, Nuevo León State (Figure 3), began the extraction of flagstones in some marl outcrops three kilometers west of this town. Shortly, the beauty of a large array of vertebrate and invertebrate fossils discovered in these rocks encouraged the local government to create a small museum, "Museo Casa de la Plomada", to exhibit these materials.

The fossil bearing strata of Vallecillo consist of a homogeneous sequence of about 4 m thick finely laminated pink marlstone interlayed by 6-3 cm thick red-brown shales. Blanco-Piñón (1998) and Blanco-Piñón *et al.* (2002) regarded this lithologic sequence as a unit within the Agua Nueva Formation, which consists of a laminated dark-gray limestone sequence bearing rich microfossils assemblage, being any vertebrate remains rather rare. However in Vallecillo, the marlstone unit underlies concordantly the Agua Nueva limestones and contrarily to what was expected, exhibits a well-preserved vertebrate assemblage, mainly fishes. The marlstone and fossiliferous unit of Vallecillo is clearly differentiable from the typical lithology of the Agua Nueva Formation; in consequence, Blanco and Covin (2003) suggested the "Vallecillo Member" for this unit, which age is Lower-Middle Turonian due the presence of the ammonites *Mammites* and *Pseudoaspidoceras*.

During 1998-2002, Blanco-Piñón leaded paleontological and geological investigations on Vallecillo supported by national and international institutions. His research includes diverse taxonomical, taphonomical and sedimentological aspects of this fossil Lagerstätte (Blanco-Piñón *et al.* 1998, 2000a-b, 2001, 2002; Blanco and

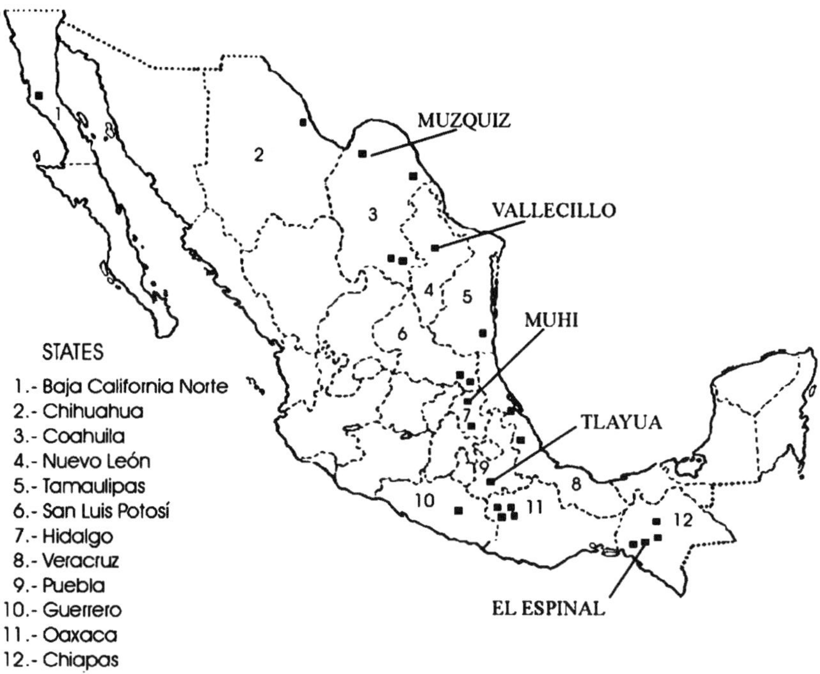

FIGURE 3. The main Mexican Mesozoic Osteichthyan localities.

Cavin, 2003; Blanco-Piñón and Frey, 2001a and b). In Vallecillo the macrofossils are relatively well known, the rare invertebrates mainly are ammonites and bivalves (inoceramids). In contrast, the diverse vertebrate assemblage that mainly involves fishes also includes turtles and other unidentified reptiles. All these vertebrate fossils are preserved in two dimensions, due to compressional forces, and show several degrees of disarticulation from isolated elements to complete specimens, total bone-calcite replacement, and rarely the soft tissues are preserved [although parts of a gill complex, belonging to an uncertain pycnodontiform, have been documented (Blanco-Piñón, 1998; Blanco *et al.* 2002)].

The Vallecillo fossiliferous marl strata were formed in a deep and restricted deposit, under relatively normal to low oxygen concentrations, which probably represent an early Turonian maximum transgressive event. This paleoenviromental model is based on some sedimentological features of this fossil bearing locality, including the fine lamination, the occurrence of micritic matrix with few to abundant bioclasts, as well as, the lack of bioturbation, added to the occurrence goethite. These conditions could have set the framework for the excellent preservation of the organisms, deposited after their death under normal-causes, rather than in a scenario of an obrutional deposit in which the organisms die intoxicated under effects of different physicochemical factors (turbiditic currents, low concentration of oxygen or high salinity in the water column, etc.).

The approximately 400 fossil fishes discovered during the first five years of paleontological investigations in Vallecillo include at least 12 shark and bony fish taxa. The sharks are represented by several lamniform vertebral centra, a trunk of a *Scyllyorhinus*-like specimen, and some *Ptychodus* teeth (Blanco-Piñón et al. 2002). About 40 % of bony fishes collected in Vallecillo belong to *Tselfatia* sp. (Blanco and Cavin, 2004). Other fish taxa found in Vallecillo are: an undescribed pycnodontiform related or belonging to *Nursallia* (Blanco and Frey, 2001a). Blanco and Cavin (2003) reported the occurrence of the genus *Araripichthys*, previously known only in remote localities in the northern part of South America and Africa, and erected two new teleosts species, *Vallecillichthys multivertebratum*, an inusual ichthyodectid with an inusual high number of vertebral centra, and *Goulmimichthys roberti,* a pachyrhizodontid with fusiform body. Blanco Piñón (1998, 2001, among others), reported some other fishes from Vallecillo, including a new species of *Rhynchodercetis* (Blanco-Piñón and Alvarado-Ortega, in prep.).

4.3. Muhi Quarry

The Muhi Quarry, discovered in 1998 by investigators of the University Autónoma del Estado de Hidalgo (UAEH), is located near Zimapán town, 120 km nortwestern Pachuca City, Hidalgo State, in central Mexico (Figure 3). Since 2002, Katia A. González-Rodríguez has been involved in the geological and paleontological studies carried on this important locality.

The Muhi Quarry represents an eight meters thick sequence of gray laminated limestones with interbeding lutites and flint nodules, which have been recognized as part of the Albian-Cenomanian age San Joaquín Facies, within El Doctor Formation (González-Rodríguez and Bravo-Cuevas, in press.).

Today, more than 700 specimens of both invertebrates and fishes from Muhi Quarry are housed into the collection of the Paleontological Museum of UAEH. Among these fossils, the invertebrates include ammonites, planctonic crinoids, equinoid spines, and crustaceans (González-Rodríguez *et al.* 2000; Vega *et al.* 2003); whereas the Muhi vertebrate assemblage, represented exclusively by fishes, includes condrichthyes and teleosts. Sharks identified are isolated teeth of *Ptychodus* sp., *Squalicorax* sp., and some vertebra and tooth remains of uncertain affinity. Even when, a large amount of the teleosts from Muhi involve today are under process of identification, indubitable taxonomical identifications involve specimens belonging to Ichthyodectiformes, Enchodontidae (being *Enchodus* sp. the most abundant fish in the quarry), Pachyrhizodontidae, Ichthyotringidae, Dercetidae, Halecidae, Beryciformes, Ostariophsyans, as well as uncertain Clupeomorphs and Elopomorphs (González-Rodríguez and Applegate, 2000; González-Rodríguez, 2002; González-Rodríguez *et al.* 2004).

4.4. Other Localities

Today, several Mexican Cretaceous fossil fish localities are known (Figure 3) in the States of Baja California Sur, Chiapas, Chihuahua, Coahuila, Durango, Guerrero, Hidalgo, Nuevo León, Oaxaca, Puebla, San Luis Potosí, Tamaulipas, and Veracruz. Barrios-Rivera (1985), González-Rodríguez (1990) and Alvarado-Ortega *et al.* (2002) catalogued all this fish assemblages; nevertheless, many detailed studies on the systematic of these fishes is needed to have a comprehensive knowledge about the Mexican Fossil fish record.

Even when only scarce and fragmentary fossil specimens from several of these localities are known, due to the fact that in most of the cases they have been visited and collected only once since their discovery; nowadays, new and old fossil fish localities are included in running geological and paleontological programs that will produce new and better samples soon.

5. Systematic Review of Mesozoic Osteichthyan Fishes of Mexico

A review of taxonomical studies on the Mesozoic Osteichthyan fishes of Mexico is provided in this section (Table 1). This updated catalogue includes all of the recent discoveries, as well as some preliminary results and advances on the study of a large part of the Mexican fossil fish record.

5.1. Sarcopterygians

Actinistia

In 1995, Grande recognized a coelacanth specimen in the unidentified fish section of the Tlayúa collection at UNAM. Espinosa-Arrubarrena *et al.* (1996) identified this specimen as *Axelrodichthys* cf. *A. araripensis*, a monospecific taxon previously collected in Aptian age outcrops of the Santana Formation, Brazil. Although, these authors are expecting to find more coelacanth specimens to improve such identification to a specific level, this first Mexican coelacanth represents the younger and northern most record of *Axelrodichthys* (also see Applegate *et al.* 2000).

Dipnoi

Aguilera (1896) reported the occurrence of *Ceratodus* in the Late Cretaceous (Turonian) El Peyote locality, into the Parras Basin, Coahuila State. Later, Aguilera (1907) confirmed this record later but the fossil material never was described or illustrated appropriately and its specific determination is still pending. According to Maldonado-Koerdell (1949:242), this material was deposited into the paleontological collection (now Colección Nacional de Paleontología) of the Instituto de Geología, UNAM.

5.2. Primitive Actinopterygians

The Mexican fossil record of the ray-finned fishes or actinopterygians involves several groups. The chondrosteans (primitive actinopterigians) are represented by a scute remain belonging to *Ancipenser* recovered in the middle Miocene outcrops of La Misión, Baja California State (Aranda-Manteca, 1994). In contrast, the Mexican Mesozoic neopterygians (also referred as higher actinopterygians) include several groups that constitute the most diverse and abundant among the Mexican Mesozoic fossil fishes.

Amiiformes

The characterization of *Otomitla speciosa* Felix, 1891, is based on a well-preserved skull and scales of a large fish collected in Cerro de la Virgen, Oaxaca State (Figure 1.C). This species is mis-spelled in Applegate (1988) as "*Otamitla speciosa*" and in Silva Barcenas (1969) as "*Otomitla speciola*". Although, originally *Otomitla* was though as part of the Saurodontidae (now in the Order Ichthyodectiformes); Hay (1929) and Applegate (1988) (among others) rejected this idea. Now, *Otomitla* is regarded as a primitive actinopterygian probably related to the amiids, such an idea that we follow here (Table 1), not without considering that a complete study of this specimen is required to determinate its affinities. Despite of Applegate´s (1987) suggestion that this material is housed at the Felix collection at Lepzing University, Germany, the actual existence of this material is questioned.

The best-preserved and well-known Mexican fossil amiids are from the Tlayúa Quarry and they have been described as *Pachyamia mexicana* Grande and Bemis 1998. Previously, Applegate and Espinosa-Arrubarrena (1982) considered these fossils belonging to *Vidalamia*, and later Applegate (1996) suggested the presence of two amid species in Tlayúa; nevertheless, according to Grande and Bemis (1998: 377) there is not evidence of any additional amiid species in Tlayúa. In addition some uncertain Caturid-like fishes have also been identified in this locality (Applegate *et al.* 2000).

The occurrence of the genus *Melvius* in Late Cretaceous sediments of Coahuila was reported by Rodríguez de la Rosa and Cevallos-Ferríz (1998), based on scarce vertebral fragments recovered in El Pelillal locality, Cerro del Pueblo Formation, Coahuila State. Additional discoveries in this formation include the recovery of a skull that supports occurrence of *Malvius* in northern Mexico (Hernández-Rivera *et al.* 2000), which represents the southern most record of this genus in America.

Additional Late Cretaceous amiid fossil specimens have been collected in outcrops belonging to the Corral de Enmedio Formation (Cabullona Group), Sonora State (Lucas and González-León, 1995), but best-preserved material and/or detailed studies are needed to define their fine taxonomic affinities.

Contrary to Blanco-Piñón *et al.* (2000a), a recent detailed review of the fossil fishes from Vallecillo (Blanco-Piñón, pers. observ., 2004) revealed the total absence of amiid remains.

Ionoscopiformes

The order Ionoscopiformes, erected by Grande and Bemis (1998), is a Triassic-Early Cretaceous actinopterygian group. Its Mexican record is restricted to only two species from the Early Cretaceous (Albian) lithographic limestones of the Tlayúa Quarry. Although, the description of *Teoichthys kallistos* Applegate 1988, is based on a few specimens, today it is known by at least 15 beautiful specimens. *Teoichthys* belongs to a group of heavy rhomboid-scaled Ionoscopiformes that includes the American genus *Macrepistius* and the European genus *Ophiopsis*.

The second Mexican ionoscopiform fish was independently identified by Grande (in Grande and Bemis, 1998:611), who regarded it as *Ionoscopus* sp. (a Jurassic-Cretaceous European genus), and Applegate *et al.* (2000), who considered it as an *Oshunia*-like form due to its apparent similarity with this South American fish. Preliminary results on the study of this specimen show that it is an *Ionoscopus-Oshunia* intermediate form (Alvarado Ortega and Espinosa-Arrubarrena, 2001;

Espinosa-Arrubarrena and Alvarado-Ortega, 2002), and its formal description is under editorial review.

Blanco-Piñón *et al.* (2004) erroneously report the occurrence of fish belonging to Ionoscopidae in the Late Cretaceous quarries, near Múzquiz, Coahuila State. Actually, as far as we know there is no evidence of any non-teleost osteichthyan fishes in these localities (Alvarado-Ortega, 2005, pers. observ.).

Lepisosteiformes

Lucas and González-León (1995) reported the occurrence of scales belonging to *Lepisosteus* or gars (Lepisosteiformes) from the Late Cretaceous strata of Packard Shale, Cabullona Group, Sonora State.

Additional *Lepisosteus* remains, including scales, vertebrae, opercles, and skull fragments, are from localities within Cerro del Pueblo Formation, Coahuila State (Rodríguez de la Rosa and Cevallos-Ferríz, 1998; Hernández-Rivera *et al.*, 2000).

Macrosemiidae

The Mexican macrosemiids represent the first record of the family in America (González-Rodríguez, 1996, 2000, 2004; González-Rodríguez and Applegate, 2001). Near 160 well-preserved macrosemiid specimens so far recovered in the Albian lithographic limestone strata of Tlayúa Quarry include *Macrosemiocotzus americanus* González-Rodríguez *et al.* 2004, *Notagogus novomundi* González-Rodríguez and Reynoso 2004, as well as other two new forms which are in process of formal descriptions. The other two macrosemiids found in the Aptian strata of El Espinal Quarry, Chiapas State, constitute the older record of this family in México. One of these specimens from El Espinal belongs to *Macrosemius founeti*, species known from European Jurassic rocks. This remains of macrosemiid found Chiapas represent the second freswater macrosemiid record known in the world (González-Rodríguez *et al.* 2002).

Pycnodontiformes

The Pycnodontiformes are a highly specialized Triassic-Eocene fish group, characterized by their deep bodies and crushing teeth battery. This group is widely represented by numerous Mexican fossils (Figure 4), including Jurassic, Cretaceous, as well as probably Eocene specimens (Applegate *et al.* 2004).

The oldest record of a Mexican pycnodontiforms is a vomer plate from the Jurassic strata near Tamán Village, San Luis Potosí State, which is identified as *Gyrodus* cf. *G. macrophthalmus* by Dunkle and Maldonado-Koerdell (1953).

Carranza-Castañeda and Applegate (1994) reported another unidentified pycnodont vomer from the Cretaceous strata of Cerro Los Mendoza, El Doctor

Formation, Hidalgo State. Later this specimen was recognized as cf. *Pycnodus* sp. by Castillo-Cerón *et al.* (1996). Neverthelles, in this paper, we prefer to report it as uncertain pycnodontiform taxa, because there are five vomerine tooth rows in *Pycnodus* (Poyato-Ariza and Wenz, 2002, character 38) and this specimen shows only three.

Mülleried (1951) suggested that the pycnodont materials from Palenque, Chiapas (firstly reported by Sapper, 1899), are Eocene remains that belong to *Pycnodus*. Recently, the journalist E. Romero (2005) reported that Kimberly Smith (from Michigan University) discovered a new Eocene pycnodont species of the genus *Coelodus* from a limestone quarry of Chiapas State. According to this journalist, this species is named in honor of Mr. Alvaro Pascual, who, eleven years before found a single small and complete pycnodont in a limestone quarry in a nearby area (which probably is the same locality of those pycnodonts found in the ruins of Palenque). Mean while, at the close of the final edition of this chapter no other manuscript reporting this recent findings has been published; nevertheless, neither a precise description of these specimens nor any detailed geological studies of their possible locality or localities have ever performed. Additional recently discovered pycnodontiform remains from this archeological site (Ovalles-Damián and Alvarado-Ortega, 2002, pers. obs.) support the idea that these fishes are older than Eocene, probably Late Cretaceous, and that they do not belong to *Pycnodus*, but the confirmation of this new interpretation must wait for detailed analyses.

Undoubtedly, the Tlayúa Quarry, have provided the largest and well-preserved record of Pycnodontiformes in Mexico. Nowadays, *Tepexichthys aranguthyorum* Applegate, 1992, is the only taxon formally described from Tlayúa Quarry; nevertheless, other numerous forms identified in this Early Cretaceous locality (Applegate, 1996, 2001; Applegate *et al.* 2000; Applagate *et al.* 2004). Applegate (2001) recognized an isolated *Neoproscinetes* vomer bone (which shows the characteristic alternation of one large and two small teeth on the main vomerine tooth row). Contrary to Applegate (2001), herein we simply referred to all the remaining undescribed forms only as pycnodontiforms (Table 1) because their description still is in process.

The first works of the fossil fish assemblage from in Vallecillo did not include the occurrence of Pycnodontifomes (Blanco-Piñón *et al.* 1998), but recently identified specimens of this locality are reported as *Nursallia* cf. *N. gotturosum* (Blanco-Piñón et al. 2000a) or simply as *Nursallia* sp. (Blanco-Piñón and Frey, 2001 b, 2001c; Blanco-Piñóna *et al.* 2002). We used the designation *Nursallia* sp. in our Mexican Mesozoic Osteichthyan Assaemblage List (Table 1), due the ambiguity of the earlier and isolated previous "specific level" report. Even when this taxonomic assignation still requires a detailed analysis, we noted that *Nursallia* is a Cenomanian-Eocene marine genus present in Brazil, Morocco, Lebanon, and Italy; whereas Vallecillo is a Late Cretacoeus (Turonian) locality localized very far from those countries. Therefore, the future confirmation of such taxonomical assignation of this pycnodontiforms from

FIGURE 4. A prognated Pycnodontiform fish from El Espinal Quarry, Chiapas.

Vallecillo, will contribute to the understanding of the evolutionary and biogeographical patterns of this group.

Actually, the formal description of the rare prognated and single complete pycnodontiform specimen from El Espinal Quarry, Chiapas (Figure 4), is in process; but it is remarkable that, as occurs with the macrosemiids from this locality, this rare fish represents the second freshwater pycnodontiform record known in the world.

Finally, some other Mexican pycnodontiforms include a complete specimens from the Late Cretaceous Temozol Quarry, near Taxco, Guerrero State deposited in a private collection; some tooth plates from a possible Late Cretaceous locality, near Ixcaquixtla, Puebla State (Alvarado-Ortega and Applegate, 2001, pers. obs.); and fragmentary specimens from El Chango Quarry, Chiapas.

Semionotiformes

The record of this primitive heavy-scaled armed fish group in Mexico only involves Early Cretaceous (Aptian) *Lepidotes* specimens from the Sates of Chihuahua and Puebla.

Luis R. Segura (Universidad Nacional Autónoma de Baja California Sur), collected an isolated body fragment showing the spine dorsal edge scales that helps to its identification as *Lepidotes* sp., in the outcrops of Rancho Consuelo, Cuchillo Formation, Chihuahua State.

Additionally, about a dozen complete and fragmentary skeletons belonging to *Lepidotes* sp. have been identified in Tlayúa, Puebla State (Figure 5). These seminotiforms are though to represent at least two different species (Applegate, 1996; Applegate *et al.* 2000). A formal description of this fishes is under preparation by Applegate and Alvarado-Ortega.

5.3. Teleostei: the Crowning Actinopterygians

Teleostei, the monophyletic groups of the modern dominate fishes, appeared during the Triassic, during the Jurassic they rapidly evolved into an array of diverse life strategies, body forms, and eventually became adapted to the most diverse aquatic habitats, progressively displacing to the primitive actinopterygian groups.

The fossils of early teleosts discovered in Mexico are classified within the following orders.

Aspidorhynchiformes

This pipefish group is a characteristic Jurassic-Paleocene marine and continental monophyletic teleost clade (Brito, 1997), which involves two genera found in the Mexican Cretaceous record, *Belenostomus* and *Vinctifer*.

The original description of the early American aspidorhynchiform known, *Belenostomus ornatus* Felix 1891, is based on Neocomian isolated remains from Cerro de la Virgen, Tlaxiaco, Oaxaca State. Recently, Brito (1997:745) noted some remarkable characters of this valid species but the scarce and poor material available prevent any further taxonomical analysis. Although, Applegate and Comas (1980) collected some few additional remains of *Belenostomus* in this area of Oaxaca State

FIGURE 5. *Lepidotes* sp. from Tlayúa Quarry.

(probably belonging to the same species), their fragmentary nature has not permited any further advance.

A large amount of well-preserved *Belonostomus* specimens in Mexico comes from the Albian lithographic limestones of Tlayúa Quarry, Puebla. Brito (1997) named some of these specimens as "*Belenostomus* sp. form 3" giving some of their peculiar characters but their formal description is still pending. At the same time, *Vinctifer* only occurs in the limestones of Tlayúa but the study of about a dozen specimens known is in progress.

Ichthyidectiformes

Mexican Ichthyodectiformes represent an abundant and diverse fossil assemblage. Alvarado-Ortega (1998a, 2005) analyzed this record concluding that all these Mexican fossil fishes are relative derived species belonging to the suborder Ichthyodectoidei. Also, he agreed with previous authors (i.e. Bardack, 1965) on the not-diagnostic nature of the fragmented fin ray from Cerro de la Virgen, Oaxaca, recognized by Felix (1891) as *Thrissops*?, and in the dubious assignation of the scale and he impression of a partial vertebral column (which shows the typical ichthyodectifom double lateral socket in the centra) from the Turonian sediments of Agua Nueva Formation, near Xilitla, San Luis Potosí State, as *Xiphactinus molossus* and *Ichthyodectes* respectively, as they were reported by Maldonado-Koerdell (1956).

From Vallecillo, Nuevo León State another Mexican Ichthyodectifrom was described, *Vallecillichthys multivertebratum* Blanco and Cavin, 2003 (Figure 6). It is characterized by its large number of vertebrae (near to a hundred) and relative small teeth.

Prymnetes longiventer Cope 1871, a monospecific and monotypic genus, is from an uncertain locality (possible of Early Cretaceous age), near Tuxtla Gutiérrez, Chiapas. This isolate report and poor record precluded the inclusion of this fish in any phylogenetic study, and often it was considerated as an *incertae sedis* form (i.e. Maisey, 1991); until Alvarado-Ortega (2004, 2005) suggested that *Prymnetes* is close related to *Cladocyclus* and other derived ichthyodectoid genera.

Alvarado-Ortega (1998b, 2004, 2005) recognized at least four new ichthyodectiform species from Tlayúa Quarry. Among them, only *Unamichthys espinosai* Alvarado-Ortega, 2004, has been described formally. It is remarkable that these specimens are differing from each other mainly by the number of vertebrae.

Recently, Garibay-Romero and Alvarado-Ortega (2002) identified the presence of a first pectoral ray that shows the characteristic "ichthyodectid" blade-shape and some isolated ichthyodectiform vertebrae in the Turonian strata exploited in the Arroyo Las Bocas Quarry, near Taxco, Guerrero. Remains (scales, tails, skull, vertebrae and other parts of the body) of large ichthyodectiforms (which total length is estimated in 1 m or more) have been collected in the Late Cretaceous outcrops near Múzquiz, Coahuila State (Blanco-Piñón *et al.* 2004). Two ichthyodectiform specimens are included in the fish assemblage recovered in the middle Cretaceous limestones of the

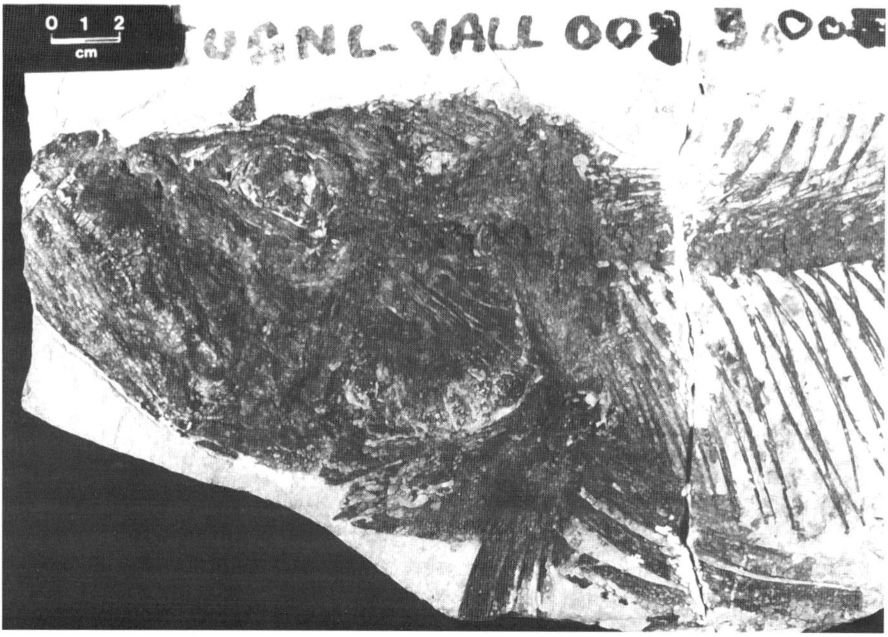

FIGURE 6. *Vallecillichthys multivertebratum* Blanco and Cavin, 2003, from Turonian outcrops of Vallecillo, Nuevo León State, Mexico.

Muhi Quarry, Hidalgo, a beautifully preserved head and a delicate specimen lacking the skull (Alvarado-Ortega, pers. observ., 2003).

Leptolepiformes

Leptolepis was a Mesozoic highly successful cosmopolitan taxon capable of living in both marine and freshwater environments. *Leptolepis tamanensis* Dunkle and Maldonado-Koerdell 1953 from the Jurassic locality in Tamán, San Luis Potosí State, is the only identified leptolepiform fish in Mexico.

Pholidophoriformes

The Pholidophoriformes are a paraphyletic group known from the Late Triassic to the Cretaceous. They represent the most primitive teleost clade (Arratia, 2000). In Mexico, their record is restricted to scarce undescribed Albian *Pholidophorid*-like specimens from Tlayúa Quarry (Applegate, 1996; and Applegate *et al.* 2000).

5.4 Teleocephala: the Most Advanced Teleosts

Teleocephala is the crown-group of teleosts (Pinna, 1996) and includes taxa with fossil and living forms, as the Osteoglossomorphs (bony-tongue fishes), Elopomorphs (tarpons, eels and relatives), and Clupeomorphs (herrings and allies), all with Jurassic species, and Euteleostei (carps, characins, catfishes and advanced spiny fishes, among others) that evolved since the Early Cretaceous. In Mexico, there is no fossil or living osteoglossomorphs known, but a large number of the other Mesozoic teleocephala forms are present in the country.

Albuliformes

Brannerion is an *incertae sedis* Albuliform genus (Fullel, 2000) that includes two Early Cretaceous (Aptian) species from the Santana Formation, Brazil. In the Tlayúa Quarry, a few specimens have been recognized and simply named *Brannerion*-like forms (Applegate *et al.* 2000). These Mexican albuliforms represent the possible latests, northern-most record of this genus.

Beryciformes

This diverse clade probably appeared in the Albian and radiated during the Late Cretaceous, becoming a successful group all over the world (Goody, 1969). In Mexico, still undiscribed bericiforms are present in two localities. The earliest record is from the Albian Tlayúa Quarry (Applegate *et al.* 2000; González-Rodríguez *et al.* 2004). Recently, fossils of these fishes were recovered in the Albian-Cenomanian sediments of the Muhi Quarry (González-Rodríguez, 2002; and González-Rodríguez and Applegate, 2000). In both cases the specimens are under study.

Previous reports about the Turonian bericiforms from Vallecillo are erroneous and certainly there is no evidence to corroborate their occurrence in this locality (Blanco-Piñón, 2004, pers. obs.).

Clupeiformes

Aguilera (1896) reported the occurrence of *Syllaemus latifrons* in Mexico, the referred material is lost and the locality was never properly recorded. As with previous authors (i.e. Maldonado-Koerdell, 1949), here we regard this record as an invalid report.

Elopiformes

Mexican elopiforms include *Megalops* sp. and *Paraelops* sp., from Tlayúa Quarry (Applegate, 1996; Applegate *et al.* 2000; Maisey, 2000). In addition, there are some indeterminate elopiforms specimens from the Arroyo las Bocas quarry, near Taxco, Guerrero (Garibay-Romero and Alvarado-Ortega, 2002), from the Muhi Quarry, Hidalgo (González-Rodríguez *et al.* 2004), and from Oaxaca (Applegate and

Comas, 1980); nevertheless, a formal study of these fossils is required.

Several beautiful well-preserved specimens belonging to the rare elopid fish, *Laminospondylus* Springer, 1957, have been collected from the Late Cretaceous (probably Turonian) quarries near Múzquiz, Coahuila State (Piñón-Blanco *et al.* 2004). In these fishes, the vertebrae have the laminar aspect of the transverse processes recognized as a diagnostic character of *Laminospondylus*. A careful anatomical study of this uncertain elopid is in process and soon the description of the specimens of the second world report of this genus will be published (Figure 7).

Ellimmichthyiformes

In Mexico, at least two known Early Cretaceous localities bear specimens that belong to this group, commonly named the double armored herrings.

Aguilera (1896) reported the occurrence of a Mexican *Diplomystus* sp. from an uncertain locality in northern Mexico. The authenticity of this report is challenged due the lack of the appropiete field data and the apparent lost of the referred material.

Applegate (1996) reported the occurrence of *Diplomistus* related forms, but in our present knowledge, these ellimmichthyiforms rather belong to *Ellimmichthys* (Chang and Grande, 1997: fig. 7e, f), although these fishes could represent a new genus (Applegate *et al.* 2000). Presently Lilia Arias is preparing her Master Degree thesis, including the description of these fossil fishes.

Gonorhynchiformes

There are no descriptive studies on Mexican Gonorhynchiformes or Milkfishes and allies; nevertheless, there are reports of these fishes from two localities.

Applegate (1996) illustrated an Albian gonorhynchiform from the Tlayúa Quarry, Puebla State (also see Grande, 1999; and Applegate *et al.* 2000). Some other uncertain gonorhynchiformes have been collected in the Muhi Quarry, Hidalgo (González-Rodríguez and Applegate, 2002).

Pachyrhizodontiformes

According to Cavin (2002), *Araripichthys, Notelops, Rhacolepis, Pachyrhizodus, Goulmichthys,* and other genera form part of this clupeocephala order. Fossils of these genera have been recovered in Mexican Cretaceous localities.

The younger and first North American specimens belonging to *Araripichthys*, a deep-bodied fish genus, were collected in the Turonian deposits of Vallecillo, Nuevo León State (Blanco and Cavin, 2002).

At least four pachyrhizodontiformes genera have been collected in Mexico. The *Rhacolepis*-like form and *Notelops* sp. specimens have been collected in the Albian limestones of Tlayúa Quarry. The Turonian *Goulmimichthys roberti* Blanco and Cavin, 2002, from Vallecillo, Nuevo León State, represents the first Mexican pachirhizodontid identified and characterized at a specific level. Other specimens reported as pachyrhyzodontids are from Cerro de la Virgen, Tlaxiaco, Oaxaca (Applegate and Comas, 1980), the Muhi Quarry, Hidalgo State (González-Rodríguez *et al.* 2004), and the quarries near Múzquiz, Coahuila State (Blanco *et al.* 2004).

Tselfatiiformes (Bananogmiiformes)

FIGURE 7. *Laminospondylus* sp. from the quarries near Múzquiz, Coahuila State, Mexico. Head and part of the trunk (above) and vertebrae showing the laminar aspect of the transverse processes (below). Scale bars equal 1 cm.

Fossils belonging to Tselfatiiformes or Bananogmiiformes conform a large Late Cretaceous marine group. Previously, these fossil fishes were collected only in North America, Europe, and Africa.

Within the last 10 years remains of Tselfatiiformes have been recovered in different Upper Cretaceous Mexican localities. Applegate (1996) reported an Albian *Bananogmius*-like fish from Tlayúa Quarry, central Mexico. Blanco-Piñón (1998) illustrated some complete Turonian specimens of *Tselfatia* sp. from Vallecillo, Nuevo León State, in northern Mexico (Figure 8). Garibay-Romero and Alvarado-Ortega (2002) identified some Turonian Tselfatiiformes remains from Arroyo las Bocas, Guerrero, in southern Mexico.

Recently Piñón-Blanco (2004) reported some pletodontid-like fishes found in the quarries near Múzquiz, Coahuila, northern Mexico; nevertheless the analysis of some complete and better preserved specimens shows that this fishes belong to *Tselfatia* (Alvarado-Ortega, 2005 pers. observ.).

Salmoniformes

Salmoniformes are a diverse group of teleosts, encompassing several subordes that radiated during the Late Cretaceous. This great radiation is evident in the Upper Cretaceous localities of México (see Table 1). The suborders Enchodontoidei, Ichthyotringoidei, Cimolichthyoidei and Halecoidei represented in México are found throughout most of the Late Cretaceous localities.

Several fossils belonging to *Enchodus*, probably representing different species, have been collected in Vallecillo, Nuevo León (Blanco-Piñón, 1998); San José de las Rusias, Tamaulipas; and Xilitla, San Luis Potosí (Maldonado-Koerdell, 1956.).

Garibay-Romero and Alvarado-Ortega (2002) identified an isolated lower jaw of *Enchodus* sp. from Arroyo las Bocas, Guerrero, and González-Rodríguez and Applegate (2002) recognized complete and fragmentary specimens belonging to this genus from the Muhi Quarry, Hidalgo. Besides, specimens referred to *Paraenchodus* sp. have been collected in the Muhi Quarry (González-Rodríguez and Applegate, 2002)). A large complete *Enchodus* specimen and several fragments of its body have been collected in the quarries near Múzquiz, Coahuila (Piñón-Blanco *et al.* 2004).

Other yet undescribed Late Cretaceous Mexican Salmoniformes involve *Dercetis, Rhynchodercetis*, cf. *Sauroramphus, Enchodus*, Halecids, Ichthyotringids from Vallecillo, Nuevo León (Blanco-Piñón, 1998, 2001, 2002, among others). Similar assemblages have been reported from the Muhi Quarry Hidalgo (González-Rodríguez *et al.* 2004) and the quarries near Múzquiz, Coahuila (Piñón-Blanco, *et al.* 2004). In addition, specimens of this genus were collected in Arroyo las Bocas, Guerrero State, by Garibay-Romero and Alvarado-Ortega (2002). Actually, Blanco-Piñón and Alvarado-Ortega are working on the description of a new *Rhynchodercetis* species from the Vallecillo Quarry.

Yabrudichthy-like specimens have been reported from the Muhi Quarry, Hidalgo State (González-Rodríguez and Applegate, 2002), and from the Tlayúa Quarry, in

Puebla (Applegate *et al.* 2002). Some other *Yabrudichthys*-like specimens have been collected in the Aptian strata of El Espinal Quarry, Chiapas (Alvarado-Ortega and Ovalles-Damián, 2004, pers. observ.).

Undeterminated Teleosts

Several and diverse remains of undeterminated Teleosts from Mexican Cretaceous localities are in process of identification, but in many cases, additional fossil and recent material is needed for adequate anatomical comparisons. It is to note the remarkable increase in the number of specimens and localities housed, properly curated and recorded in paleontological Museums and Universities, mainly those reported from Coahuila, Chiapas, Guerrero and Hidalgo.

6. Paleobiogeographical implications

The scarce ocurrence of Mexican Jurassic fossil fish assemblage precludes any significant paleobiogeographical analysis except for a brief commentary. In recent and comprehensive phylogenetic approaches on Pycnodontiformes, the central papilla in vomerine and prearticular teeth has been considered as an autapomorphic and diagnotic character for *Gyrodus* (Poyato-Ariza and Wenz, 2002: 218); based on this condition, the identification of the vomer from the Jurassic strata, near Tamán, San Luis Potosí, described by Dunckle and Maldonado-Koerdell (1953), as *Gyrodus*

FIGURE 8. *Tselfatia* sp. from Vallecillo, Nuevo León

TABLE 1. Mesozoic Bony fishes of Mexico. Abbreviations – ERCT = Escuela Regional de Ciencias de la Tierra, Universidad Autónoma de Guerrero, Taxco, Guerrero, México. FCMP = Museo de Paleontología, Facultad de Ciencias, UNAM. FCT = Facultad de Ciencias de la Tierra, Universidad Autónoma de Nuevo León, Linares, Nuevo León. FMNH = Field Museum of Natural History, Chicago, Illinois, USA. IGLUNAM Instituto de Geología, UNAM. IGM = Colección Nacional de Paleontología, IGLUNAM; INAH-DP = Departamento de Prehistoria, Instituto Nacional de Antropología e Historia, D.F., México. MDC (informal abbreviation) = Museo del Desierto, Coahuila, México. MNHN = Museo Nacional de Historia Natural, D.F., México. MEPA (informal abbreviation) = Museo Eliseo Palacios Aguilera, Chiapas, México. PASAC = Colección de la Sociedad Civil "Paleontólogos Asociados de Sabinas A. C." UAHMP = Museo de Paleontología, Universidad Autónoma de Hidalgo, México. USNM = United States National Museum, Washington, D. C., USA. Personal observations are supported by unpublished photographs. Specimens under study are denoted by underlined abbreviations (meaning that their official incorporations to a collection is in process or that these will be deposited in the referred collections after publication). Italic abbreviations denote that duplicates and/or peels of the referred specimens are housed in the respective collections.

TAXA	Locality, State, and (Collection)
	SARCOPTERYGII
Axelrodichthys cf. *A. araripensis*	Tlayúa, Albian, Puebla (IGM)
Ceratodus sp.	Parras, Turonian?, Coahuila (IGM)
	ACTINOPTERYGII
Amiiformes	
Melvius sp.	Cerro del Pueblo Fm., Coahuila (IGM and MDC)
Otomitla speciosa	Cerro de la Virgen, Oaxaca (Collection Felix, Leipzig University?)
Pachyamia mexicana	Tlayúa, Puebla (IGM and FMNH)
Amiids	Corral de Enmedio Fm., Sonora (IGM).
Caturid-like	Tlayúa, Puebla (IGM)
Ionoscopiformes	
Teoichthys kallistos	Tlayúa, Puebla (IGM)
Ionoscopid new gen. and sp.	Tlayúa, Puebla (IGM)
Lepisosteiformes	
Lepisosteus	Cerro del Pueblo Fm., Coahuila (MDC)
Lepisosteids (*Lepisosteus*)	Cerro del Pueblo Fm., Coahuila (IGM and MDC); Packard Shale, Sonora, (IGM)
Macrosemidae	
Notagogus novomundi	Tlayúa, Puebla (IGM)
Mcrosemius founeti	El Espinal, Chiapas, (MEPA)
Macrosemiocotzus americanus	Tlayúa, Puebla, IGM
Macrosemiocotzus americanus	Tlayúa, Puebla (IGM)
Macrosemids	

TABLE 1. Continued.

TAXA	Locality, State, and (Collection)
Pycnodontiformes	? (see Romero, 2005)
Coelodus n. sp.	Near Tamán, San Luis Potosí (USNM)
Gyrodus cf. *G. macrophthalmus*	Vallecillo, Nuevo León (FCT)
Nursallia sp.	Tlayúa, Puebla (IGM)
Neoproscinetes	Palenque, Chiapas (INAH-DP or INAH)
Pycnodus sp.	Tlayúa, Puebla (IGM)
Tepexichthys aranguthyorum	Tlayúa, Puebla (IGM); Vallecillo, Nuevo León (FCT);
Pycnodontids	El Espinal, Chiapas (FCP); Cerro Los Mendoza, Hidalgo (IGM); and Temozol quarry, near Taxco, Guerrero (Alvarado-Ortega, 2002, pers. obs.); Ixcaquixtla, Puebla (Applegate and Alvarado-Ortega, 2000, pers. obs.)
Semionotiformes	Tlayúa, Puebla (IGM); Rancho Consuelo, Chihuahua (?)
Lepidotes sp.	
	Teleostei
Aspidorhynchiformes	
Belenostomus ornatus	Cerro de la Virgen, Oaxaca (Collection Felix, Leipzig University?)
Belonostomus sp.	Tlayúa, Puebla (IGM); Muhi, Hidalgo (UAHMP); Cerro de la Virgen, Oaxaca (FCMP).
Vinctifer sp.	Tlayúa, Puebla (IGM)
Pholidophoriformes	
Pholidophorid-like	Tlayúa, Puebla (IGM)
Leptolepiformes	
Leptolepis tamanensis	Near Tamán, San Luis Potosí (USNM)
Ichthyodectiformes	
Ichthyodectes	Xilitla, San Luis Potosí (INAH-DP)
(cf. *Ichthyodectes)*	
Prymnetes longiventer	Near Tuxtla Gutiérrez, Chiapas (USNM, IGM, MEPA)
¿Thrissops? sp.	Cerro de la Virgen, Oaxaca (Collection Felix, Leipzig University?)
Unamichthys espinosai	Tlayúa, Puebla (IGM)
Vallecillichthys multivertebratum	Vallecillo, Nuevo León (FCT)
Xiphactinus audax (molossus)	Xilitla, San Luis Potosí (Maldonado-Koerdell, 1956, personal observ and draw)
Ichthyodectids or Ichthyodectiformes	Quarries near Múzquiz, Coahuila (PASAC); Tlayúa, Puebla (IGM); Arroyo las Bocas, Guerrero (IGM); Muhi Quarry, Hidalgo (UAHMP)
	Teleocephala
Albuliformes	
Brannerion-like form	Tlayúa, Puebla (IGM)
Beryciformes	
Beryciform undescribed froms	Tlayúa, Puebla (IGM); Vallecillo, Nuevo León (FCT); Muhi Quarry, Hidalgo (UAHMP)

TABLE 1. Continued.

TAXA	Locality, State, and (Collection)
Clupeomorpha	
Syllaemus ¿latifrons?	Unknown (Aguilera, 1896)
Ellimmichthyiformes	
Ellimychthys sp.	Tlayúa, Puebla (IGM)
¿Dyplomystus? sp.	Unknown (Aguilera, 1896)
Paraclupeids	El Espinal, Chiapas (MEPA)
Elopiformes	
Megalops sp.	Tlayúa, Puebla (IGM)
Paraelops sp.	Tlayúa, Puebla (IGM)
Laminospondylus sp.	Quarries near Múzquiz, Coahuila (PASAC)
Elopidae	Muhi Quarry, Hidalgo (UAHMP)
Gonorhynchiformes	
Gonorhynchiformes	Tlayúa, Puebla (IGM); Muhi, Hidalgo (UAHMP)
Pachyrhizodontiformes	
Araripichthys sp.	Vallecillo, Nuevo León (FCT)
Goulmimichthys roberti	Vallecillo, Nuevo León (FCT)
Notelops sp.	Tlayúa, Puebla (IGM)
Rhacolepis –like form	Tlayúa, Puebla (IGM)
Pachyrhyzodontids	Near Cerro de la Virgen, Tlaxiaco, Oaxaca (FCMP); Vallecillo, Nuevo León (FCT); Muhi Quarry, Hidalgo (UAHMP); quarries near Múzquiz, Coahuila (PASAC)
Tselfatiiformes	
Tselfatia sp.	Arroyo las Bocas, Guerrero (IGM); quarries near Múzquiz, Coahuila (PASAC); Vallecillo, Nuevo León (FCT)
Bananogmius-like	Tlayúa, Puebla (IGM)
Salmoniformes	
Enchodus sp.	Vallecillo, Nuevo León (FCT); Xilitla, San Luis Potosí (pers. obs.); Arroyo las Bocas, Guerrero (IGM); Muhi, Hidalgo (UAHMP); San José de las Rusias, Tamaulipas (INAH-DP)
Dercetis sp.	Vallecillo, Nuevo León (FCT)
Paraenchodus sp.	Muhi, Hidalgo (UAHMP)
Rhynchodercetis sp.	Vallecillo, Nuevo León (FCT); Arroyo las Bocas, Guerrero (IGM)
cf. *Sauroramphus*	Vallecillo, Nuevo León (FCT)
Yabrudichthy-like	Tlayúa, Puebla IGM; Muhi, Hidalgo (UAHMP); El Espinal Quarry, Chiapas (MEPA)
	Other undiscribed forms
Cimolichtids	Vallecillo, Nuevo León (FCT)
Enchodontids	Tlayúa, Puebla (IGM); Vallecillo, Nuevo León (FCT); Muhi, Hidalgo (UAHMP); Quarries near Múzquiz, Coahuila (PASAC)
Dercetids	Vallecillo, Nuevo León (FCT); Muhi, Hidalgo (UAHMP) Quarries near Múzquiz, Coahuila (PASAC);
Halecids	quarries near Múzquiz, Coahuila (PASAC)

TABLE 1. Continued.

TAXA	Locality, State, and (Collection)
	Vallecillo, Nuevo León (FCT); Muhi, Hidalgo (UAHMP)
Ichthyotringids	Vallecillo, Nuevo León (FCT); Muhi, Hidalgo (UAHMP)
Aulopiformes	Vallecillo, Nuevo León (FCT); Muhi, Hidalgo (UAHMP)
Elophomorps undescribed forms	Tlayúa, Puebla (IGM); Arroyo las Bocas, Guerrero (IGM); near Cerro de la Virgen, Tlaxiaco, Oaxaca (FCMP)
Clupeomorphs undescribed	Tlayúa, Puebla (IGM); Arroyo las Bocas, Guerrero (IGM); quarries near Múzquiz, Coahuila (PASAC)
Salmoniformes	Muhi, Hidalgo (UAHMP)
Undeterminated Osteichthyans and Teleosts	Tlayúa, Puebla (IGM); Muhi, Hidalgo (UAHMP); Vallecillo, Nuevo León (FCT); Near Tula Hidalgo (MNHN), Arroyo las Bocas, Guerrero (ERCT); Chiapas (MEPA)

cf. *G. macroptalmus*, is confirmed. Unfortunately, due to the lack of better specimens, neither this Mexican record nor those three additional American (two from Texas and one from Cuba) *Gyrodus* records (Romer, 1945; Gregory, 1923) were included in Poyato-Ariza and Wenz´s (2002) monographic review of Pycnodontiformes.

According to Matos (1992), the Tethys Sea evolved from the east to the west, forming part of the Mediterranean Sea during the Triassic and part of the Atlantic Ocean during the Early Cretaceous (Aptian-Albian). Geological evidence suggests that the formation of the Gulf of Mexico started in the Late Triassic and concluded at the end of the Jurassic (Ortega-Gutiérrez *et al.* 2000). During Jurassic-Cretaceous time, shallow tropical marine waters, associated to development of islands and reefal communities occupied the continental platform of North America, Mexico, and Caribbean. Based on evidence derived from fossil marine reptiles, Gasparini (1985) suggested a Jurassic seaway connection between the Gulf of Mexico and Europe. The North American, South American, and European Jurassic occurrence of *Gyrodus*, including the Mexican report discussed above, supports this conclusion about such Jurassic seaway. Later, Ferrusquía-Villafranca and Comas-Rodríguez (1988), also based on fossil marine reptiles, proposed a similar connection but for the Early Cretaceous. The Cuban and European distribution of other Jurassic fishes [as *Caturus eunatoides*, and probably *Leptolepis* and *Sauropsis* (Thies, 1989)] provides additional support to these ideas, as well as, the occurrence of additional well-known European Jurassic fish forms now recorded in the Mexican Cretaceous assemblages. This the case of the *Macrosemius founeti*, the specimen collected in El Espinal González-Rodríguez *et al.* 2002) and the serveral macrosemiid species known from the Tlayúa Quarry (González-Rodríguez 1989, 1996, 2004; and González-Rodríguez *et al.* 2004) reported for the first time in the New World.

On the other hand, Maisey (2000) analyzed the association of the geographical distribution of fishes in Africa, South America, and the Caribbean (including Mexico) within the context of the aperture of the Atlantic Ocean throughout the Early Cretaceous (Aptian-Abian); an episode that involves the continental break up of Western Gondwana (BWG), including the Western Africa and Eastern South American areas. Based on the present review of the Mexican Mesozoic fishes, some notes can be added to Maisey´s (2000) revision, although this exercise will produce a more complex process to explain the distribution of some of the Early Cretaceous species. Maisey (2000) did not report Jurassic fishes in the BWG area, but the taxonomical affinities of some Early Cretaceous fishes from Central and Southern Mexico clearly are closely related to clades known in the Jurassic-Early Cretaceous Western Tehys realm. These Mexican fossil fishes (see Table 1) are Macrosemiidae, Ionoscopiformes, Aspidorhynchiformes, Leptolepiformes, Semionotiformes, and Amiiformes. The occurrences of these fishes probably show that they arrived to the Mexico during Jurassic times, or at the beginning of the Early Cretaceous, and probably they evolved with some independence in Mexico since then.

Besides that, Blanco and Cavin (2003) and González-Rodríguez and Bravo Cuevas (in press) recognized that several Late Cretaceous Mexican fishes show affinities with other Tethyan assemblages located in North America, South America, Africa, and Europe. These authors based their observations on the presence of several fish clades as *Rhynchodercetis, Tselfatia, Araripichthys, Yabrudichthys*, and *Paraenchodus*, that accentuate the importance of the Mexican Mesozoic fish record, in the understanding of the paleobiogeographic patterns that osteichthyans had during the Mesozoic.

7. Final Remarks

An exciting future is waiting for the small Mexican Paleoichthyological Community that is facing the challenge of studying a comparatively large Mexican Mesozoic fish record. Many potential new species had been recognized and their number is constantly increasing, so several years of intense work are required. Nevertheless, we have just started learning some of the implications that this record has within our knowledge of Mexican Paleontology. Surely, these incoming efforts are going to change our understanding of the systematics, evolution, and biogeography of many Mesozoic fish groups.

References

Aguilera, J. G., 1896, Bosquejo Geológico de México, *Bol. Inst. Geol. Méx.*, 4-6:1-267.

Aguilera, J. G., 1906, Les Gissements carbonifères de Coahuila. Guide *Géologique du Mexique,* 28, 17 p.

Aguilera, J. G., 1907, Aperçu sur la Géologie du Mexique pour servir d'explication á la carte géologique de l´Amérique du Nord, *Congreso Geológico Internacional*, Xéme Session, Mexico, I:227-248.

Aguillón-Martínez, M. C., Espinosa, B., and Kirkland, J. I., 1998, Rostro de Pez Sierra, *Ischyrhyza mira* (Familia Schlerorhinchidae) de la Formación Lutita Parras (Campaniano Tardío, Cretácico Superior)

Coahuila, México, *VI Congreso Nacional de Paleontología, Soc. Mex. Paleont.*, Libro de Resúmenes, p. 2.

Alvarado-Ortega, J., 1998a, Descripción preliminar de un Género Nuevo Ichthyodectiforme (Neopterygii: Teleostei) de la Cantera Tlayúa (Cretácico Temprano) en Tepexi de Rodríguez, Puebla, México. Unpublished Professional thesis, Universidad Nacional Autónoma de México.

Alvarado-Ortega, J., 1998b, Ichthyodectifromes mexicanos: Un análisis de su diversidad. Avances en Investigación: Paleontología de Vertebrados, *Univ. Autón. Edo. Hidalgo, Inst. Inv. Cienc. Tierra,* Pub. Esp. 1: 109-119.

Alvarado-Ortega, J., 1998c, Avances en el Estudio de los Ichthyodectiformes de la Cantera Tlayúa, Tepexi de Rodríguez, Puebla, *VI Congreso Nacional de Paleontología, Soc. Mex. Paloent.,* Resúmenes, p. 3-4.

Alvarado-Ortega, J., 2000, The characterization of the ichthyodectiforms of the Tlayua quarry, in Puebla State, Mexico, *80th Annual Meeting, Soc. Ichth. Herp.*, Abstracts, p. 171.

Alvarado-Ortega, J., 2004, Description and relationships of a new Ichthyodectiform fish from the Tlayúa Formation (Early Cretaceous: Albian), Puebla, Mexico, *J. Vert. Paleont.*, 24:502-813.

Alvarado-Ortega, J., 2005. Sistemática de los peces Ichthyodectiformes de la Cantera Tlayúa, Puebla, México, Unpublished PhD Thesis, Universidad Nacional autónoma de México, 302 p.

Alvarado-Ortega, J., and Applegate, S. P., 1999, The fossil conservation on the Tlayúa quarries, Tepexi de Rodríguez, Puebla, México, *33rd Annual Meeting, Geol. Soc. Am., South-Central Section*, Abstracts with Programs, 31:A1.

Alvarado-Ortega, J., and Espinosa-Arrubarrena, L., 2001, Los ionoscopidos de la Cantera Tlayúa, Tepexi de Rodríguez, Puebla, Reunión Anual 2001, Un. Geof. Mex. and XII *Reunión Nacional SELPER*, Programa y Resúmenes, p. A3.

Alvarado-Ortega, J., and Carranza, O., 2002a, The fossil fishes from Pliocene localities near Tula de Allende, Hidalgo, México, *62nd Annual Meeting Society of Vertebrate Paleontology, J. Vert. Paleont.*, 22(Supplement to Number 3): 32A.

Alvarado-Ortega, J., and Carranza, O., 2002b, Evidencias fósiles de la Presencia de un Lago durante el Plioceno en las cercanías de Tula, Hidalgo, *III Reunión Nacional de Ciencias de la Tierra (IIIRNCT), Un. Geof. Mex.*, Resúmenes, p. 56.

Alvarado-Ortega, J., and Carranza, O., 2002c, Los peces fósiles de la región de Tula, Hidalgo. *VIII Congreso Nacional de Paleontología, Soc. Mex. Paleont.*, Ponencias, p. 25.

Alvarado-Ortega, J., Espinosa-Arrubarrena, L., and Guzmán, F., 2002, Peces, in: *Fósiles de México: Coahuila, una ventana a través del tiempo. Gobierno del Estado de Coahuila, Coahuila, México* (González-González, A. H. and A. De Stéfano-Farías, eds.), p. 107-127.

Alvarado-Ortega, J., Garibay-Romero, L. M., and Vega, F., 2004, Rayas fósiles del Cretácico Superior del Estado de Guerrero, *IX Congreso Nacional de Paleontología, Soc. Mex. Paleont.*, Libro de Resúmenes, p. 68.

Alvarez, J., 1966, Contribución al conocimiento de los bagres fósiles de Chapala y Zacoalco, Jalisco, México, *Inst. Nal. Antropol. Hist., Dept. Prehist. Paleoecol.*, 1:26.

Alvarez, J., 1974, Contribución al conociemiento de los peces fósiles de Chapala y Zacoalco (Aterínidos y Ciprínidos), *An. Inst. Nal. Antropol. Hist., 7ª época*, 4:191-209.

Alvarez, J., and J. Arriola, 1972, Primer goodeido fósil procedente del Plioceno jaliciense (Pisces, Teleostomi), *Bol. Soc. Cienc. Nat. Jalisco,* 6:6-15.

Alvarez, J., and M. E. Moncayo, 1976, Contribución a la paleoictiología de la Cuenca de México, *An. Inst. Nal. Antropol. Hist., 7ª época*, 6:191-242 p.

Applegate, S. P., 1978, Phyletic studies; Part I; Tiger Shark, *Univ. Nal. Autón. México, Inst. Geol., Rev.*, 2: 55-64.

Applegate, S. P., 1979, El primer registro de una selacifauna miocénica de Michoacán, *III Congreso Nacional de Zoología, Soc. Zool. Mex.*, Resúmenes, p. 41.

Applegate, S. P. 1986, The El Cien Formation, strata of Oligocene and Early Miocene age in Baja California Sur, *Univ. Nal. Autón. México, Inst. Geol., Rev.*, 6:145-162.

Applegate, S. P., 1987, A preliminary study of the Tlayúa Quarry near Tepexi de Rodríguez, Puebla, *Soc. Mex. Paleont., Rev.,* 1:40-50.

Applegate, S. P., 1988, A new genus and species of a holostean belonging to the family Ophiopsidae, *Teoichthys kallistos*, from the Cretaceous, near Tepexi de Rodríguez, Puebla, *Univ. Nal. Autón. Méx., Inst. Geol., Rev.*, 7:200-205.

Applegate, S. P., 1989, Peces, in: *Fósiles Tipo Mexicanos. Publication to celebrate the centenary of the Instituto de Geología 1886-1986* (Carreño, A. L., M. C. Perrilliat, C. González-Arreola, S. P. Applegate, O. Carranza-Castañeda, and E. Martínez-Hernández, eds), *Univ. Nal. Autón. Méx., Inst. Geol.*

Applegate, S. P, 1992, A new genus and species of pycnodont, *Tepexichthys aranguthyorum* from the Cretaceous (Albian) of Central Mexico, Tepexi de Rodríguez, Puebla, *Univ. Nal. Autón. Méx., Inst. Geol., Rev.*, 10:164-178.

Applegate, S. P., 1993, The use of tooth sets in Mexican species of the genus *Carcharhinus* with comments on the fossil history of this genus in Mexico, *Am. Elasmobranch Soc., Univ. Texas at Austin,* Abstracts Book, p. 70.

Applegate, S. P., 1996, An overview of the Cretaceous fishes of the quarries near Tepexi de Rodríguez, Puebla, México, in: *Mesozoic Fishes: Systematics and Paleoecology* (Arratia, G. and Viohl, G., eds.), München, p. 529-538.

Applagate, S. P., 1999, Recent advances in our knowledge of te Lagerstatën that Occurs in the Tlayua quarries near Tepexi de Rodriguez, Puebla, Mexico, *Third International Symposium on Litographic Limestones, Revista del Museo Civico di Scienze Naturali "Rnrico Caffi", suplemmento al Volumen 20.*

Applegate, S. P., 2001, The Pycnodont fauna from the Lower Cretaceous of Tepexi de Rodríguez (Puebla, México), *III International Meeting on Mesozoic Fishes: Systematics, Paleoenviroments and Biodiversity,* Abstracts book, p. 3.

Applegate, S. P., and Alvarado-Ortega, J., 1999, Update of the Tlayúa Model, *33rd Annual Meeting, Geol. Soc. Am., South-Central Section*, Abstracts with Programs, 31:A2.

Applegate, S. P., Alvarado-Ortega, J., García-Barrera, P., Ovalles-Damián E., and Espinosa-Arrubarrena, L., 2004, Análisis de la diversidad de peces pycnodontiformes de México, *IX Congreso Nacional de Paleontología, Soc. Mex. Paleont.*, Libro de Resúmenes, p. 17.

Applegate, S. P., Buitrón, B. E., and López-Neri, P., 1982, La Cantera Tlayúa en la región de Tepxi de Rodríguez, Puebla, El Solnhofen de México, *VI Congreso Nacional de Zoología, Soc. Zool. Mex.*, Resumenes (not paginated).

Applegate, S. P., and Comas, O., 1980, Primera ictiofauna del Cretácico Inferior de México, estado de Oaxaca, *V Congreso Geológico Nacional, Soc. Geol. Mex.*, Resúmenes, p 10.

Applegate, S. P. and Espinosa-Arrubarrena, L. 1981. The geology and selachian paleontology of Loma del Tirabuzón (Corkscrew Hill), Santa Rosalía, B. C. S., in: *Geology of Northwestern Mexico and Southern Arizona. Field trip for theGeological Society of America* (Ortlieb, L. and Roldán, J., eds.), *Univ. Nal. Autón. Méx., Inst. Geol.*, p. 257-263.

Applegate, S. P., and Espinosa-Arrubarrena, L., 1982, Litographic limestone like deposits in Tepexi de Rodríguez, Puebla, Mexico, *42th Annual Meeting, Society of Vertebrate Paleontology,* Field Guide for the excursion, 39 p.

Applegate, S. P., Ferrusquia-Villafranca, I., and Espinosa-Arrubarrena, L., 1979, Preliminary observations on the Geology and Paleontology of the Arroyo Tiburón area, Bahía Asunción, Baja California Sur, México, in: *Baja Califronia Geology, Field guides and Papers* (Abbott, P. A., and R. Gordon Gastil, eds.), *Geol. Soc. Am. Annual Meeting,* p. 113-115.

Applegate, S. P., González-Rodríguez, K., and Alvarado-Ortega, J., 2000, Fish fauna of the Tlayua quarries; in: *Paleontological highlights of the Mixteca Poblana in Central Mexico. Field Guide* (Espinosa-Arrubarena, L., Montellano-Ballesteros, M., and Applegate, S. P., eds.), *60th Annual Meeting of Society of Vertebrate Paleontology*, p. 95-105.

Applegate, S. P., and Wilson, E. C., 1976, Correlation of fossiliferous upper Oligocene or Lower Miocene sections at San Telmo Point and Arroyo Guadalupe, Baja California Sur, Mexico and a possible new phosphate source, *Congreso Lationoamericano de Geología,* Resúmenes, 3:6.

Aranda-Manteca, F. J., 1994, Vertebrados marinos fósiles de la Misión, Baja Califronia, México, Unpublushied M.S. thesis, Universidad Autónoma de Baja Califronia.

Aranda-Manteca, F. J., 2002, Tafonomía de Tiburones Fósiles del Género *Isurus* en La Misión, Baja California, *VIII Congreso Nacional de Paleontología, Soc. Mex. Paleont.*, Ponencias, p. 30.

Aranda Manteca, F. J., 2004, Tafofacies de la Formación Rosario Beach, Mioceno medio, Baja California, México, *IX Congreso Nacional de Paleontología, Soc. Mex. Paleont.*, Libro de Resúmenes, p. 18.

Arratia, G., 2000, The Sister-Group of Teleosteo: Consensus and Disagreements. *J. Vert. Paleont.* , 21:767–773.

Ashby, J. R., 1987, Vertebrate Paleoecology of the Late Pliocene Arroyo Salada Local Fauna, Baja California Sur, México. Paleoecología de vertebrados de la fauna local del Arroyo Salada del Plioceno tardío, Baja California Sur, México, *Ciencias Marinas,* 13:23-30.

Avendaño-Gil, M. J., Coutiño-José M. A., and Pimentel-Zepeda F., 1998, Rescate del patrimonio Paleontológico de la Carretera Ocozocuautla-Cosoleacaque, Estado de Chiapas, México, *VI Congreso Nacional de Paleontología, Soc. Mex. Paleont.*, Resúmenes, p. 7-8.

Avendaño-Gil, J., F. Pimentel, and Coutiño, M., 1999, Cretaceous vertebrates from Chiapas, Southern Mexico, *33rd Annual Meeting, Geol. Soc. Am., South-Central Section,* Abstracts with Programs, 31: A2.

Barbour, C. D., 1973, A biogeographical history of *Chirostoma* (Pisces: Atherinidae): A species flock from the Mexican plateau, *Copeia*, 533-556.

Bardack, D., 1965, Anatomy and evolution of Chirocentrid fishes, *Univ. Kansas Paleont. Contr., Vert. article,* 10, 88 p.

Barrios-Rivera, M., 1985, Estudio Analítico del registro paleovertebradológico de México. Unpublished Professional thesis, Universidad Nacional Autónoma de México.

Becerra-Martínez, C. A., 2003, Estudio anatómico de las aletas impares de los Goodeidos fósiles procedentes de Sanctorum (Formación Atotonilco el Grande), Hidalgo. Unpublished Bachelor's degree Thesis, Universidad Nacional Autónoma de México, Facultad de Estudios Superiores de Zaragoza, 37 p.

Benammi, M, Urrutia-Fucugauchi, J., and Alvarado-Ortega, J.,. 2003, Magnetic polarity stratigraphy of the Tlayúa Formation near Tepexi de Rodríguez, State of Puebla, Mexico, *Reunión Anual 2003 de la Unión Geofísica Mexicana*, Libro de Resúmenes, p. 56.

Benammi, M, Urrutia-Fucugauchi, J., and Alvarado-Ortega, J., 2004, Magnetic polarity stratigraphy of the "Mexican Solnhofen": The Tlayúa Formation near Tepexi de Rodríguez, State of Puebla, Mexico, *European Geosciences Union 2004, Geoph. Res. Abstr.,* vol 6, p. 56.

Blanco, A., and Cavin, L., 2003, New Teleostei from the agua Nueva Formation (Turonian), Vallecillo (NE Mexico), *C. R. Paleoevol.*, 2(2003):299-306.

Blanco-Piñón, A., 1998, Vallecillo, Nuevo León: Yacimiento fosilífero del Noreste de México, Unpublished M.S. thesis, Universidad Autónoma de Nuevo León.

Blanco-Piñón, A., 2001, Late Cretaceous (Turonian) fish assemblage from Vallecillo, northeastern Mexico, *Colloque sur le Cénomanien, Bull. Soc. Et. Sci. Nat.*, p. 19.

Blanco-Piñón. A., 2002, Primer registro de *Araripichthys* Silva Santos, 1985, (Telesotei: Araripichthydae) en sedimentos Cretácicos (Turoniano) del Noreste de México, *VIII Congreso Nacional de Paleontología, Soc. Mex. Paleont.*, Ponencias, p. 34.

Blanco-Piñón, A., and Frey, E., 2001a, A new Ichthyodectoidei from the Agua Nueva Formation (Late Cretaceous), Vallecillo, NE Mexico, *III International Meeting on Mesozoic Fishes: Systematics, Paleonviroments and Biodiversity,* Abstract book, p. 9.

Blanco Piñón, A., and Frey, E., 2001b, First occurrence of *Nursallia* sp. (Pycnodontoidei) in marine sediments of the Agua Nueva Formation (Turonian), Vallecillo, NE Mexico and its significance, *VI European Workshop on Vertebrate Paleoontology,* Abstract volume, p. 20.

Blanco Piñón, A., and Frey, E., 2001c, First occurrence of Nursallia sp. (Pycnodontoidei) in marine sediments of the Agua Nueva Formation (Turonian), Vallecillo, NE Mexico and its significance, *III International Meeting on Mesozoic Fishes: Systematics, Paleonviroments and Biodiversity,* Abstract book, p. 10.

Blanco-Piñón, A., Frey, E., and Stinnesbeck, W., 2000a, A new vertebrate fauna from Vallecillo N. L.:

Fossil-Lagerstätte in the Upper Cretaceous of NE Mexico, *V European Workshop on Vertebrate Paleoontology,* Abstracts, p. 17.

Blanco-Piñón, A., Stinnesbeck, W., and López-Oliva, J. G., 2000b. Ictiofauna de Vallecillo, Nuevo León, México, *VII Congreso Nacional de Paleontología, Soc. Mex. Paleont., Libro de Resúmenes*, p. 14-15.

Blanco-Piñón, A., Frey, E., Stinnesbeck, W., and López-Oliva, J. G., 2002., Late Cretaceous (Turonian) fish assemblage from Vallecillo, Northeastern Mexico, *N. Jb. Geol. Paläont. Abh.*, 225:39-54.

Blanco-Piñón, A., López-Oliva, J. G., Stinnesbeck, W., Addete, T., and González, A. H., 1998, Vallecillo, Nuevo León, "Fossil Lagerstäte", en el Noreste de México, *VI Congreso Nacional de Paleontología, Soc. Mex. Paloent.*, Resúmenes, p. 10-11.

Blanco-Piñón, A., Porras, H., Vega-Vera, F., González-Rodríguez, K. A., and Alvarado-Ortega, J., 2004, Múzquiz, Coahuila: a new fossiliferous locality, northeastern Mexico, *IX Congreso Nacional de Paleontología, Soc. Mex. Paleont.*, Libro de Resúmenes, p. 23.

Blanco-Piñón, A., Stinnesbeck, W., López-Oliva, J. G., Frey, E., Adatte, T., and González, A. H., 2001, Vallecillo, Nuevo León: una nueva localidad fosilífera del Cretácico Tardío en el noreste de México, *Rev. Mex. Cienc. Geol.*, 18:186-199.

Block-Iturriaga, C., 1963, Contribución al estudios de los peces fósiles del Valle de México, Unpublished Professional thesis, Universidad Nacional Autónoma de México.

Böse, E., 1913, Algunas faunas del Cretácico Superior de Coahuila y regiones limítrofes, *Inst. Geol. Méx., Bol.*, 30, 56 p.

Böse, E., and Cavins, O., 1928, The Cretaceous and Tertiary of southern Texas and northern Mexico, *Univ. Texas Bull.*, 2748, 142 p.

Bradbury, P. J., 1971, Paleolimnology of Lake Texcoco, Mexico. Evidence from diatoms, *Limn. Oceanogr.*, 16:180-200.

Bridges, L. W. and DeFord, R. K., 1962, Pre-Carboniferous Paleozooic Rocks in central Chihuahua, Mexico, *Am. Assoc. Petr. Geol., Bull.,* 45:98-104.

Brito, P. M., 1997, Révision des Aspidorhynchidae (Pisces, Actinopterygii) du Mésozoïque: ostéologie, relations phylogénétiques, données enviromentales et biogéographiques, *Geodiversitas*, 19:681-772.

Brown, B., 1912, *Brachyostracon* a new genus of Glyptodonts from Mexico, *Am. Mus. Nat. Hist., Bull.*, 31:67-177.

Carranza-Castañeda, O., and Applegate, S. P., 1994, Primer regsitro de peces picnodóntidos en el Estado de Hidalgo, *Univ. Autón. Edo. Hidalgo, Pancromo 23:*56.

Carranza-Castañeda, O., and Miller, E. W., 2000, Selected late Cenozoic vertebrate localities in the State of Hidalgo and Guanajuato, Mexico, *Field Guide, 60th Annual Meeting of Society of Vertebrate Paleontology*, p. 3-47.

Carranza-Castañeda, O., Miller, E. W., and Martínez, J. 1982. Guidebook of Guanajuato, *42nd Ann. Meet., Soc. Vert. Paleont.*, 75 p.

Castañeada-Posadas, C., 2004, Identificación de maderas terciarias de Panotla, Tlaxcala, México, y sus implicaciones paleobiológicas. Unpublished Professional thesis, Universidad Nacional Autónoma de México.

Castillo-Cerón, J., Cabral-Perdomo, M. A., and Carranza-Castañeda, O. 1996. Vertebrados Fósiles del Estado de Hidalgo (Fossil Vertebrates from the State of Hidalgo), *Univ. Autón. Edo. Hidalgo,* 127 p.

Cavender, M. T., and Miller, R. R., 1982, *Salmo australis*, a new species of fossil salmonoid from southwestern Mexico, *Univ. Michigan, Contrib. Mus. Paleontol.*, V. 26, 457 p.

Cavin, L., 2001, *Goulmichthys arambourgi* Cavin, 1995 and the phylogenetic relationships pf the Cretaceous Pachyrhizodontidae, *III Internacional meeting on Meswozoic fishes- Systematics, Paleoenviroments and Biodiversity*, Abstract book, p. 16.

Chang, M. M., and Grande, L., 1977, Redescription of *Paraclupea chetungensis*, an early clupeomorph from the Lower Cretaceous of Southern China, *Fieldiana Geol., n. ser.*, 37:1-19.

Cope, E. D., 1871, On two extinct forms of Physostomi of the Neotropical region, *Proc. Am. Phil. Soc.*, 12:52-55.

De Cserna, Z., 1972, Memorial to Manuel Maldonado-Koerdell 1908-1972, *Geol. Soc. Am.,* 7 p.

Dickerson, R. E., and Kew, W. S. W., 1917, Tertiary Mollusks and Echinoderms of Tuxpan, México, *Geol. Soc. Am., Bull.,* 28:224-225.

Downs, T., 1958, Fossil vertebrates from Lago de Chapala, Jalisco, Mexico, *Congreso Geológico Internacional, XX^{a} session, Sec. VII -Paleontología, Taxonomia y Evolución,* p. 75-77.

Dunkle, D. H., and Maldonado-Koerdell, M., 1953, Notes on some Mesozoic fossil fish remains from Mexico, *J. Wash. Acad. Sci.*, 43:311-317.

Durham, W. J., Applegate S. P., and Espinosa-Arrubarena, L., 1981, Onshore marine Cenozoic along southwest Pacific coast of Mexico, *Geol. Soc. Am., Bull.*, 92:384-394.

Espinosa-Arrubarena, L., 1979, Los tiburones fósiles (Lamniformes) del Rancho Algodones, Baja California Sur, México. Unpublished Professional thesis, Universidad Nacional Autónoma de México.

Espinosa-Arrubarena, L., 1987, Neogene species of the genus *Isurus* (Elasmobraanchii, Lamnidae) in southern California, U.S.A., and Baja California Sur, Mexico. Unpublished M.S. thesis, California State University.

Espinosa-Arrubarena, L., and Alvarado-Ortega, J., 2002, Las afinidades de los ionocóspidos de la Cantera Tlayúa, Puebla, *VIII Congreso Nacional de Paleontología. Soc. Mex. Paleont.*, Ponencias, p. 44.

Espinosa-Arrubarena, L., and Applegate, S. P., 1996, A possible model for the paleoecology of the vertebrate bearing beds in the Tlayúa quarries, near Tepexi de Rodríguez, Puebla, México, in: *Mesozoic Fishes-Systematics and Paleoecology* (Arratia, G. and Viohl, G., eds.), München, p. 539-550.

Espinosa-Arrubarena, L., and Applegate, S. P., 2000, Revisión de los tiburones de la Formación El Cien, Baja California Sur, México, V*II Congreso Nacional de Paleontología, Soc. Mex. Paleont.,* Resúmenes, p. 35-36.

Espinosa-Arrubarena, L., Applegate, S. P., and González- Rodríguez, K., 1996, The first mexican record of a coelacanth (Osteichthyes : Sarcopterygii) from the Tlayúa quarries near Tepexi de Rodríguez, Puebla, with a discussion on the importance of this fossil, *Sixth North American Paleontological Convention, Paleont. Soc. Spec. Pub.*, 8: 116.

Espinosa-Pérez, H., González-Rodríguez, K. and Applegate, S. P., 1991, Nuevos Registros de Peces Dulceacuícolas Fósiles en México, *II Congreso Nacional de Ictiología*, Resúmenes, p. 33.

Feldmann, R. M., Vega, F. J., Applegate, S. P., and Bishop, G. A. 1998. Early Cretaceous arthropods from the Tlayúa Formation at Tepexi de Rodríguez, Puebla, Mexico, *J. Paleont.*, 72:79-90.

Felix, J., 1891, Versteinerungen aus der mexicanischen Jura und Kreide Formation, *Palaeontographica* 37:140-194.

Ferrusquía-Villafranca, I., 1978, Distribution of Cenozoic vertebrate faunas in Middle America and problems of migration between North and South America, in: *Conexiones terrestres entre Norte y Sudamérica* (Ferrusquía-Villafranca I., ed.), *Univ. Nal. Autón. Méx., Inst. Geol., Bol.,* 101:193-321.

Ferrusquía-Villafranca, I., Applegate, S. P., and Espinosa-Arrubarrena, L., 1999, First Paleogene selachifauna of the Middle American-Caribbean-Antillean region, La Mesa de Copoya, west central Chiapas, Mexico-Systematics and Paleontological significance, *Rev. Mex. Cienc. Geol.*, 16:155-174.

Fierstine, H. L., Applegate, S. P., González-Barba, G., Schwennicke, T., and Espinosa-Arrubarrena, L., 2001, A fossil blue marlin (*Makaira nigricans* Lacepede) from the middle facies of the Trinidad Formation (upper Miocene to upper Pliocene), San José del Cabo Basin, Baja California Sur, Mexico, *South Calif. Acad. Sci., Bull.* 100:57-73.

Filleul, A., 2000, *Baugeichthys caeruleus,* Gen. et Sp. nov. , a new albuliform fish from the Hauterivian of Massif des Bauges (France), *J. Vert. Paleont.*, 20:637–644.

Fitch, E. J., 1969, Fossil Lanternfish otoliths of California, with notes fossil Myctophidae of North America, *Los Angeles County Mus. Contrib. Sci.*, 173:1-20.

Garibay-Romero, L. M., and Alvarado-Ortega, J., 2002, El hallazgo de peces fósiles en el Arroyo Las Bocas y la Cantera Temazol, Cretácico Superior, en la región Norte del Estado de Guerrero, *VIII Congreso Nacional de Paleontología. Soc. Mex. Paleont.*, Ponencias, p. 104.

Gasparini, Z. B., 1985, Los reptiles marinos jurásicos de América del Sur, *Ameghiniana*, 22:23-34.

Gío-Arágez, F., and Rodríguez-Arévalo, H. E., 2003, Panorama de la paleontología Mexicana, *Ciencia ergo sum*, 10:1-20.

González-Barba, G., 1995, Fauna de Elasmobranquios del Eoceno medio de la Formación Tepetate, Baja California Sur, México, *III Reunión Internacional sobre Geología de la Península de Baja Califronia, Univ. Autón. Baja California Sur,* Abstracts, p. 66-67.

González-Barba, G., 2000a, Cenozoic Chondrichthyans assemblages from Baja California Peninsula, *80th Annual Meeting, Am. Soc. Ichthyol. Herpetol.*, Abstracts, p. 171.

González-Barba, G., 2000b, Faunal succession of the Cretaceous and Tertiary chondrichthyans from Baja California Peninsula, Mexico, *60th Annual Meeting, Soc. Vert. Paleont.* Abstracts, p. 45A.

González-Barba, G., Coutiño-Jose, M. A., and Ovalles-Damián, E. 2002, *Carcharocles angustidens* Agassiz, 1843 (Chondrichthyes: Otodontidae) del Oligoceno Tardío de la Formación La Quinta, Grupo Simojovel, Localidad los Pocitos, Simojovel, Chiapas, *VIII Congreso Nacional de Paleontología. Soc. Mex. Paleont.*, Ponencias, p. 107.

González-Barba, G., Coutiño, M. A., and Ovalles-Damián, E. 2004, Adiciones a la fauna de selacios (tiburones y rayas) del Eoceno medio de la Formación San Juan, Chiapas, México, *IX Congreso Nacional de Paleontología, Soc. Mex. Paleont.*, Libro de Resúmenes, p. 39.

González-Barba, G., Herrera-Gil, L. A., and Rosales-López, A., 1998, Catálogo de localidades de la Cuenca Buenavista-San José del Cabo, Baja California Sur, México, *VI Congreso Nacional de Paleontología, Soc. Mex. Paleont.*, Resumenes, p. 25.

González-Rodríguez, K. A., 1989, Descripción preliminar de un nuevo holósteo (Macrosemiidae) del Cretácico Temprano (Albiano) de la Cantera Tlayúa en Tepexi de Rodríguez, Puebla, Unpublished Professional thesis, Universidad Nacional Autónoma de México.

González-Rodríguez, K. A., 1990, Los peces Cretácicos de México, *Univ. Nal. Autón. Méx., Inst. Geol.*, Unpublished report, 11 p.

González-Rodríguez, K. A., 1996, Taxonomía de la Familia Macrosemiidae (Osteichthyes:Holostei) de la Cantera Tlayúa (Cretácico Temprano) Tepexi de Trodríguez, Puebla, México. Unpublished M.S. thesis, Universidad Nacional Autónoma de México.

González-Rodríguez, K. A., 2000, Paleoichthyofauna of the Tlayua quarry, Puebla in central Mexico (State of Puebla), *80th Annual Meeting, Am. Soc. Ichthyol. Herpetol.*, Abstracts, p. 171.

González-Rodríguez, K. A., 2002, Paleoictiofauna de la Cantera Muhi, en Zimapán, Hidalgo, *VIII Congreso Nacional de Paleontología, Soc. Mex. Paleontol.*, Ponencias, p. 109.

González-Rodríguez, K. A., 2004, Descripción y Sistemática de la Familia Macrosemiidae (Neopterygii-Halecostomi) de la Cantera Tlayúa, Puebla, significancia paleogeográfica y paleontoecológica, Unpublished Ph.D. thesis, Universidad Nacional Autónoma de México.

González-Rodríguez, K. A., and Applegate, S. P., 2000, Muhi quarry, a new Cretaceous fish locality in Central Mexico, *60th Annual Meeting, Soc. Vert. Paleont.*, Abstracts, p. 45A.

González-Rodríguez, K. A., and Applegate, S. P., 2001, The presence of the Macrosemiidae (Halecostomi) in the Tlayúa Quarries, Central Mexico, *III International Meeting on Mesozoic fishes*, Abstracts, p. 34.

González-Rodríguez, K., and Reynoso, V. H., 2004, A new *Notagogus* (Macrosemiidae-Halecostomi) species from the Albian Tlayúa Quarry, Central Mexico, in: *Mesozoic Fishes 3 – Systematics, Paleoenviroments and Biodiversity* (G. Arratia and A. Tintori, eds.), Verlag Dr. Friedrich Pfeil, München, p. 265-278.

González-Rodríguez, K. A., Applegate, S. P., and Espinosa-Arrubarena, L., 2004, A new Macrosemiid (Pisces: Neopterygii-Alecostomi) from the Albian of Mexico, *J. Vert. Paleont.*, 24:281-289.

González-Rodríguez, K. A., Fielitz, K. C., and Bravo-Cuevas, V., 2004, Los peces aulopiformes de la Cantera Muhi de Zimapán, Hidalgo, *IX Congreso Nacional de Paleontología, Soc. Mex. Paleont.*, Libro de Resúmenes, p. 40.

González-Rodríguez, K. A., García-Barrera, P., and Avendaño, J., . 2002, Un nuevo registro de la Familia Macrosemiidae (Neopterygii: Halecostomi) en Chiapas, Sureste de México, *XVI Convención Geológica Nacional and III Reunión de Ciencias de la Tierra, Soc. Geol. Mex.*, Abstract book p. 303.

González-Rodríguez, K. A., Moreno-Tovar, R., and Vega-Vera, F., 2000, Zimapán, una nueva localidad cretácica marina en el estado de Hidalgo, México, *VII Congreso Nacional de Paleontología, Soc. Mex. Paleont.*, Resúmenes, p. 35-36.

Grande, L., and Bemis, W. E., 1998, A Comprehensive Phylogenetic Study of Amiid fishes (Amiidae) based on comparative skeletal anatomy. An Empirical Search for Interconnected patterns on Natural History, *Soc. Vert. Paleont., Mem. 4, Suppl. J. Vert. Paleont.*, 18, p. 690.

Grande, L., 1999, Distribution patterns and historical biogeography of gonorynchiform fishes (Euteleostei: Ostariophysi), in: *Mesozoic Fishes - Systematics and Stratigraphy* (Arratia, G. and H.-P. Schultze, eds.), Verlag Pfeil. München, p. 425-444.

Guerrero-Arenas, R., 2004, Icnofósiles de invertebrados de la Formación Tlayúa en Tepexi de Rodríguez, Puebla, México, Unpublished M.S. thesis, Universidad Nacional Autónoma de México.

Guzmán, A. F., Stinnesbeck, W., Robles-Camacho, J., and Polaco, O. J., 1998, El paleolago de Amatitlán, Jalisco: Estratigrafía, Sedimentología, Paleontología de la localidad tipo de Tapatia occidentalis (Osteichthyes: Goodeidae), *Soc. Mex. Paleont., Rev.*, 8:127-134.

Hanna, G. D., and Hertlein, L. G., 1927, Expedition of the California Academy of Sciences to the Gulf of California in 1921. Geology and Paleontology, *Proc. California Acad. Sci.*, 4:137-157.

Hay, P. O., 1929, Second bibliography and catalogue of the fossil vertebrate of North America, *Carnegie Inst. Washington, Publ.*, 390:1-916.

Hernández-Junquera, A., 1977, Fauna Local Laguna Media Luna, Pleistoceno Tardío, Municipio de Río Verde, San Luis Potosí, México, Unpublished Professional thesis, Universidad Nacional Autónoma de México.

Hernández-Rivera, R., 1998, Las playas del Cretácico en Coahuila, *VI Congreso Nacional de Paleontología, Soc. Mex. Paleont.,* Libro de Resúmenes, p. 33.

Hernándea-Rivera, R., Kirkland, J. I., and Gómez-Núñez, R. 2000, Cretaceous Dinosaurs from the State of Coahuila, Mexico, in: Espinosa-Arrubarrena, L., Montellano-Ballesteros, M., and Applegate, S. P. (coords.) *Field Guide of the 60th Ann. Meet., Soc. Vert. Paleont.*, p. 131-172.

Hertlein, L. G., 1966, Pliocene fossils from Rancho El Refugio, Baja California, and Cerralvo Island, Mexico, *Proc. California Acad. Sci.*, 30:265-284.

Hertlein, L. G., and Jordan, E. K., 1927, Paleontology of the Miocene of Lower California, *Proc. California Acad. Sci.,* 4:605-647.

Jordan, D. S., and Gilbert, J. Z., 1919, *Fossil Fishes of Southern California,* Stanford University Publications, University Series, p. 13-64.

Jordan, E. K., and Hertlein, L. G., 1926, XIV Expedition to Revillagigedo Islands, Mexico, in: 1925. VII-Contributions to the Geology and Paleontology of the Tertiary of Cedros Islands and adjacent parts of lower California, *Proc. California Acad. Sci.* XV:409-464.

Kashiyama, Y., Fastovsky, D. E., Rutherford, S., King, J., and Montellano, M. 2004. Genesis of a locality of exceptional fossil preservation: paleoenviroments of Tepexi de Rodríguez (mid-Cretaceous, Puebla, Mexico), *Cret. Res.*, 24:407-431].

Kirkland, J. I., and Aguillón-Martínez, M. C., 1995, *Schizorhiza* from Late Cretaceous of northern Mexico: the ultimate saw fish, *J. Vert. Paleont.*, 15(suppl. 3), p. 29A.

Kirkland, J. I., Hernández-Rivera, R., Aguillón-Martinez, M. C., Delgado de Jesús, C. R., Gómez-Nuñez, R., and Vallejo, I., 2000, The Late Cretaceous Difunta Group of the Parras Basin, Coahuila, Mexico, and its Vertebrate fauna, in: *Cretaceous Dinosaurs from the State of Coahuila, Mexico* (H. Hernández-Rivera, J. I. Kirkland, and R. Gómez-Nuñez, eds.), *Field Guide, 60th Annual Meeting of Society of Vertebrate Paleontology, Univ. Autón. Edo. Hidalgo, Avances en Investigación, special publication,* p. 133-172.

Kruckow, T., 1957, Die stratigraphische und palaeogeographische Bedeutung der miozänen Elasmobranchier-fauna von Baja California, Mexico, *N. Jahrb. Geol. Paläont., Monatch Jarhg,* 1957:444-449.

Kruckow, T., 1959a, Das Miozän der Golfküstenebene von Mexico, *N. Jahrb. Geol. Paläont. Abh.* 109:130-146.

Kruckow, T., 1959b, Beitrag zur miozänen Elasmobranchier-Fauna von Baja California, Mexico, *Bol. Asoc. Mex. Geol. Petrol.*, 50 p.

Kruckow, T., 1961, Fossiluche in Süd-Mexico, *Sonderdruck as "Der Aufschlus"* 1961(2):34-39.

Kruckow, T., and Thies, D., 1990, Die Neoselachier der Paläokaribik, *Courier Forsch. Senckenberg,* 119, 58 p.

Leriche, M., 1938, Contribution a étude des poissons fossiles des pays riverains de la Mediterranée Americane (Venezuela, Trinité, Antilles, Mexique), *Mém. Soc. Paleontol. Suiss.*, 61:1-42.

Lillegraven, A. J., 1972, Preliminary report on the Late Cretaceous mammals from the El Gallo Formation,

Baja Califronia Norte, México, *Los Angeles County Mus., Contrib. Sci.*, 232:1-11.

López-Ramos, E., 1981, Paleogeografía y Tectónica del Mesozoico en México, *Univ. Nal. Autón. Méx., Inst. Geol., Rev.*, 5:158-177.

Lucas, S., and González-León, C., 1995, Paleontology of the Upper Cretaceous Cabullona Group, Northeastern Sonora, in: *Studies on the Mesozoic of Sonora and Adjacent Areas* (Jacques-Ayala, C., C. M. González-León, and J. Roldán Quintana, eds.), *Geol. Soc. Am. Spec. Pap.*, 301:143-165.

Maldonado-Koerdell, M., 1948a, Peces Fósiles de México, I. Elasmobranquios, *Rev. Soc. Mex. Hist. Nat.*, 9:127-133.

Maldonado-Koerdell, M., 1948b, Adiciones a "Peces Fósiles de México, I. Elasmobranquios", *Rev. Soc. Mex. Hist. Nat.*, 9:295-300.

Maldonado-Koerdell, M., 1949, Peces fósiles de México II. Dipnoos, Ganoides y Teleosteos, *Rev. Soc. Mex. Hist. Nat.*, 10:241-245.

Maldonado-Koerdell, M., 1956, Peces fósiles de México III. Nota preliminar sobre peces del Turoniano superior de Xilitla, San Luis Potosí, México, *Ciencia*, 16:31-36.

Maisey, J., 2000, Continental breack up and the distribution of fishes of Western Gondwana during the Early Cretaceous, *Cret. Res.*, 21:281-314.

Matos, R. M. D., 1992, The north-east Brazilian rifts system, *Tectonics*, 11:766-791.

Meyer, H. von, 1840, Uber Uhde´s Sammlung Mexikanischen Antiquitäten, Mineralien und Petrefakten. *N. Jarhb.Miner., Geol. Paläont.*, p. 574-587.

Meyer, H. von, 1867, Studien über das Genus *Mastodon, Paleontographica,* 18:1-72.

Miller, R. R., 1965, Quaternary freshwater fishes of North America, in: Wright E. H. and Frey G. D. (eds.)*The Quaternary of the United States* , Princenton Univ. Press, New Jersey, p. 569-581.

Miller, R. R., 1973, Report of Committee on Research, *Am. Phil. Soc., Philadelphia, Yearbook,* p. 336.

Minch, A. J., Schulte C. K., and Hofman G., 1970, A Middle Miocene age for Rosarito Beach Formation in nortwestern Baja Califronia, Mexico, *Geol. Soc. Am. Bull.*, 81:3149-3154.

Müllerried, F. K., 1939, Apuntes Geológicos y Estratigráficos sobre el Valle del Mezquital, Estado de Hidalgo, *An. Esc. Nal. Cienc. Biol.*, 1:225-254.

Müllerried, F. K., 1945, El Edéstido *Helicoprion*, encontrado por primera vez en México, en el Estado de Coahuila, *Ciencia*, 6:208-212.

Mullerried, F. K., 1951, Algunos fósiles marinos del Terciario Inferior y Medio de Palenque, Chiapas. *Rev. Soc. Mex. Hist. Nat.*, 12:209-227.

Ortega-Gutiérrez, F., Mitre-Salazar, L. M., Roldán-Quintana, J., Morán-Zenteno, D., Alaniz-Alvarez, S. A., and Nieto-Samaniego, A. F. 1992, Texto explicativo de la Quinta Edición de la Carta Geológica de la República Mexicana Escala 1:2´000,000. Universidad Nacional Autónoma de México, Instituto de Geología, and Consejo de Recursos Naturales, Secretaría de Energía, Minas e Industria Paraestatal; Mapa con texto explicativo, escala 1:2´000,000, 74 p.

Ortega-Gutiérrez, F., Sedlock, R. L., and Speed, R. C., 1992, Evolución Tectónica de México durante el Fanerozoico, in: Llorente-Bouquets, J. E., E. González-Soriano, and N. Papavero (eds.) *Biodiversidad, Taxonomía y Biodiversidad de Artrópodos de México: Hacia una Síntesis de su conocimiento , Vol. II, Univ. Nal. Autón. Méx., Fac. Cienc., CONABIO, and Bayer, México,* p. 3-59.

Ovalles-Damián, E., 2004, Determinación Taxonómica de un Paraclupeido (Teleostei: Clupeomorpha) fósil de la Cantera El Espinal, Ocozocoautla de Espinosa, Chiapas, Unpublished Professional thesis, Universidad de Ciencias y Artes de Chiapas.

Ovalles-Damián E., and Alvarado-Ortega, J., 2002, Los paraclupeidos de la Cantera El Espinal, Ocozocoautla de Espinosa, Chiapas, *VIII Congreso Nacional de Paleontología, Soc. Mex. Paleont.*, Ponencias, p. 120.

Pantoja-Alor, J., 1992, Geología y paleoambientes de la Cantera Tlayúa, Tepexi de Rodríguez, estado de Puebla, *Univ. Nal. Autón. Méx., Inst. Geol., Rev.*, 9:156-176.

Pinna, M. de, 1996, Teleostean monophyly, in: M. L. J. Stiassny, L. R. Parenti, and G. D. Johnson (eds.) *Interrelationships of Fishes*, Academic Press, San Diego, p. 147–162.

Rodríguez de la Rosa, R., and Cevallos-Ferríz, S. R. S., 1998, Vertebrates of the El Pelillal Locality (Campanian, Cerro del Pueblo Formation), Southeastern Coahuila, Mexico, *J. Vert. Paleont.*, 18:751-764.

Romero, E., 2005, Descubren fósil de pez no conocido, Tabasco Hoy, Martes 8 de Marzo de 2005 (Sección Sureste), p. 49.

Sapper, K., 1899, Ueber Gebirgsbau und Boden des noerdlichen Mittelamerika, *Peterm. Mitt., Erg. Heft*, 127.

Schwennicke, T., and González-Barba, G., 1998, Fósiles alóctonos de las capas de fosforita granular del Miembro San Juan (Formación el Cien), Baja California Sur: Su importancia para el Modelo sedimentológico, *VI Congreso Nacional de Paleontología, Soc. Mex. Paloent.*, Resúmenes, p. 68-69.

Seibertz, E., and Buitrón, B. E., 1987, Paleontología y Estratigrafía de los *Neohibolites* del Albiano de Tepexi de Rodríguez, Estado de Puebla (Cretácico Medio, Mexico), *Rev. Soc. Mex. Paleont.,* 1:285-299.

Seibertz, E., and Spaeth, C., 2002, Cretaceous belemnites of Mexico III. The Albian *Neo-* and *Mesohibolites* of the "Mexican Solnhofen" Tepexi de Rodríguez (State of Puebla) and their biostratinomy (Lower Cretaceous), *N. Jb. Geol. Paläont. Abh.*, 225:55-74.

Shaw, A. C., 1982, La fauna regional "El Golfo" del Pleistoceno Medio proveniente del Noreste de Sonora, Mexico, *42th Annual Meeting, Soc. Vert. Paleont.,* Abstracts (not paginated).

Silva-Bárcena, A., 1969, Localidades de vertebrados fósiles de la República Mexicana, *Univ. Nal. Autón. Méx., Inst. Geol., Paleont. Mex.* 28:1-34.

Smith, L. M., 1980, The evolution and ecological history of the fish fauna of the Río Lerma Basin, Mexico, Unpublished Ph.D. Dissertation, University of Michigan.

Smith, L. M., 1981, Late Cenozoic fishes in the warm desert adaptations: in: *Fishes in North America Deserts* (Naiman J. R., and Soltz L. D., eds.), Jonh Wiley and Sons, New York, p. 11-38.

Smith, L. M., Cavender, M. T., and Miller R. R., 1975, Climatic and Biogeographic significance of a fish fauna from the Late Pliocene - Early Pleistocene of the Lake Chapala Basin (Jalisco, Mexico). *Studies on Cenozoic Paleontology and Stratigraphy in honor Claude W. Hibbard, Univ. Michigan, Pap. Paleont.*, 12:29-38.

Sour-Tovar, F., Quiroz-Barroso, S. A., and Applegate, S. P., 2000, Presence of *Heliocoprion* (Chondrichthyes, Elasmobranchii) in the Permian Patlanoaya Formation, Puebla, Mexico, *J. Paleont.,* 74: 363-366.

Springer, V. G., 1957, A new genus and species of Elopid fish (*Laminspondylus transverses*) from the Upper Cretaceous of Texas, *Copeia*, 2:135-140.

Thies, D., 1989, *Lepidotes gloriae*, sp. nov. (Actinopteryggii: Semionotiformes) from the Late Jurassic of Cuba, *J. Vert. Paleont.*, 9:18-40.

Vega, F. J., García-Barrera, P., Coutiño, M., Nyborg, T., Cifuentes-Ruiz, P., González-Rodríguez, K., Martens, A., Delgado, C. R., and Carbot, G. 2003, Early Cretaceous Arthopods from Plattenkalk facies in Mexico, *Contr. Zool.*, 72:187-189.

Villada, M., 1897, Catálogo de la Colección de Fósiles de México, *An. Mus. Nac., México, Imprenta del Museo Nacional, México,* 79 p.

Weiler, W., 1959, Miozäne Fisch-Otolithen aus der Bohrung S. pablo-2 im Beckein von Veracruz in Mexiko, *N. Jarhb. Geol. Paläon., Abh.*, 109:147-172.

Willer, R. R., and Smith, M. L., 1985, Origin and geography of the fishes of Central Mexico. Cap. 14,. in: *The zoogeography of North American freshwater fishes* (Hocutt, Ch. H., and Wiley. E. O., eds.), Wiley Interscience Publication, p. 487-517.

Wittich, E., 1913, Restos de Selacios del Terciario de la División del Norte de la Baja California, Actas Sesión Veracruzana, *Soc. Geol. Mex.*, p. 1-16.

Chapter 10

Research on Fossil amphibians and reptiles in Mexico, from 1869 to early 2004 (including marine forms but excluding pterosaurs, dinosaurs, and obviously, birds)

VÍCTOR-HUGO REYNOSO

Instituto de Biología, UNAM. Circuito Exterior. Ciudad Universitaria 04510, Coyoacán, México, D.F. vreynoso@ibunam.ibiologia.unam.mx

1. Abstract

The amphibian and reptile fossil record in Mexico is scarce. There have been reported 255 unique taxon/locality/age records from 85 localities. Most fossils are fragmentary and their taxonomic status is difficult to assess; however, some are beautifully preserved, permitting even good cladistic and biogeographical analysis. Among amphibians there are records of caudates and anuras, restricted to the Neogene and Quaternary. Reptiles include turtles (which are the most abundant), crocodiles, ichthyosaurs, sauropterygians, a basal lepidosauromorph, sphenodontians, a basal squamate, true squamate lizards, and snakes, from the Late Triassic to the Holocene. Localities are distributed in 21 Mexican states from which Sonora bears most records. Nineteen species based on Mexican fossils have been erected. Three have been considered

F. Vega et al. (eds.), Studies on Mexican Paleontology, 209–231.

invalid, one has changed taxonomic status, and at least two more are questioned. Most of the type specimens designated are housed in Mexican institutions. There have been 99 main non-redundant published documents that mention, list, or describe fossil amphibians and reptiles, including scientific papers, books and book chapters of international and national circulation, meeting abstracts, guidebooks and newsletters. Of all papers published, only 51 are formal publications in either national or international press and 40 were published in international journals or editorials. Mexicans have published only 10 international papers. However, Mexican paleoherpetology has significantly increased in the last two decades, although most information is hidden in gray literature of difficult access to the international community. It is imperative that Mexican workers document their discoveries in formal publications.

2. Introduction

As with most countries of the world, the amphibian and reptile fossil record in Mexico is scarce and fragmentary. The lack of good descriptive fossil material has lead Mexican paleontologists to overlook these groups, and no specialists were ever formed until recently, when the well-preserved reptile fossils from the Tlayua Formation, in Tepexi de Rodríguez, Puebla, Mexico, were first described. Since then, the number of researchers, and particularly students, interested in paleoherpetology has increased.

Up until late 1990s most literature referring to fossil amphibians and reptiles was restricted to listings of approximate taxonomic names from fossil sites well known mainly for their mammals, a grouping which research is very active. Papers which their main subject were amphibians, reptiles or both were practically non-existent, and even today only a small portion of the specimens reported has been reviewed and the true identity of most is still uncertain.

Mexicans are proud of their wildlife diversity. Amphibians and reptiles are noteworthy because Mexico bears the second highest diversity worldwide, following Australia. In fact, Mexico has the highest diversity per area unit when is taken into account that Australia has an area 3/4ths larger. The great wildlife diversity of Mexico has been explained by an extremely complex physiography associated with its transitional position between two important biogeographic realms: the Neotropical and the Nearctic. The complex physiography results in the generation of a greater number of habitats and barriers that promotes speciation. Mexican landscape is not a modern occurrence, since most of the mountain chains were formed during the Tertiary. Even in the Cretaceous and earlier, Mexico was subjected to a series of oceanic transgressions and regressions, which resulted in the appearance and disappearance of islands, a phenomenon that also promotes speciation. Mexican complex geological history predicts the presence of a large amphibian and reptile biodiversity in the past.

This paper intends to provide a revision of the literature concerning fossil amphibians and reptiles in Mexico from 1869 to early 2004. Taxonomic groups, areas, and time periods were analyzed separately, as well as the work of Mexican paleontologists on the subject. Many localities and taxonomic names reported in Mexican literature, which are not readily accesible are also introduced to the international community.

This work is largely based on Barrios-Rivera (1985), professional thesis where he lists many reports of fossil amphibian and reptile occurrences in Mexico.

3. Historical Background

The first report of a fossil reptile from Mexico was by José Aguilera (1869) in his "Bosquejo Geológico de México". In this paper Aguilera describes the presence of plates of a Late Cretaceous (Early Turonian) soft-shelled turtle *Trionyx* sp.? from the Peyotes Division at Parras, Coahuila, in the Río Bravo. After this first discovery, nothing else is mentioned in the literature until the early 1900s.

Papers restricted to the first half of the 20th century are those of Aguilera (1907) who mentions for the second time the *Trionyx* finding; Wieland (1910) describing a fragmentary lower jaw of the marine reptile, *Plesiosaurus mexicanus* from the early Cretaceous deposits near Tlaxiaco, Oaxaca, the first new species of a fossil reptile from Mexico; and Müllerried (1931; 1943) mentioning two partial mosasaurs from the Late Cretaceous of Tamaulipas, and an internal mold of a turtle from the Tertiary of Chiapas, respectively.

During the second half of the 20th century and the early beginnings of the 21st, fossil amphibian and reptile research increased significantly. During this time frame most of the main documents concerned with or referring to fossil amphibians and reptiles were published. Of these publications, it is worth mentioning the descriptions of beautifully preserved holotypes of turtles, sphenodontians, basal lepidosaurs and lizards. These fossils are the skull of *Paradipsosaurus mexicanus* (Fries *et al.*, 1955) from the late Eocene deposits of Marfil, Guanajuato; the almost complete skeleton of *Gopherus auffenbergi* (Mooser, 1972) from the Illinoian Cedazo local fauna in Aguascalientes; one of the few skulls of *Gopherus* associated with its shell and postcranial skeleton from which *G. donlaloi*, a giant turtle from the Rancholabrean deposits of Villagrán, northwest Tamaulipas was recently described (Reynoso and Montellano, 2004); the exquisitely well preserved basal lepidosauromorphs and sphenodontians from the mid-Jurassic red beds of La Boca Formation at Huizachal Canyon, Tamaulipas (Clark and Hernández, 1994; Reynoso, 1996, 2003, in press; Reynoso and Clark, 1998); the beautifully preserved complete skeletons of sphenodontians, basal squamates and true lizards from the Early Cretaceous of the Tlayúa Formation, near Tepexi de Rodríguez, Puebla (Reynoso, 1997, 1998, 1999, 2000; Reynoso and Callison, 2000; Reynoso *et al.*, 2000); and the skull of the new crocodile species *Geosaurus vignaudi* (Frey *et al.*, 2002) from the middle Tithonian deposits of La Pimienta Formation, near Mazatepec, Puebla.

Other important fossils that have been recently described are remains of the largest known marine carnivore *Liopleurodon ferox* (Buchy *et al.*, 2003) from the Kimmeridgian La Casita Formation. Also outstanding are the not yet described complete skeletons, including the frogs and salamanders from the Pliocene fossiliferous region of Santa María Amajac, Atotonilco el Grande, Hidalgo (Fuentes-Guerra and Velasco-de León, 2002), the turtles and crocodiles from Tlayúa (Reynoso *et al.*, 2000), and an almost complete skull of a goniolophid crocodile from the Maastrichtian Ocozocoautla Formation of Chiapas (Carbot-Chanona and Coutiño-José, 2000).

Remaining described or listed taxa are represented by incomplete skeletons and skulls, isolated osteological elements and fragmentary material from many sites broadly distributed through 21 states of the Mexican Republic. New genera and/ or species described from fragmentary material are: the early Cretaceous marine reptile *Plesiosaurus mexicanus* (Wieland, 1910) from Tlaxiaco, Oaxaca known from the fragmentary upper and lower jaws; the Hemphillian *Bufo campi* (Brattstrom, 1955a) from Yepómera, Chihuahua, only known from its left tibia and fibula; the late Pleistocene lizards *Ctenosaura premaxillaris* and *Lepidophyma arizeloglyphus* (Langebartel, 1953) from the Spukil Cave, Yucatan, known from the premaxilla and right dentary respectively; the Wisconsian *Phrynosoma josecitensis* (Brattstrom, 1955b) from Nuevo León, known from a skull fragment; and the turtle *Gopherus pargensis* (Mooser, 1980) only known from a fragmentary shell from the Illinoian Cedazo local fauna in Aguascalientes. A controversy on the validity of these taxa has been raised mainly because their descriptions were based on non-diagnostic features.

Additional indirect evidences of the presence of fossil reptiles in Mexico are an isolated braincase of a crocodile (J. Clark, pers. com.) from the late Campanian Cerro del Pueblo Formation, near Saltillo, Coahuila, originally thought to be a chasmosaur braincase (Hernández-Rivera and Delgado-de Jesús, 1999); an ammonite shell with probable mosasaur bites from the Maastrichtian found at the Arroyo de Santa Catarina, El Rosario, Baja California (Saul, 1976); and a turtle ignite from the Late Cretaceous of Coahuila (Rodríguez-de la Rosa and Dávila-Rodríguez, 1998).

4. Taxa Analysis

Taxa and biogeographical analyses were performed by gathering unique reports of fossil amphibians and reptiles mentioned, listed or described in literature, considered independently from their taxonomic rank. A unique record is defined as taxa present in a different locality and stratigraphic level (even within the same locality). Taxa reports from the same locality and age mentioned in different papers, were considered redundant and were not used. To avoid admission errors (e. g. the duplication of taxa from different reports) redundant taxa were merged from each other into a single record as much as possible. In such cases, the valid name used was that of the most recent citation of the taxa, or the name used in the most important publication. In a few published documents it was impossible to distinguish whether authors were referring to different taxa or to the same one. These cases were considered in the analyses but carefully discussed since they might be increasing richness artificially. Special case are documents produced by Castillo–Cerón's working team, where what seems to be the same fossil specimen, is mentioned repeatedly at different taxonomic ranks or identities. For example, it is uncertain if the Miocene-late Pleistocene Tehuichila-Zacualtipan amphibians referred by Castillo-Cerón *et al.* (1996), include those of the Miocene Tehuichila-Zacualtipan amphibians referred by Castillo-Cerón (1994), Castillo-Cerón and Goyenechea (1997), Castillo-Cerón *et al.* (1998) and Castillo-Cerón *et al.* (2002); although these last papers certainly do not include the late Pleistocene amphibians that have been never reported in a formal paper. For all calculations, undetermined various amphibians and reptiles reported by Miller and Carranza-Castañeda (2001) and the record of cf. *Platecarpus* (Frey *et al.*, 2001) were excluded, since they did not provide collecting data in their reports.

Up to now there have been reported 43 amphibians and 212 reptiles (Table 1) from Mexico. Among amphibians, five are undetermined reports, five caudates, and 33 anurans. Among reptiles, 10 are undetermined, 89 turtles (2 pleurodires), one basal lepidosauromorph, seven sphenodontians, one basal squamate, 67 true squamates (39 lizards and 28 snakes), 26 crocodiles and 11 non-squamate marine reptiles (Figure 1).

Unidentified amphibians and reptiles are usually reported when introducing non-lower vertebrate faunal assemblages, being uncertain if these reports refer to a single or several taxa. *Ambystoma* is the only genus of caudates properly identified (Brattstrom, 1955a; Álvarez and Huerta, 1975). In addition, *Bufo, Gastrophryne, Hyla, Hylactophrine (=Eleutherodactylus), Leptodactylus, Pternohyla, Rana, Rhinophrynus, Scaphiopus, and Syrrhophus,* have been reported for Anura.

Reptiles are considerably more abundant than amphibians. There maybe more unidentified reptiles not considered in this paper. For example, Cope *et al.* (2003) mentioned a Campanian reptile microfauna for the Cañón Santa Elena National Park, Chihuahua, without specifying the number or identity of the taxa. Similar cases were considered as a single report causing omission errors.

Among reptiles, turtles are the most abundant. Of 89 turtle records, 87 are cryptodirans and two pleurodirans. The most abundant identified genus is *Kinosternon* with 18 records, followed by *Gopherus* with 12. The remaining taxa are only reported one or few times. These include A*docus, Aspideretes, Compsemys, Geochelone* and/or *Heperotestudo*, cf. *Hoplochelys*, similar to *Neurankylus, Naomichelis*, cf. *Protochelydra, Terrapene, Testudo, Trachemys* (reported as *Chrysemys*, *Pseudemys* or *Trachemys*), *Trionyx*, and the pleurodiran *Bothremys*.

Crocodiles are scarce in Mexican fossil record, although very diverse. Many taxa have been reported, although material is usually not well preserved for good taxonomic identification. Primitive crocodiles include goniolophids, thallatosuchids, metriorhynchines, teleosaurids, and unidentified metasuchians and mesosuchians. Modern crocodiles are all referred to the genus *Crocodylus*. Miller (1980) identified a lower jaw from the Blancan Las Tunas Local Fauna in Baja California Sur as *Crocodylus moreletti.*

TABLE 1. Number of undetermined, proposed, and described new species, and taxonomic level of identification of amphibians and reptiles reported for Mexico.

	records	**proposed new species**	**described new species**	**undetermined species**	**records referred to the suprageneric level**	**records referred to the generic level**	**records referred to the species level**
AMPHIBIANS	43	2	1	11	0	10	22
%		4.7	2.3	25.6	0.0	23.3	51.2
undet. amphibians	5	0	0	5	0	0	0
%				100			
urodeles	5	0	0	2	0	2	1
%				40		40	20
anurans	33	2	1	4	0	8	21
%		6.1	3.0	12.1		24.2	63.6
REPTILES	212	5	18	50	35	59	68
%		2.4	8.5	23.6	16.5	27.8	32.1
indet. reptiles	10	0	0	10	0	0	0
%				100			
turtles	89	4	3	26	9	34	20
%		4.5	3.4	29.2	10.1	38.2	22.5
crocodiles	26	1	1	7	12	4	3
%		3.8	3.8	26.9	46.2	15.4	11.5
lepidosauromorphs	76	0	13	7	8	20	41
%			17.1	9.2	10.5	26.3	53.9
"euryapsids"	11		1		6	1	4
%			9.1		54.5	9.1	36.4
TOTAL	255	7	19	61	35	69	90

There have been reported 76 lepidosauromorphs. The Mesozoic forms are the best known, particularly those of the mid-Jurassic La Boca Formation and the early Cretaceous Tlayúa Formation. From Huizachal are: the basal lepidosauromorph *Tamaulipasaurus morenoi* (Clark and Hernández, 1994) known from a skull and a basicranium; the sphenodontians *Cynosphenodon huizachalensis* (Reynoso, 1996; Reynoso 2003), the most abundant local form known from several dental elements; the dwarf *Zapatadon ejidoensis* (Reynoso and Clark, 1998), known by a single almost complete skull; a venomous species (Reynoso, in press); and two not yet formally described forms (Reynoso, 1992). From Tlayúa are the aquatic sphenodontian *Ankylosphenodon pachyostosus* (Reynoso, 2000) and the beaded sphenodontian *Pamizinsaurus tlayuaensis* (Reynoso, 1997). World wide important forms include the basal squamate *Huehuecuetzpalli mixtecus* (Reynoso, 1998), sister group of all known squamates, and the basal scincomorph *Tepexisaurus tepexii* (Reynoso and Callison, 2000). All Tlayúa reptile skeletons are among the best preserved.

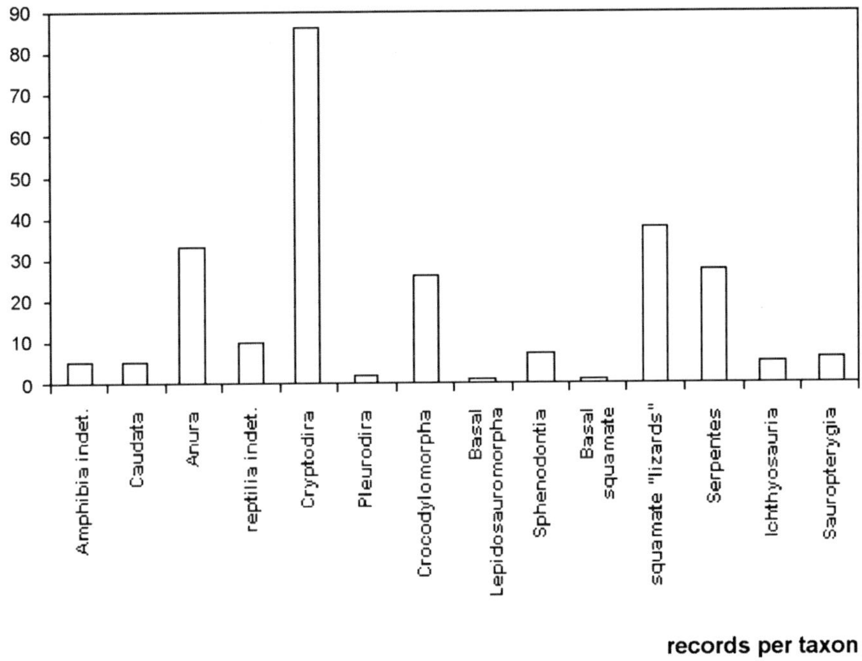

FIGURE 1. Frequency of unique taxon/locality/age records of amphibians and reptiles per high rank taxonomic group.

There are 39 reports of true lizards (and marine mosasaurs) and 28 snakes. These includes the modern genera *Ameiva, Barisia, Callisaurus, Cnemidophorus (=Aspidoscelis), Ctenosaura, Sceloporus* (the genus with the most reports), *?Laemanctus, Lepidophyma*, and *Phrynosoma*. A varanid lizard, *Chamops, Polyglyphanodon*, and the mosasaurs *Clidastes*, cf. *Mosasaurus*, and cf. *Platecarpus* are Cretaceous (Lucas *et al.*, 1995; Nydam, 1999, Everth *et al.*, 2003; Frey *et al.*, 2001).

All reported snakes are from the late Cenozoic with the exception of an isolated vertebra referred as *Coniophis* sp. by Aguillón (2000) from the Late Cretaceous Cerro del Pueblo Formation, Coahuila; however, this identification needs to be verified. Late Cenozoic snakes are *Boa, Crotalus* and/or *Bothrops, Drymarchon, Drymobius, Elaphe, Hypsiglena, Lampropeltis, Masticophis?, Pituophis, Salvadora, Spilotes, Thamnophis* (the most abundant genus) and *Trimorphodon*. Most reports are from the Spukil Cave (Langebartel, 1953), Rancho La Brisca, Sonora (Van Devender *et al.*, 1985), and Baja California Sur (Miller, 1980).

For the purpose of this paper non-squamate marine reptiles are grouped as "euryapsids". These are the less abundant taxa among Mexican reptiles and only five ichthyosaurs and six sauropterygian have been reported. Most fossils were collected in the Mesozoic deposits of northern Mexico, with the exception of the holotype of *Pliosaurus mexicanus* and another ?Pliosauridae from Oaxaca (Wieland, 1910; Ferrusquía-Villafranca and Comas, 1980). Only *Shashtasaurus altispinus* and *Toretocnemus californicus* from Sonora (Lucas and González-León, 1995), and the Aramberri monster *Liopleurodon ferox* (Buchy *et al.*, 2003) were complete enough to be identified to the species level.

5. Amphibians and Reptiles in Time and Space

Amphibians and reptiles are broadly distributed throughout the Mexican territory having records from the Triassic to the Holocene. Twenty-one of the 31 states have fossil herpetofauna in 85 localities. The most important states listed according to their richness are Sonora with 40 reports, Coahuila with 22, Tamaulipas and Yucatán with 19, Nuevo León with 18 and Estado de México, Hidalgo, Jalisco, and Puebla, all with 17 (Figure 2). The remaining states have less than 15 reports. All reports of Yucatán are of Pleistocene-Holocene subfossils from local caves, contrasting with Coahuila, where most fossils are from the Mesozoic. Of all unique records, 129 are from the Quaternary, 47 from the Late Cretaceous, 45 from the Neogene, 11 from the late Jurassic, 10 from the Early Cretaceous, eight from the middle Jurassic, three from the Paleocene, and two from the Triassic (Figure 3). There are no reports earlier than the late Triassic.

The earliest known records are the ichthyosaurs *Shashtasaurus altispinus* and *Toretocnemus californicus* from the late Triassic deposits at Mina el Antimonium at Sierra del Alamo, Sonora (Lucas and González-León, 1995). The Huizachal Canyon red beds of La Boca Formation is the only known Middle Jurassic locality. Additional taxa from Huizachal are several partial skulls of Metasuchia (Clark *et al.*, 1994b), and a primitive crocodylomorph skull fragment (Clark *et al.*, 1994a). Late Jurassic deposits are from Coahuila, Nuevo León and a single record from Puebla, where the crocodile *Geosaurus vignaudi* (Frey *et al.*, 2002) was described. The most important late Jurassic deposits are the series of localities near Aramberri, Nuevo León (Buchy *et al.*, in press; Aranda-Manteca and Stinnesbeck, 1993; Frey *et al.*, 2001), comprising deposits of La Casita and La Caja Formations at Asunción, Galeana, San Lucas, Río Pablillo, Cieneguillas, Hidalgo e Iturbide, Zaragoza, La Angostura, and Sierra el Montelongo Pedregoso, among others. From these localities, there have been described partial skulls and skeletons of crocodiles (thalattosuchians, a metriorhynchine, and a teleosaurid), an euichthyopterygian ichthyosaur, an elasmosaurid, pliosaurid, and the giant *Liopleurodon ferox* (Buchy *et al.*, 2003; Buchy *et al.*, in press). From the type Section of La Casita Formation, at Sierra de Parras, Coahuila, south of General Cepeda, only some vertebrae and plates of *Ichthyosaurus*? (Peterson-Rodríguez *et al.,* 2000) of Kimmeridgian or early Tithonian age, are mentioned.

Most of the known Early Cretaceous lower vertebrates in Mexico are from the Tlayúa Quarry, near Tepexi de Rodríguez, Puebla. In addition to the previously listed lepidosaurs (Reynoso, 1997, 1998, 2000; Reynoso and Callison, 2000) this locality

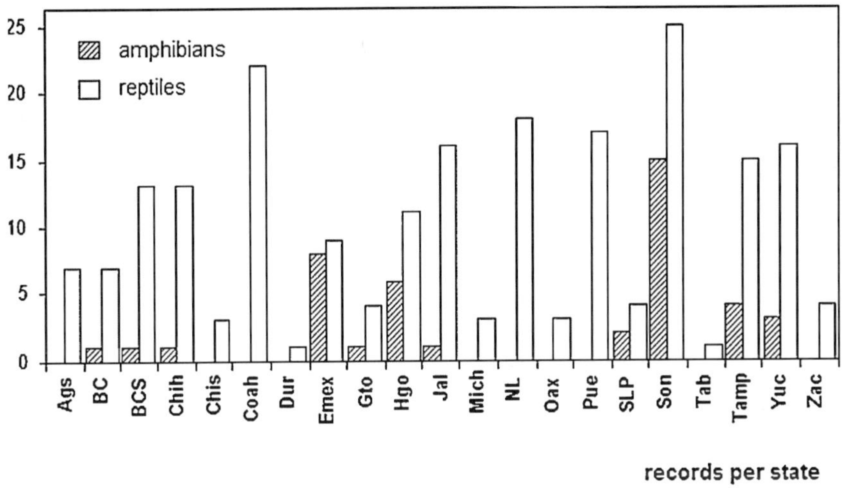

FIGURE 2. Frequency of unique taxon/locality/age records of amphibians and reptiles per state. Abbreviations: Ags, Aguascalientes; BC, Baja California; BCS, Baja California Sur; Chih, Chihuahua; Chis, Chiapas; Coah, Coahuila; Dur, Durango; Emex, Estado de México; Gto, Guanajuato; Hgo, Hidalgo; Jal, Jalisco; Mich, Michoacán; NL, Nuevo León; Oax, Oaxaca; Pue, Puebla; SLP, San Luis Potosí; Son, Sonora; Tab, Tabasco; Tamp, Tamaulipas; Yuc, Yucatán; Zac, Zacatecas.

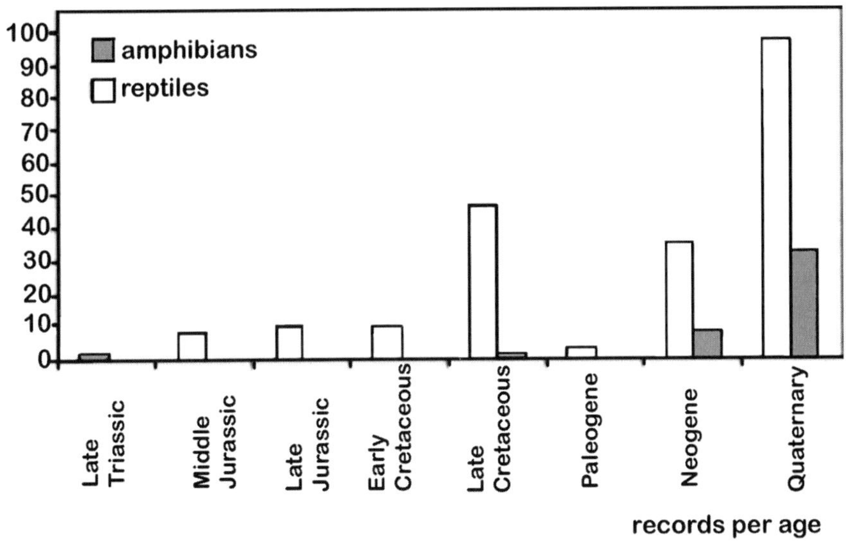

FIGURE 3. Frequency of unique taxa/locality/age records of amphibians and reptiles per age.

has produced several complete and partial skeletons of two turtle taxa, and a crocodile (Reynoso *et al.*, 2000). Other early Cretaceous localities are found in Oaxaca, in the Tlaxiaco area and near Huajuapan de León (Wieland, 1910; Ferrusquía-Villafranca and Comas, 1980). From Tlaxiaco, the type specimen of *Plesiosaurus mexicanus* was recovered (Wieland, 1910), and from the late Berriasian-early Hauterivian of Huajuapan de León, Oaxaca, an ichthiosaurid centra and a Pliosauridae? dentary associated with a scapular fragment (Ferrusquía-Villafranca and Comas, 1980) were also recovered.

Late Cretaceous deposits bearing lower vertebrate are abundant. Most localities are either Campanian or Campanian-Maastrichtian in age and are basically distributed through the northern states of the republic: Baja California, Sonora, Chihuahua, Coahuila, and Nuevo León. Only two localities have been reported to be Turonian: the Peyotes División at Parras on the Río Bravo from the north Presidium to Piedras Negras, Coahuila (Aguilera, 1896; 1907), and Vallecillo, Nuevo León (Blanco-Piñón *et al.*, 1998) where a soft shelled turtle, *Trionyx* sp., and undetermined chelonians have been reported respectively.

The oldest Late Cretaceous record of a marine reptile is that of Müllerried (1931) who reported two partial individuals of a mosasaur from Rayón, north of the Yamesí River at 60 kms NW of Tampico, Tamaulipas. These specimens were deposited in the Museo Nacional de Historia Natural, in Mexico City, but are now probably lost.

The Cerro del Pueblo Formation area, near Saltillo, bears the most important Late Cretaceous (late Campanian) reptile fauna in Mexico; however, opposite to the fossils from the Jurassic and Early Cretaceous deposits, the fossils from Coahuila are preserved to be used for good taxonomy and systematics. These sites have produced 18 different records including: the turtles *Compsemys*, cf. *Hoplochelys*, Kinosternoidea, cf. *Protochelydra*, Trionichidae, *Adocus*, and a form similar to *Neurankylus* (Rodríguez de la Rosa and Cevallos-Ferriz, 1998; Everth *et al.*, 2003); the southern most record of the pleurodire *Bothremys* (García and Reynoso, 2002); a varanid lizard (Everth *et al.*, 2003); the snake questionably referred to as *Coniophis* (Aguillón, 2000); the mosasaurs *Clidastes*, probably cf. *Platecarpus*; an undetermined Mosasauria (Everth *et al.*, 2003; Frey *et al.*, 2001); and longirostrine crocodiles, probably Goniopholididae (Rodríguez-de la Rosa and Cevallos-Ferriz, 1998). The lack of good material makes identifications to the generic and particularly to the specific level dubious, so, the recovery of more complete material is needed before a final faunal list can be produced.

The possible Campanian-Maastrichtian Aguja, Javelina and San Carlos formations, located on the Chihuahua-Coahuila border, are the second most abundant Late Cretaceous deposits in Mexico. These have been recently studied by M. Montellano-Ballesteros and J. Westgate's teams (Montellano-Ballesteros *et al.*, 2000, Westgate *et al.*, 2000; 2002a, b, Cope *et al.*, 2003). Unfortunately, up to now no formal lists have been produced. The presence of a reptile microfauna; the turtle *Aspideretes*;soft-shelled turtles and other chelonians; and crocodiles, including goniolophids and a short snouted form have been reported.

In Baja California, all known Late Cretaceous lower tetrapods are from the late Campanian El Gallo Formation. This site bears the oldest known fossil amphibian from Mexico (Lillegraven, 1976) although no taxonomic identity has been given. Reptile fossils have only been referred to higher taxonomic ranks (e.g. turtle, lizards, and alligator (sic) (Lillegraven, 1976), with the exception of the recently reported *Naomichelis* (Rodríguez-de la Rosa and Aranda-Manteca, 2000), known from fragments of costal and plastral plates and the lizard *Polyglyphanodon bajaensis* (Nydam, 1999) known from several isolated jaw fragments.

Sonoran Late Cretaceous deposits are from the Campanian-Maastrichtian Corral de Enmedio, Lomas Coloradas, and Packard Shale formations from the Cabullona Group (Lucas *et al.*, 1995). There have been reported isolated plates of trionichid turtles, isolated eusuchian crocodile vertebrae, and the right dentary of the lizard *Chamops segnis.*

No records of Late Cretaceous amphibians and reptiles are known from the central part of Mexico, but Chiapas, located at the southern most part of the country, has produced remains of an almost complete skull of a goniolophid crocodile associated with some postcranial elements, several isolated crocodile remains of a larger crocodile, and several turtle elements (Avendaño-Gil *et al.,* 1998; Carbot-Chanona and Coutiño-José, 2000). These fossils were collected in the Maastrichtian Ocozocoautla Formation that outcrops in several road cuts at the Tuxtla Gutiérrez area, Chiapas.

Early Cenozoic remains are very scarce. From the early Oligocene (Chadronian) a carapace fragment of a turtle (Ferrusquía-Villafranca, 1967) from Rancho Gaitán, Prietos Formation, Chihuahua, was reported; and from the late Eocene the lizard *Paradipsosaurus mexicanus* (Fries *et al.,* 1955) from Marfil, Guanajuato was described.

Neogene deposits are among the best prospected within Mexico for the establishment of faunal correlations between locality based fossil mammals. Within this research model, the knowledge of amphibians and reptiles has been left out although their presence has been documented thoroughly. Most Neogene deposits are located in central Mexico and have been extensively collected by O. Carranza-Castañeda and W. Miller's team, who have discovered many new fossil sites. These localities have been dated either as Hemphillian (late Miocene-early Pliocene) or Blancan (late Pliocene-early Pleistocene). Neogene localities are found in Aguascalientes, Baja California, Baja California Sur, Chihuahua, Guanajuato, Hidalgo, Jalisco, Michoacán, and Nuevo León. Miller and Carranza-Castañeda (2001) mention new Hemphillian to Blancan localities at Zacatecas and Estado de México although they do not specify where these fossil amphibians or reptiles were collected. Most Neogene reports are from Hidalgo with 15 unique records, followed by Baja California Sur with 11, and Jalisco with six. The most abundant records are from turtles including the genera: *Kinosternon* from Zietla near Zacualtipan and the fossiliferous region of Tula, both in Hidalgo, and La Goleta, Michoacán; *Trachemys*, also from Zietla and an uncertain locality from Jalisco; *Geochelone* from the San José del Cabo basin and Bahía Tortugas, Baja California Sur; *Testudo* from La Goleta; and cf. *Gopherus* from near Rincón de la Concha, Chihuahua (Brattstrom, 1955a, Castillo-Cerón *et al.*, 1996; Flores Villela, 1998; Miller and Carranza-Castañeda, 2002). Crocodiles are only known from six records of *Crocodylus* from Zietla, the Tecolotlán basin and the Cantera Jocotepec, Jalisco, and Las Tunas Baja California Sur (Smith, 1980; Miller, 1980; Rodríguez-Huerta and Guzmán-Gutiérrez, 1991; Castillo-Cerón *et al.*, 1996; Miller and Carranza-Castañeda, 1998; García-Pérez and Reynoso, 2002). Lizards are known from the genera *Phynosoma* and *Sceloporus* at Zietla (Castillo-Cerón, 1994; Castillo-Cerón *et al.*, 1996) and Yepómera (Lindsay, 1984); snakes from cf. *Boa*, cf. *Crotalus*, and ?*Pituophis* from Las Tunas, Baja California Sur (Miller, 1980); and amphibians from *Bufo campi* from Yepómera, Chihuahua, and *Hyla* and *Rana* from Zietla and the fossiliferous region of Santa María Amajac, near Atotonilco El Grande, Hidalgo (Brattstrom, 1955a, Castillo-Cerón and Goyenechea, 1997; Fuentes-Guerra and Velasco-de León, 2002). The remaining reports are either undetermined or identified to a higher taxonomic rank.

Quaternary deposits preserve a greater diversity of amphibians and reptiles in Mexico. Without taking into account early Pleistocene (Blancan) localities, here included with the Neogene records, there are 129 unique reports in 15 states: 33 amphibians and 97 reptiles. Most records are from Yucatán, Estado de México, Jalisco, and Tamaulipas. Important Pleistocene localities with amphibians and reptiles are: the Late Pleistocene-Recent Cueva de Spukil near Calcehtok and Cueva de Lara, near Yokat and Ticul, Yucatán (Langebartel, 1953), the late Pleistocene Cerro de Tlapacoya south of Tlapacoya, and Zumpango de Ocampo, Estado de México (Brattstrom, 1955a; Álvarez and Huerta, 1975); the Wisconsian San Josecito Cave, near Aramberri, Nuevo León (Brattstrom, 1955b; Mead *et al.,* 1999); the Illinoian Toba Aguascalientes in the Cedazo and San Francisco Creek near Aguascalientes city, (Mooser, 1958; 1980); the Late Pleistocene Cueva del Abra, near Villa Antigua Morelos, Tamaulipas (Holman, 1969); the late Sangamon-Wisconsian Jocotepec area, Jalisco (Smith, 1980); the Irvingtonian (?end Kansian) El Golfo de Santa Clara, Sonora (Shaw, 1982); the Late Pleistocene (Sangamon) Rancho La Brisca, Sonora (Van Devender *et al.* 1985); and the end of the Pleistocene Los Tanques, Zacatecas (Guzmán and Polaco, 1988; 2000).

As with most of the fossil record, turtles are abundant. Forty six turtles have been reported including the genera *Geochelone, Gopherus, Kinosternon, Terrapene, Testudo,* and *Trachemys* (Langebartel, 1953; Hibbard, 1955; Flannery, 1967; García-Cook, 1968; Guenther, 1968; Cabrera-Castro, 1972; Álvarez and Huerta, 1975; Hernández-Junquera, 1977; Moodie and Van Devender, 1979; Ferrusquía-Villafranca and Torres-Roldán, 1980; Mooser, 1980; Smith, 1980; Lorenzo and Mirambell, 1981; Polaco-Ramos, 1981; Shaw, 1982; Van Devender *et al.* 1985; Guzmán and Polaco, 1988, 2000; Guillette *et al.* 2000; Bravo-Cuevas, 2001; Reynoso and Montellano, 2004). Lizards includes *Ameiva, Barisia, Callisaurus, Aspidoscelis,*

Ctenosaura, ?Laemanctus, Lepidophyma, Phynosoma, and *Sceloporus* (Langebartel, 1953; Brattstrom, 1955a, 1955b; Flannery, 1967; Holman, 1969; Lorenzo and Mirambell, 1981; Van Devender *et al.* 1985; Castillo-Cerón *et al.* 1996; Mead *et al.* 1999). Castillo-Cerón *et al.* (1996) referred to *Barisia* fragmentary maxillas and jaws from Tula, however, Mead *et al.* (1999) considered these to be a robust form of phrynosomatid (e.g. *Sceloporus*) due to the lack of adhering osteoderms. Snakes include *Boa, Crotalus* and/or *Bothrops, Drymarchon, Drymobius, Elaphe, Hypsiglena, Lampropeltis, Masticophis, Thamnophis, Trimorphodon, Salvadora,* and *Spilotes* (Langebartel, 1953; Brattstrom, 1954, 1955a; Álvarez and Huerta, 1975; Smith, 1980; Van Devender *et al.* 1985). Pleistocene crocodiles are only represented by two reports of *Crocodylus* from Sonora and Jalisco (Miller and Carranza-Castañeda, 1998; Mead and Báez, 2003).

Amphibians are also abundant, genus *Ambystoma* is known from the Estado de México (Brattstrom, 1955a; Álvarez and Huerta, 1975), and the anurans *Bufo, Eleutherodactylus, Gastrophrine, Leptodactylus, Pternohyla, Rana, Rhinophrynus, Scaphiopus* and *Syrrhophus* from Sonora, Estado de México, Tamaulipas, and Yucatán (Langebartel, 1953; Brattstrom, 1955a; Holman, 1969; Álvarez and Huerta, 1975; Shaw, 1982; Van Devender *et al.,* 1985). Many other Pleistocene reports are unidentified or known only up to the suprageneric level.

6. The Validity of Taxa Erected with Mexican Fossil Material

Of the 255 unique records of amphibians and reptiles, 19 were described as new genera and/or species. Many of these taxa are not valid anymore, mainly because their original descriptions were based on incomplete type material, or taxa were erected using invalid criteria. For example, *Deltatmena premaxillaris* (Langebartel, 1953) description was based on a premaxilla with three teeth, was later referred to *Ctenosaura* by Estes (1983), indicating that the species *Ctenosaura premaxillaris* is not definable with the available material and thus should be a *nomen dubium*; *Impesodens arizeloglyphus* (Langebartel, 1953), which description was based on a right dentary is certainly a *Lepidophyma* (Hecht, 1956; Estes, 1983), however, *L. arizeloglyphus* should still be considered a valid species differing only from other lepidophymas in the relatively larger and more robust teeth; the partial snout into which *Plesiosaurus mexicanus* (Wieland, 1910) was described, has been said to be undiagnostic (Welles, 1962), and that the species should be considered a *nomen dubium*; and, although *Gopherus auffenbergi* (Mooser, 1972) was based on an almost complete skeleton, it is now considered to be a *G. berlandieri* (Bramble, 1982; Morafka 1988) due to the absence of true diagnostic features.

Other taxa that are probably based on undiagnostic holotype specimens are *Bufo campi* (Brattstrom, 1955a), *Gopherus pargensis* (Mooser, 1980), and *Phrynosoma josecitensis* (Brattstorm, 1955b). Although literature questioning their status has not been found, the validity of these species needs revision. *Bufo campi* is based on a left tibia and fibula, *G. pargensis* on a posterior fragment of a carapace, and *P. josecitensis* on a temporal bone with four spines. On the other hand, *Paradipsosaurus mexicanus* description (Fries *et al.* 1955) was based on a nice complete cranium associated to its lower jaw, and has not been questioned. Estes (1983) considered it valid. The recently described *Polyglyphanodon bajaensis* from Baja California (Nydam, 1999) is also based upon fragmentary dental material.

Mexican authors have for the most part erected recently described new species. These are based on beautifully preserved complete holotype skeletons or skulls from the Tlayúa and Huizachal deposits (Reynoso, 1996; 1997; 1998; 2000; in press; Reynoso and Clark, 1998; Reynoso and Callison, 2000). Two other new taxa based on well preserved holotypes are *Geosaurus vignaudi* (Frey *et al.* 2002) and *Gopherus donlaloi* (Reynoso and Montellano, 2004).

7. Museums and Institutions Housed Types and Other Material

Most holotypes of species first described with material collected within the Mexican territory are now housed in Mexican museums or institutions. Eight holotypes, plus several paratypes and hypodigm material from Huizachal and Tlayúa are housed in the Museo de Paleontología of the Instituto de Geología, UNAM. These specimens are the types of: *Tamaulipasaurus morenoi; Cynosphenodon huizachalensis, Zapatadon ejidoensis,* a new genus and species of sphenodontian, *Pamizinsaurus tlayuaensis; Ankylosphenodon pachyostosus; Huehuecuetzpalli mixtecus;* and *Tepexisaurus tepexii.* Also in this same museum, are housed the type specimen of *Polyglyphanodon bajaensis* from Baja California, and *Gopherus donlaloi* from near Villagrán, Tamaulipas, the same as a referred specimen of *Paradipsosaurus mexicanus*. Other Mexican Museums housing type specimens are: the Museo Regional de Historia de Aguascalientes, in the City of Aguascalientes, Aguascalientes, where the type specimen of *Gopherus auffenbergi* is still in exhibition; and the Colección Paleontológica of the Facultad de Ciencias de la Tierra, Universidad Autónoma de Nuevo León, which keeps the recently described *Geosaurus vignaudi*. Other holotype specimens are in Los Angeles County Museum, Texas Memorial Museum, U.S. National Museum, and American Museum of Natural History. The recent location of the holotype of *Pliosaurus mexicanus* remains uncertain.

It is not well documented where most non-holotype specimens are held, particularly those mentioned in abstracts or field guides where the final placement might not be considered important to mention. Other material is kept in many other museums. Among Mexican museums are the Estación Regional del Noroeste, UNAM; Museo de Paleontología y Geología of the Facultad de Ciencias, UNAM; Museo de Paleontología of the Instituto de Geología, UNAM, including almost all of the material originally placed in the Instituto Geológico de México; Colección Paleontológica of the Instituto de Historia Natural, Tuxtla Gutiérrez, Chiapas; Departamento de Prehistoria of the Instituto Nacional de Antropología; Museo

Nacional de Historia Natural, Mexico; Museo de Paleontología de Guadalajara; Museo Regional de Historia de Aguascalientes y Laboratorio de Paleontología y Arqueología del Centro INAH, Aguascalientes, including partially FC material of O. Mooser; Secretaría de Educación Pública, Coahuila (now at the Museo del Desierto, Saltillo); Museo de Paleontología of the Instituto de Investigaciones de Ciencias de la Tierra, Universidad Autónoma del Estado de Hidalgo, Pachuca; and Colección Paleontológica of the Facultad de Ciencias de la Tierra Universidad Autónoma de Nuevo León, Linares. Foreign Museums are mostly North American and include the American Museum of Natural History, New York; Los Angeles County Museum, Los Angeles, California; Midwestern University, Texas; Texas Memorial Museum, Austin, including remaining FC material of O. Mooser; University of California, Los Angeles, California; Museum of Paleontology, University of Michigan, Ann Arbor, Michigan; the U. S. National Museum, Washington, D. C.

8. Analysis of the Documents Produced Concerning with amphibians and Reptiles in Mexico

This section attempts to provide a panorama of the research on amphibians and reptiles of Mexico. Analysis was performed on all documents that mention, list, or describe any amphibian or reptile found in national and international literature. The term document is used here to refer to all printed papers that display information about an amphibian or a reptile. There are three categories of documents according to their importance and range of distribution: a) scientific papers, books, and book chapters produced in international journals or editorials; b) scientific papers, books, and book chapters produced in national journals or editorials; and c) theses, abstracts in national and international meetings and congresses, field guides, and newsletters. To decide whether a paper was written by a national or an international author only the nationality of the first author was considered.

Results indicate that there have been 99 published non redundant main documents that mention, list or describe amphibians and reptiles, giving a total of 255 records with a unique locality and age (Table 2). Of all these documents, only 51 are formal publications in either national or international journals, books or book chapters, number that represents barely the 51.5% of all non-redundant documents. This means that the remaining 48.5% of the information can be considered as gray literature, not available or of difficult access to the international paleontological community. This problem is stressed in crocodiles, where 70% of the reports are presented in informal literature. Mexican researchers have produced only 50.5% of all formal publications as first authors.

Considering all non-redundant documents, 40.4% were published in international journals, books and books chapters, that is, 40 papers. Of these papers, Mexican researchers wrote only 10 in the whole history of Mexican paleoherpetology: three are faunal listings (Flores-Villela, 1998; Rodríguez-de la Rosa and Cevallos-Ferriz,

1998; Polaco Ramos, 1981), and the remaining are formal descriptions of new taxa including cladistic and in some cases biogeographical analyses in the modern sense (Reynoso, 1996, 1997, 1998, in press; Reynoso and Clark, 1998; Reynoso and Callison, 2000; Reynoso and Montellano, 2004). These last analytical papers are unique among Mexican authors.

Of the 255 unique records of amphibians and reptiles, Mexican researchers mention, list or describe only 102. This is only 40%, a low percentage when scaled to the 50.5% documents produced by Mexicans. Recent papers have increased the number of holotypes designated by Mexicans, and now national authors have designated nine of the 19 holotypes of species described from material collected in Mexico.

Historically, the production of documents mentioning, listing, or describing amphibians and reptiles has increased exponentially since the 1950s (Figure 4). In the last four years there were 30 documents produced, two more than in the 1990s decade. The increase in the production of documents is co-opted with a substantial growth of gray literature (abstracts, theses, field guides, and newsletters) towards the last decades (Figure 5), but the number of formal papers does not show an important growth, since the papers produced per decade have remainded almost stable since the 1950s. During the 1990s, formal scientific production appears to have improved, but this should only be reflected with a larger number of expected papers to be published during the first decade of the 21st century. The increase of gray literature denotes a growth in the number of research groups in the whole country that has paid more attention to reporting the presence of amphibians and reptiles at their localities. However, it also denotes the lack of professional groups capable of publishing their findings in broadly distributed journals, either national or international.

9. Conclusions

Mexico is a country in which fossil amphibians and reptiles are well distributed over a broad range of time, from the late Triassic to the Holocene. Taxonomically, Mexico is also very diverse, although it cannot be compared to other countries with a broader time range and more fossil localities. Mexico lacks amphibian and reptile bearing deposits areas from the Paleozoic to the mid Triassic, which are well represented in the United States and Canada. This does demonstrate that Mexican vertebrate bearing faunal outcrops are always younger.

The absence of paleontologists interested in the study of amphibians and reptiles has kept Mexico from a complete understanding of their vertebrate faunas. The amphibian and reptile fossil record is quite large and still barely known. Many reports coming mainly from the Cenozoic and the Late Cretaceous need to be carefully reviewed, since they are based on fragmentary and not well preserved material that could lead to false identifications. In addition, much of the unidentified material mentioned to be present in several localities, but set aside by the authors, also needs to be studied. It is particularly interesting that fossils from the Neogene of central

TABLE 2. Analysis of amphibian and reptile records in documents produced considering national and international journals, books, and book chapters, and gray literature including abstracts, theses, field guides, and newsletters.

	Main references	International publication	National publication	Abstracts/thesis/field guides/ /newsletters	% of formal papers	% of documents written by Mexicans	% of formal documents written by Mexicans	% of international papers written by Mexicans	number of taxa referred	% of taxa referred by Mexicans
amphibians	17	9	1	7	58.8	47.1	10	-	43	32.6
undet. reptiles	7	-	-	7	-	71.4	-	-	10	80.0
turtles	52	18	11	23	55.8	53.8	41.4	22.2	89	50.6
cocodiles	20	6	-	14	30.0	45.0	16.7	16.7	26	42.3
lepidosauromorphs	37	21	5	11	70.3	40.5	34.6	28.6	76	23.7
"euryspsids"	6	3	1	2	66.7	16.7	-	-	11	18.2
Total	139	57	18	64	54.0	47.5	30.7	19.3	255	38.4
Documents	99	40	11	48	48.5	50.5	31.4	25	255	40

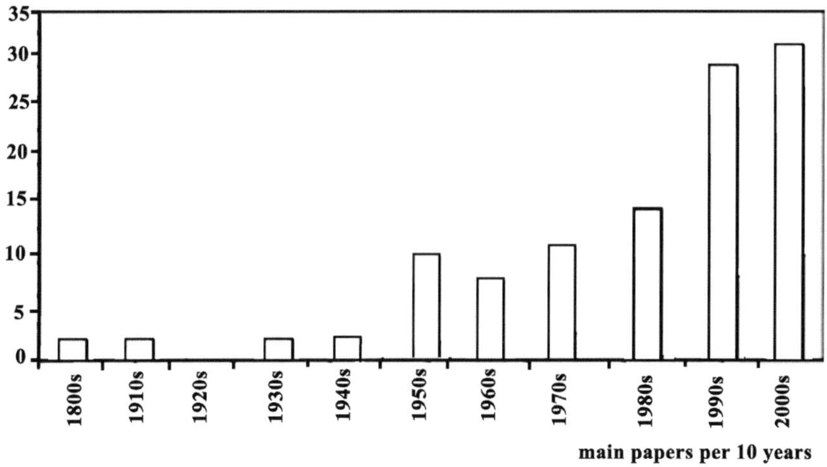

FIGURE 4. Frequency of main documents that mention, list or describe amphibians and reptiles produced in 10 year classes during the 1900s and 2000s. In the 19th century, only a single paper was published.

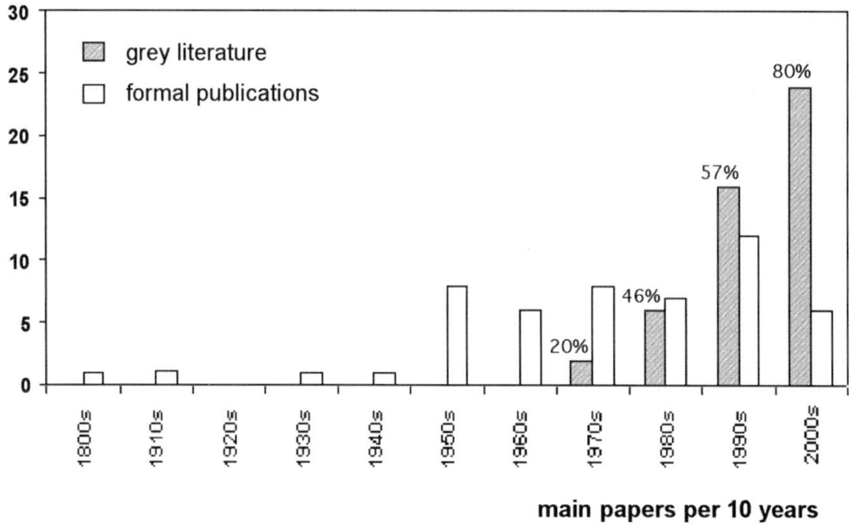

FIGURE 5. Frequency and percentage of gray literature documents that mentions, lists, or describes amphibians and reptiles, outlined as a proportion of formal publications in national or international journals, books, or book chapters.

Mexico that were reported in informal documents are abundant.

The paleoherpetological research production in Mexico is far from being satisfactory. In the whole history of this discipline, we have produced as first authors, only 16 formal papers of which only 10 are of broad international distribution. Although it is clear that research groups are more and more interested in the study of amphibians and reptiles, there is a need to formalize their discoveries by publishing their results in either national or international journals. It is imperative to include in every research team program their findings available to the national and international community. Otherwise, all the information we have produced to date will remain domestic and not available to the international scientific community.

Acknowledgments

I would like to thank Rosa Elena Tovar for providing me with some literature I was not aware of, and Pilar Melgarejo and Eugenia Zarza for reviewing the grammar.

References

Aguilera, J., 1896, Bosquejo Geológico de México, *Bol. Inst. Geol. Méx.*, 4-6, 270 p.

Aguilera, J., 1907, Carte Geológique du l'Amerique du Nord, *Congr. Geol. Inter. Compte Rendous du la Xeme session* 1906: 227-248.

Aguillón, M., 2000, Snake trunk vertebrae (*Coniophis* sp.) of Campanian age, from Coahuila, México, *J. Vert. Paleont.*, 20:25A.

Álvarez, T., and Huerta, P., 1975, Restos óseos de anfibios y reptiles en Tlapacoya (Tlapacoya) IV, México,

Bol. Inst. Nal. Antrop. Hist. Méx., 11:37-42

Aranda-Manteca, F., and Stinnesbeck, W., 1993, Reptiles marinos del Jurásico superior (Kimmeridgiano) del Noreste de México, *VI Congr. Nal. Paleont., Soc. Mex. Paleont., Mem.*, p. 15-16.

Avendaño-Gil, M. J., Coutiño-José, M. A., and Pimentel-Zepeda, F., 1998, Rescate del patrimonio paleontológico de la carretera Ocozocuautla-Cosoleacaque, Estado de Chiapas, *VI Congr. Nal. Paleont., Soc. Mex. Paleont., Mem.*, p. 7-8.

Barrios-Rivera, H., 1985, Estudio analítico del registro paleovertebradológico de México, Unpublished Professional thesis, Universidad Nacional Autónoma de México.

Blanco-Piñón, A., López-Oliva, J. G., Stinnesbeck, W., Adatte, T., and González, A., 1998, Vallecillo, Nuevo León, "fossil lagerstätte" en el Noreste de México, *VI Congr. Nal. Paleont., Soc. Mex. Paleont., Mem.*, p. 10-11.

Bramble, D. M., 1982, *Scaptochelys*: generic revision and evolution of gopher tortoises, *Copeia*, 1982:852-867.

Brattstrom, H. B., 1954, The fossil pit-vipers (Reptilia: Crotalidae) of North America, *Trans. San Diego Soc. Nat. Hist.*, 12:31-46.

Brattstorm, H. B., 1955a, Records of some Pliocene and Pleistocene reptiles and amphibians from Mexico, *Bull. South Cal. Acad. Sci.*, 54:1-4.

Brattstorm, H. B., 1955b, Pleistocene lizards from San Josecito Cavern, Mexico, with description of new species, *Copeia*, 1955:133-134.

Bravo-Cuevas, V. M., 2001, Pleistocene faunal assemblage from south-central region of the state of Hidalgo, central Mexico, *J. Vert. Paleont.*, 21:35A.

Buchy, M.-C., Frey, E., Stinnesbeck, W., and López-Oliva, J. G., 2003, First occurrence of a gigantic pliosaurid plesiosaur in the late Jurassic (Kimmeridgian) of México, *Bull. Soc. Geol. France,* 174:271-278.

Buchy, M.-C., Frey, E., Stinnesbeck, W., and López-Oliva, J. G., in press. An annotated catalogue of the upper Jurassic (Kimmeridgian and Tithonian) marine reptiles in the collections of the Universidad Autónoma de Nuevo León, Facultad de Ciencias de la Tierra, Linares, Mexico, *Oryctos*.

Cabrera-Castro, R., 1972, Fauna fósil pleistocénica en Jocotepec, Jalisco, *Bol. Inst. Nal. Antrop. Hist. Méx.*, 2a. época (oct-dic): 37-44.

Carbot-Chanona, G. F., and Coutiño-José, M. A., 2000, Cocodrilos fósiles del Cretácico Superior (Maastrichtiano) provenientes de la Formación Ocozocoautla, Chiapas, VII Congr. Nal. Paleont., *Soc. Mex. Paleont., Mem.*, p. 92-93.

Castillo-Cerón, J. M., 1994, Bioestratigrafía de los sedimentos continentales del área de Zacualtipan (Mioceno tardío), Estado de Hidalgo, México, Unpublished M.S. thesis, Universidad Nacional Autónoma de México.

Castillo-Cerón, J. M., and Goyenechea, I., 1997, Fossil anurans from Hidalgo, *J. Vert. Paleont.*, 17:36A.

Castillo-Cerón, J. M., Cabral-Perdomo, M. A., and Carranza-Castañeda, O., 1996, *Vertebrados Fósiles del Estado de Hidalgo,* Univ. Autón. Edo. Hidalgo, Spec. Pub., 127 p.

Castillo-Cerón, J. M., Carranza-Castañeda, O., and Cabral-Perdomo, M. A., 1998, Vertebrados cenozoicos del estado de Hidalgo, *VI Cong. Nal. Paleont., Soc. Mex. Paleont., Mem.*, p. 17-18.

Castillo-Cerón, J. M., Cabral-Perdomo, M. A., and Bravo-Cuevas, V. M., 2002, Microvertebrados del Mioceno tardío del Estado de Hidalgo, *VII Cong. Nal. Paleont., Soc. Mex. Paleont., Mem.*, 40.

Clark, J. M., and Hernández, R., 1994, A new burrowing diapsid from the Jurassic La Boca Formation of Tamaulipas, Mexico, *J. Vert. Paleont.,* 14:180-195.

Clark, J. M., Fastovsky, D. E., Montellano, M., Hopson, J. A., and Hernández, R., 1994a, Additions to the middle Jurassic vertebrates fauna of Huizachal Canyon, Tamaulipas, Mexico, *J. Vert. Paleont.,* 14:21A.

Clark, J. M., Montellano, M., Hopson, J. A., Hernández, R., and Fastovsky, D. E., 1994b, An early or middle Jurassic tetrapod assemblage from La Boca formation, North East Mexico, in: *In the Shadow of the Dinosaurs* (N. C., Fraser, and H. D. Sues, eds.), Cambridge University Press. New York, p. 295-302

Cope, D., England, B., Brown, R. B., Westgate, J., and Pittman, J., 2003, A Late Cretaceous (Campanian) marine vertebrate microfauna from the Aguja Formation of Chihuahua, Mexico, *J. Vert. Paleont.,* 23:42A.

Everth, D., Sampson, S., Rodríguez-de la Rosa, R. A., Aguillón-Martínez M. C., Brinkman D. B., and López-Espinoza. J., 2003, Las Aguilas: An unusually rich Campanian-age vertebrate local in Southern Coahuila, Mexico, *J. Vert. Paleont.*, 23:47A.

Estes, R., 1983, *Handbuch der Paläherpetologie*. Gustav Fischer Verlag, Stuttgart. 249 p.

Ferrusquía-Villafranca, I., 1967, (1969), Rancho Gaitán local Fauna, Early Chadronian. Northeastern Chihuahua, *Bol. Soc. Geol. Mex.*, 30:99-138.

Ferrusquía-Villafranca, I., and Comas, O, 1980, Reptiles marinos mesozoicos en el Sureste de México y su significación geológica-paleontológica, *V Conv. Geol. Nal., Soc. Geol. Mex.*, Resúmenes, p. 104-105.

Ferrusquía-Villafranca, I., and Torres-Roldán V., 1980 (1982), El registro de mamíferos terrestres del Mesozoico y Cenozoico de Baja California, *Univ. Nal. Auton, Mex., Inst. Geol., Rev.*, 4:56-62.

Flannery, K. V., 1967, Vertebrate fauna and hunting patterns, in: *The prehistory of the Tehuacan Valley, Vol. 1: Environtment and Subsistence* (Byers, D. S. ed.), University of Texas Press, Austin, p. 132-177.

Flores-Villela, O., 1993, Herpetofauna of Mexico: Distribution and endemism, in: *Biological Diversity of Mexico: Origins and Distribution* (T. P. Rammoorthy, R. Bye, A. Lot, and J. Fa, eds.), Oxford University Press, New York, p. 251-278.

Frey, E., Buchy, M. C., and Stinnesbeck W., 2001, The monster of Aramberri and friends: New finds of marine reptiles in the Mesozoic of northeastern México, *6th European Workshop on Vertebrate Palaeontology, Abstr.*, p. 30.

Frey, E., Buchy, M. C., Stinnesbeck, W., and López-Oliva, J. G., 2002, *Geosaurus vignaudin*. sp. (Crocodyliformes: Thalattosuchia), first evidence of metriorhynchid crocodilians in the Late Jurassic (Tithonian) of central-east Mexico (state of Puebla), *Can. J. Earth Sci.*, 39:1467-1483.

Fries, C. Jr, Hibbard, W. C., and Dunkle, H. D., 1955, Early Cenozoic Vertebrates in the Red Conglomerate at Guanajuato, Mexico, *Smith. Misc. Coll.*, 123:1-25.

Fuentes-Guerra, E. P. and Velasco-de León, M. P., 2002, Anuros de la Formación Atotonilco el Grande, México, *VIII Congr. Nal. Paleont., Mem.*, p. 99.

García-Pérez, L. and Reynoso, V. H., 2002, Los fósiles del género *Crocodylus* en México y su historia biogeográfica, *VIII Congr. Nal. Paleont., Soc. Mex Paleont., Mem.*, p. 52.

García, R., and Reynoso, V. H., 2002, The southernmost record of the turtle *Bothremys* (Testudines: Pleurodira) in the Cerro del Pueblo Formation, near Saltillo, Coahuila, México, *105th Ann. Meet. Texas Acad. Sci., Texas A&M Int. Univ., Laredo, Texas*, p. 95.

García-Cook, A., 1968, Chimalhuacán: un artefacto asociado a megafauna, *Inst. Nal. Antropol. Hist., Méx.*, 25:22-27.

Goyenechea M. G., I., and Castillo-Cerón, J. M., 2002. Historia geológica de las serpientes, *Bol. Soc. Herpetol. Mex.*, 10:35-42.

Guenther, E. W., 1968, Untersuchungen zur jungeiszeitlichen und nacheiszeitlichen geologischen und paläontologischen Geschichte, in: *Das Mexiko - Project der Deseschen Forschingsgemeinschaft - Berichte über beyonnene und Geplante arbeiten* (F. Tichy, ed.) Franz Steiner Verlage GMBH-Wiesbaden, p. 32-37,

Guillette, D., Ferrusquía-Villafranca, I, Jiménez-Hidalgo, E., Gutiérrez-Limón O., and Gunnel, G. F., 2000, Preliminary report on a new Pleistocene Faunal site in southern Puebla, Mexico, *J. Vert. Paleont.*, 20:44A.

Guzmán, A. F., and Polaco, O. J., 1998, Tortugas fósiles de Durango y Zacatecas, *VI Congr. Nal. Paleont., Soc. Mex. Paleont., Mem.*, p. 28-29.

Guzmán, A. F., and Polaco, O. J., 2000, Vertebrados pleistocénicos de los Tanques, Zacatecas, *VII Congr. Nal. Paleont., Soc. Mex. Paleont., Mem.*, p. 100-101.

Hecht, M., 1956, A new xantusiids lizards from the Eocene of Wyoming, *Am. Mus. Novitates*, 1774:1-8.

Hernández-Junquera, A., 1977, Fauna local Laguna Media Luna Pleistoceno Tardío, Municipio de Río Verde, San Luis Potosí, México. Unpublished Professional thesis. Universidad Nacional Autónoma de México.

Hernández-Rivera, R., and Delgado-de Jesús, C., 1999, An endocast of a ceratopsian brain cavity from the Cerro del Pueblo Formation (Latest Campanian, Upper Cretaceous), Coahuila, Mexico, *J. Vert.*

Paleont., 19:50A-51A.
Hibbard, W. C., 1955, Pleistocene vertebrates from the Upper Becerra (Becerra Superior) Formation, Valley of Tequixquiac, Mexico, with notes on other Pleistocene forms. *Univ. Michigan, Contrib. Mus. Paleont.*, 2:47-96.
Holman, A. J., 1969 (1970), A small Pleistocene herpetofauna from Tamaulipas, *Quart. J. Florida Acad. Sci.*, 32:153-158.
Langebartel, S. D., 1953, The reptiles and amphibians., in: *Faunal and Archeological researches in Yucatan Caves (T. R., Hatt, ed.), Cranbrook Inst. Sci. Bull.*, 33:91-108.
Lillegraven, A. J., 1976, A new genus of therian mammal from the Late Cretaceous "El Gallo Formation", Baja California, Mexico, *J. Paleont.*, 50:437-443.
Lindsay, E. M., 1984, Late Cenozoic mammals from Northwetern Mexico, *J. Vert. Paleont.*, 4:208-215.
Lorenzo, L. J., and Mirambell, L., 1981, El Cedral, S.L.P., México: Un sitio con presencia humana de más de 30,000 años, *X Cong. Unión Inter. Cienc. Prehist Protohist. México, D. F. Oct. 19-24, 1981, Comisión XII, Coloquio El poblamiento de América: Evidencia arqueológica de ocupación humana en América anterior a 11,500 años a. p.* p. 112-124.
Lucas, S. G., Kues B. S., and González-León, C. M., 1995, Paleontology of the Upper Cretaceous Cabullona Group, Northeastern Sonora, in: *Studies on the Mesozoic of Sonora and Adjacent Areas.* (C. Jacques-Ayala, C. M., González-León, and J. Roldán-Quintana, eds.), *Geol. Soc. Am. Spec. Pap.*, 301:143-165.
Lucas, S. G., and González-León, C. M., 1995, Ichthyosaurs from the Upper Triassic of Sonora and the biochronology of Triassic ichthyosaurs, in: *Studies on the Mesozoic of Sonora and Adjacent Areas. (C. Jacques-Ayala, C. M., González-León, and J. Roldán-Quintana, eds.), Geol. Soc. Am. Spec. Pap.*, 301:17-20.
Mead, J. I., and Baez, A., 2003, Crocodilian remains from the Pleistocene of Northeastern Sonora, Mexico, *Croc. Spec. Gr., Newsl.*, 22:19-21.
Mead, J. I., Arroyo-Cabrales, J., and Johnson, E., 1999, Pleistocene lizards (Reptilia: Squamata) from San Josecito Cave, Nuevo León, México, *Copeia* 1999:163-173.
Miller, W., 1980, The late Pliocene Las Tunas local Fauna from Southernmost Baja California, México, *J. Paleont.*, 54:762-805.
Miller, W., and Carranza-Castañeda, O., 1998, The Tecolotlán basin, Central México: A new area for late Cenozoic terrestrial vertebrates, *J. Vert. Paleont.*, 14:21A.
Miller, W., and Carranza-Castañeda, O., 2001, New faunal discoveries in the Hemphillian-Blancan basins of central Mexico, *J. Vert. Paleont.*, 21:81A.
Miller, W., and Carranza-Castañeda, O., 2002, Late Tertiary vertebrates and sedimentation in the San José del Cabo Basin, Southern Baja California, *J. Vert. Paleont.*, 22:88A.
Montellano-Ballesteros, M., Hernández-Rivera, R., Álvarez-Reyes, G., Andrade-Ramos, P., and Martín-Medrano, L., 2000, Discovery of Late Cretaceous vertebrate local faunas in North México, *J. Vert. Paleont.*, 20:58A-59A.
Moodie, B. K., and Van Devender, R. T., 1979, Extinct and extirpation in the herpetofauna of the southern high plains with emphasis on *Geochelone wilsoni* (Testudinidae), *Herpetologica*, 35:198-206.
Mooser, O., 1958, La fauna "Cedazo" del Pleistoceno de Aguascalientes, *Univ. Nal. Auton. Méx., Inst. Biol., An.*, 28:359-363.
Mooser, O., 1972, A new species of Pleistocene fossil tortoise, genus *Gopherus*, from Aguascalientes, Aguascalientes, Mexico, *Southwest. Natur.*, 17:61-65.
Mooser, O., 1980 (1982), Pleistocene fossil turtles from Aguascalientes, *Univ. Nal. Autón. Méx., Inst. Geol.*, 4:63-66.
Morafka, D. J., 1988, Historical Biogeography of the Bolson Tortoise, *Ann. Carnegie Mus. Nat. Hist.*, 57:47-72.
Mullerried, G. K. F., 1931, Un reptil y algunos invertebrados fósiles de Rayón, Estado de Tamaulipas, *Univ. Nal. Autón. Méx., Inst. Biol., An.*, 2:171-178.
Mullerried, G. K. F., 1943, Fósiles raros de México. III. Una tortuga fósil del estado de Chiapas. *An. Inst. Biol., Univ. Nal. Autón. Méx.*, 14:623-624.
Nydam, R. L., 1999, Polyglyphanondotinae (Squamata: Teiidae) from the medial and Late Cretaceous: new records from Utah, U.S.A. and Baja California del Norte, Mexico, in: *Vertebrate Paleontology*

in Utah (D. D. Gillette, ed.), Utah Geological Survey, Salt Lake City, p. 303–317.

Peterson-Rodríguez, R. H., Aguirre-Garza, S. J., Velasco-Segura, J. A., and Delgado-de Jesús, C. R., 2000, Evidencias de un reptil marino del Jurásico Superior en la Sierra de Parras, Coahuila, México, *VII Congr. Nal. Paleont., Soc. Mex. Paleont.*, p. 115.

Polaco-Ramos, O., 1981, Restos fósiles de *Glossotherium y Eremotherium* (Edentata) en México, *Anais II Congr. Latin. Paleont.*, 2:819-833.

Reynoso, V. H., 1992, Descripción de los esfenodontes (Sphenodontia, Reptilia) de la Fauna del Cañón del Huizachal (Jurásico Temprano-Medio), Tamaulipas, México, Unpublished M.S. thesis, Universidad Nacional Autónoma de México.

Reynoso, V. H., 1996, A middle Jurassic *Sphenodon*-like sphenodontian (Diapsida: Lepidosauria) from Huizachal Canyon, Tamaulipas, México, *J. Vert. Paleont.,* 16:210-221.

Reynoso, V. H., 1997, A "beaded" sphenodontian (Diapsida: Lepidosauria) from the Early Cretaceous of Central México, *J. Vert. Paleont.*, 17:52-59.

Reynoso, V. H., 1998, *Huehuecuetzpalli mixtecus*, gen. et sp. nov. a basal squamate (Reptilia) from the Early Cretaceous of Tepexi de Rodríguez, Central Mexico, *Phil. Trans. Roy. Soc. London, Ser.* B 353:477-500.

Reynoso, V. H., 1999, A new lizard from the Early Cretaceous *Lagerstatten* of Tepexi de Rodríguez, Puebla, *Riv. Mus. Civ. Scien. Nat. "Enrico Caffi"* 20 (1999 supl.):131-132.

Reynoso, V. H., 2000, An unusual aquatic sphenodontian from the Tlayua Formation (Albian), Central México, *J. Paleont.*, 74:133-148.

Reynoso, V. H., 2003, Growth patterns and ontogenetic variation of teeth and jaws of the middle Jurassic sphenodontian *Cynosphenodon huizachalensis* (Reptilia: Rhynchocephalia), *Can. J. Earth Sci.*, 40:609-619.

Reynoso, V. H., in press, Possible evidence of a venom apparatus in a middle Jurassic sphenodontian, *J. Vert. Paleont.*

Reynoso, V. H., and Callison, G., 2000, A new scincomorph from the early Cretaceous of Puebla, México, *Zool. J. Linn. Soc.*, 130:183-212.

Reynoso, V. H., and Clark, J. M., 1998, A dwarf sphenodontian from the Huizachal Formation of Tamaulipas, México, *J. Vert. Paleont.*, 18:333-339.

Reynoso, V. H., and Montellano, M., 2004, A new giant turtle of the genus *Gopherus* (Chelonia: Testudines) from the Pleistocene of Tamaulipas, México and its bearing on the phylogeny and biogeography of gopher turtles, *J. Vert. Paleont.*, 24:822-837.

Reynoso, V. H., Cabral Perdomo, M. A., and Clark, J. M., 2000, The reptiles of the Tlayúa Formation, in: *Guide Book of the Field Trips of the 60th Meeting of the Society of Vertebrate Paleontology,* (L. Espinosa-Arrubarrena, M. Montellano-Ballesteros and S. P. Applegate, eds.) *Avances en Investigación, Univ. Autón, Edo. Hgo., Spec. Publ.*, p. 106- 110.

Rodríguez-de la Rosa, R., and Aranda-Manteca, F. J., 2000, Nuevos hallazgos de vertebrados de la Formación El Gallo, (Campaniano Tardío) del estado de Baja California, *Mem. VII Congr. Nal. Paleont., Soc. Mex. Paleont.*, p. 65.

Rodríguez-de la Rosa, R. A., and Cevallos-Ferriz, S. R. S., 1998, Vertebrates of the El Pelillal Locality (Campanian, Cerro del Pueblo Formation), Southeastern Coahuila, México, *J. Vert. Paleont.* 18: 751-764.

Rodríguez-de la Rosa, R., and Dávila-Rodríguez, J. R., 1998, Una icnita de quelonio del Cretácico Superior de Coahuila, México, Mem. *VI Congr. Nal. Paleont., Soc. Mex. Paleont.*, p. 62-63.

Rodríguez-Huerta, M., and Guzmán-Gutiérrez, R., 1991, Primer registro del género *Crocodylus* del Plioceno de Tecolotlán, Jalisco, México. *III Congr. Nal. Paleontol., Soc. Mex. Paleont., Mem.*, p. 125.

Saul, L. R., 1979, A hollo spined *Anapachydiscus penninsularis* with possible mosasaur bite impressions, *Los Angeles Co. Mus. Contrib. Sci.*, 304: 1-8.

Shaw, A. C., 1982, La Fauna regional "El Golfo" del Pleistoceno medio proveniente del Noroeste de Sonora, México. *Soc. Vert. Paleont. 42th Ann. Meeting, México, D.F.*

Smith, L. M., 1980, The evolution and ecological history of the fish fauna of the Río Lerma Basin, Mexico, Unpublished Ph.D. Dissertation, University of Michigan.

Van Devender, T. R., Rea, A. M., and Smith, M. L., 1985, The Sangamon interglacial vertebrate fauna from Rancho La Brisca, Sonora, México, *Trans. San Diego Soc. Nat. Hist.*, 21: 23-55.

Welles, S. P., 1962, A new species of elasmosaur from the Aptian of Colombia and a review of Cretaceous plesiosaurs, *Univ. California Publ., Geol. Sci.*, 44:1-96.

Westgate, J. W., Brown, R. B., Cope, D., and Pittman, J., 2000, A Late Cretaceous dinosaur-bearing community from coastal deposits in Chihuahua, Mexico, *J. Vert. Paleont.*, 20:78A.

Westgate, J. W., Brown, R. B., Cope, D., and Pittman, J., 2002a, Discovery of dinosaur remains in coastal deposits near Ojinaga, Mexico, *J. Vert. Paleont.,* 22:118-119A.

Westgate, J. W., Pittman, J., Brown, R. B., and Cope, D., 2002b, Continued excavation of the first dinosaur community from Chihuahua, Mexico, *J. Vert. Paleont.*, 22:118A.

Wieland, R. G., 1910, *Plesiosaurus (Polyptychodon) mexicanus,* Inst. Geol., México, *Parergones* 3: 359-365.

Chapter 11

A Review of the Dinosaurian Record from Mexico

HÉCTOR E. RIVERA-SYLVA, RUBÉN RODRÍGUEZ-DE LA ROSA and JORGE A. ORTIZ-MENDIETA

Departamento de Paleontología, Museo del Desierto, A.P. 307, C.P. 25000, Saltillo, Coahuila, México.

1. Abstract

Several dinosaur localities have been reported from Mexico. In 1926 Werner Janensch reported the first dinosaur bone remains as *Monoclonius* from La Soledad, western Coahuila. Since 1926, more dinosaur remains have been reported from the northern Mexican states of Baja California, Sonora, Chihuahua, Coahuila and Tamaulipas. Several dinosaur footprint localities have also been reported from the northern portion of Mexico. From the central and southern Mexican states of Michoacán, Puebla and Chiapas dinosaur remains have also been reported. Recently, fragments of *Heterodontosaurus* and *Syntarsus* were found in Tamaulipas. Re-evaluation of dinosaur remains from Puebla indicate that they belong to a tetrapod. In Baja California there are recent findings of *Aublysodon* teeth. In Chihuahua fragments of a member of Titanosauridae have been found. In Michoacán hadrosaur bones have been reported. The ichnological record has increased in recent years with the discovery of new localities in Coahuila and Puebla, where two localities include tracks of tetrapods, sauropods and ornitopods. Ichnites from one ornitopod family and from several tetrapod families have been reported from Michoacán. Since the discovery of dinosaur remains about 80 years ago, dinosaur discoveries in Mexico have been

F. Vega et al. (eds.), Studies on Mexican Paleontology, 233–248.

significant however the field is still in its infancy with much still to be learned.

2. Introduction

Mexico has several Mesozoic outcrops that have yielded fossil vertebrates including dinosaur remains. The first described dinosaur remains were reported from the La Soledad Beds in the western part of Coahuila, in northern Mexico by the German paleontologist Werner Janensch, who in 1926 commented on the discovery of several ceratopsian skeletal elements, attributed to *Monoclonius* (Janensch, 1926). Just a few years later, more dinosaur remains were reported from northern Mexico, in particular from the northern part of the state of Sonora (Langston and Oakes, 1954; Lull and Wright, 1942).

Throughout the 1950´s to 1970´s paleontologists from the Los Angeles County Museum of Natural History conducted dinosaur research in Mexico (Langston and Oakes, 1954; Morris, 1967; 1971; 1972; 1973a; 1973b; 1981; 1982; 1983). The LACM group collected hadrosaurian dinosaurs from Baja California and geological work done in southeastern Coahuila yielded evidence of hadrosaurian and ceratopsian dinosaurs (Murray *et al.* 1960).

At the end of the 1970´s, the first dinosaur tracks from Mexico were recorded from Michoacán (Ferrusquía-Villafranca *et al.* 1978a; 1978b). Since then the dinosaurian paleoichnological record of Mexico has grown considerably.

During the first part of the 1980´s, a systematic description of dinosaur remains from Coahuila was undertaken by a team frrom the Royal Ontario Museum; however no full description of this material has yet been published. Also in the 1980´s another paleontological effort was undertaken by the Instituto de Geología from the Universidad Nacional Autónoma de México (UNAM) resulting in the discovery and mounting of a gryposaur hadrosaur skeleton (Hernández, 1996; 1997).

Since the early 1990´s, a survey focused on the Late Cretaceous of southeastern Coahuila has yielded a diverse assemblage of dinosaur material (Hernández, 1992; Hernández, and Kirkland, 1993; Rodríguez-de la Rosa, and Cevallos-Ferriz, 1998; Brinkman *et al.* 2002; Eberth *et al.* 2003; Rodríguez-de la Rosa *et al.* 2003). During the last decade, Coahuila government has contributed to the study of dinosaurian paleontology, reinforced in 1999 when the Museo del Desierto was founded in Saltillo, Mexico.

Dinosaurs from Mexico have been mainly collected from the Mexican northern states of Baja California, Sonora, Chihuahua, Coahuila and Tamaulipas. The Central portion of Mexico has yielded dinosaur evidence from Puebla. Dinosaur ichnology is currently known from several track sites in central and northern Mexico.

The dinosaurian record of Mexico can be approached from two points of view; the osteological and paleoichnological perspectives. This paper reviews the dinosaur studies from Mexico taking into account the present knowledge of both perspectives.

In addition the nature and implications of the Mexican dinosaur record is also discussed.

3. Dinosaur Evidence of Mexico

3.1. Osteological Record

The oldest dinosaur remains in Mexico were collected from rocks of the Middle Jurassic La Boca Formation at Huizachal Canyon in Tamaulipas, northeastern Mexico. Vertebrate fossils in this locality occur in sediments deposited as water-lain debris flows (Clark *et al.* 1994; Fastovsky *et al.* 1995).

Among the dinosaurian remains collected at Huizachal Canyon are ornithischian teeth similar to those of *Heterodontosaurus* (Fastovsky *et al.* 1995). Theropod remains include laterally compressed, serrated teeth as well as an articulated pelvis, sacrum and a partial skull of two indeterminate dinosaurian specimens. In regard to the indeterminate dinosaurian specimens, the pelvis resembles the South African theropod *Syntarsus* whereas the fragmentary skull suggests affinities with the Maniraptora or a *Ceratosaurus*-Abelisauridae clade (Munter, 1999). The larger bone fragments found within the same deposit may also represent sauropods (Fastovsky *et al.* 1995).

In 1987 the first dinosaur remains from Central Mexico were collected near San Felipe Otlaltepec, southwest of Puebla from Late Jurassic (Tithonian) or Early Cretaceous (Berriasian) rocks (Ortega-Guerrero, 1989). In 1990 this material was considered to belong to a sauropod (Hernández and Espinosa-Arubarrena, 1990) however recent work done by Rivera-Sylva instead describe it as belonging to a medium-size theropod dinosaur (Rivera-Sylva, 2003).

One of the first studied dinosaurian faunas was from Late Cretaceous rocks of Baja California. Two different formations have yielded these dinosaur remains; the Campanian La Bocana Roja and the Late Campanian El Gallo Formation.

The La Bocana Roja Formation has yielded a partial skeleton of the large theropod dinosaur *Labocania anomala* (Molnar, 1974). The few known cranial bones share similarities with *Chilantaisaurs* and *Indosaurs* and also with tyrannosaurs, however they have some peculiarities of their own (Molnar, 1974). For instance, pubes and ischia are typical tyrannosaurian, yet, the short backward-sloping, jaw-supporting quadrate is classic allosauroid. In addition, the front teeth are not quite D-cross-sectioned (Paul, 1988).

The Late Campanian El Gallo Formation has yielded abundant hadrosaurian material (Langston and Oakes, 1954; Morris, 1967; 1971; 1972; 1973a; 1973b; 1981; 1982; 1983), as well as remains of theropod dinosaurs. Among other hadrosaurian material that remains unidentified, *Lambeosaururs laticaudus* Morris 1981 was described on the basis of a partial skeleton (Morris, 1973a; Weishampel, 1990) (Figure 1). An isolated osteoderm suggests the presence of an ankylosaur (Morris, 1967; 1971), tentatively referred to *Euoplocephalus*, and other ornithischian remains

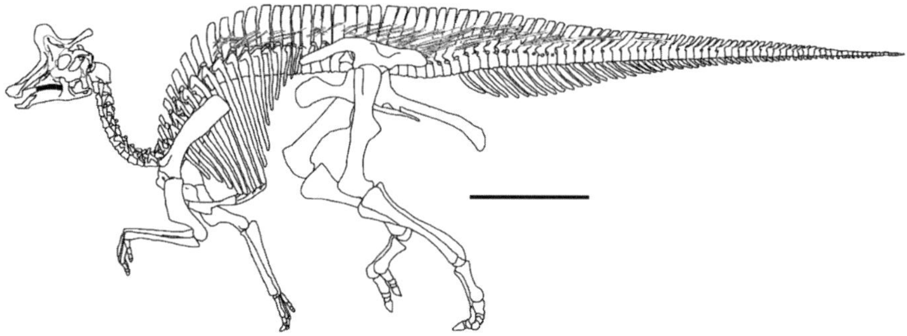

FIGURE 1. Reconstruction of *Lambeosaurus laticaudus* with the skull of *L. lambei*. Scale bar = 2 m.

have been attributed to ceratopsians (Weishampel, 1990).

To date, the remains related to Theropoda include bones attributed to Troodontidae, Dromaeosauridae, Ornithomimidae and Tyrannosauridae (Rodríguez-de la Rosa and Aranda-Manteca, 1999; Weishampel, 1990). Among the theropodan material an isolated tooth preserves an unusual posterior carina groove. This groove structure, although controversial, has been interpreted as a specialization to a kind of envenomation among theropod dinosaurs (Rodríguez-de la Rosa and Aranda-Manteca, 2000a; 2000b). Additional theropod material includes isolated teeth referred to the enigmatic genus *Aublysodon* (Paul, 1988).

The Late Cretaceous Cabullona Group in north-eastern Sonora has yielded dinosaurian remains belonging to Hadrosauridae, Ceratopsidae and Tyrannosauridae (cf. *Albertosaurus*). Additional vertebrate fossils include fishes, turtles, crocodilians and lizards (Lucas and González-León, 1993). The Cabullona Group has been dated as Late Campanian-Early Maastrichtian and the inferred paleoenvironment is that of an inland extensive fluvial and lacustrine systems (Lucas and González-León, 1990; 1993; 1996; Taliaferro, 1933).

Late Cretaceous dinosaurs from the Aguja Formation have been collected from the northeastern corner of Chihuahua. The most abundant material is that of hadrosaurian dinosaurs, but also chasmosaurine ceratopsian elements have recently been identified (Andrade-Ramos *et al.* 2002; Montellano-Ballesteros *et al.* 2000), and small to large theropods have also been recorded. Other vertebrates include amiid and lepisosteid fishes, and trionychids among other turtles (Westgate *et al.* 2000). The occurrence of titanosaurid sauropod caudal vertebrae in northeastern Chihuahua have been recently documented (Montellano-Ballesteros, 2003).

As noted in the introduction, the first dinosaur remains reported from Mexico come from Coahuila in northern Mexico (Janensch, 1926). Relatively recent extensive surveying done in the southeastern portion of Coahuila has yielded additional ornithischian and saurischian dinosaur remains.

Ornithischian dinosaurs are known from several partial skeletons of hadrosaurian dinosaurs, of both hadrosaurine and lambeosaurine subfamilies, as well as partial skeletons of centrosaurine and chasmosaurine ceratopsians (Hernández and Kirkland, 1993; Kirkland *et al.* 2000; Brinkman *et al.* 2002; Eberth *et al.* 2003). Other ornithischian remains include isolated osteoderms related to Ankylosauridae. Saurischian remains belong mainly to Theropoda and record the presence of Troodontidae, Dromaeosauridae, Ornithomimidae (cf. *Ornithomimus*) and Tyrannosauridae (Rodríguez-de la Rosa and Cevallos-Ferriz, 1998; Weishampel, 1990; Aguillón-Martínez pers. com. 2004).

The paleoenvironment in which the dinosaurs from Coahuila lived was that of an extensive low-gradient and mud-dominated lower coastal plains. An interesting observation is that these environments were subject to frequent storm-generated flooding events (Eberth *et al.* 2005). The climate was hot and humid with abundant rainfall and little opportunity for extensive subaerial exposure and drying. In fact almost all of the tracks were made in very shallow water no more than 50 centimeters in depth (Eberth *et al.* 2005).

Additional, but undescribed, dinosaur remains from Mexico are known to occur in Chiapas and Michoacán. In Chiapas, Early or Middle Jurassic remains have been collected from Juiquipilas, Cintalapa and La Concordia localities (Pimentel pers. com., 2003). Also in Chiapas fragmentary dental elements from the Maastrichtian Ocozocoautla Formation found near Jaltenango River have been discovered (Pimentel pers. com., 2003). In Michoacán, hadrosaurian dinosaur remains are known to occur in Late Cretaceous sediments; however, the geological unit is still unknown (Aguillón-Martínez pers. com., 2004).

Available information suggests that the southern plains dinosaurian fauna of the United States extended, with relatively little change, into northern Mexico. Current research is yielding extensive paleoenvironment and paleogeographical aspects of dinosaurs from these areas of northern Mexico (Eberth *et al.* 2003). A preliminary cluster analysis has revealed only three different paleogeographical regions where Cretaceous dinosaurs from Mexico lived. The first is in the northern portion of Mexico, the second in what is now Baja California Peninsula and the third is in central Mexico.

3.2. Ichnological Record

Ichnofossils record the activity of living organisms. The fossils usually considered under the scope of ichnology include eggs, nests, eggshells, skin impressions, stomach contents and associated evidence of feeding behavior (e.g., tooth-marked bones, gastroliths, etc.), coprolites, and obviously tracks or footprints (Gillete and Lockley, 1989; Lockley, 1991).

Dinosaur eggshells have been described from the Late Cretaceous (Campanian) El Gallo Formation of Baja California. These eggshells belong to the typical prolatospherulitic morphotype attributed to ornithopod dinosaurs (mainly hadrosaurs) and are associated with some neognathous avian eggshells (Rodríguez-de la Rosa,

1998a). Abundant dinosaurian eggshell fragments belonging to at least three different morphotypes were recently collected from the late Campanian Cerro del Pueblo Formation in southeastern Coahuila (Aguillón-Martínez pers. com. 2004).

Dinosaur skin impressions are known from the states of Baja California and Coahuila. Interestingly, skin patches are frequently found in association with hadrosaur carcasses (Morris, 1981; Hernández and Delgado-de Jesús, 2000).

Bromalites are fossils related to feeding behavior (Hunt, 1992), including stomach contents or cololites, regurgitalites and coprolites or fossil faeces. Up until now, neither dinosaur stomach contents, nor gastroliths have been recorded from Mexico; however, vertebrate coprolites related to dinosaurs have been described from the Late Cretaceous of Coahuila in northern Mexico (Rodríguez-de la Rosa, 2003).

Dinosaur tracks are known from several localities in the central and northern parts of Mexico. Middle Jurassic dinosaur tracks attributed to theropod and sauropod dinosaurs are known from the Xochixtlapilco area in Oaxaca (Ferrusquía-Villafranca *et al.,* 1995). Late Jurassic theropod and ornithopod dinosaur tracks are known from an area near the town of Chuta now called Playa Azul, Michoacán (Ferrusquía-Villafranca *et al.,* 1978a; 1978b). The tracks in Chuta are significant because not only do the tracks represent the first record of footprints described from Mexico but they also represent the first record of footprints described from Latin America. In addition, a slab bearing theropod tracks of probable Jurassic age, but of unknown provenence is reported from Durango (Rodríguez-de la Rosa, 1998b; Rodríguez-de la Rosa, *et al.* 2004).

Early Cretaceous ornithopod, theropod and sauropod tracks were recently found in southern Puebla, in central Mexico. These tracks are distributed within six track-bearing layers within the San Juan Raya Formation (Carrillo-Montiel *et al.* in press; Rodríguez-de la Rosa *et al.* 2004).

Late Cretaceous (Maastrichtian) tracks attributed to Hadrosauridae and Sauropoda are known from the Mexcala Formation near Mitepec, within the southwestern part of Puebla (Ferrusquía-Villafranca *et al.* 1993; Ferrusquía-Villafranca *et al.* 1995). These sauropod impressions, attributed to young individuals, are of the same size as sauropod footprints from the Late Cretaceous Jingdong Formation of Korea (Lim, Yang and Lockley, 1989). There is also the possibility that the producer of these tracks was an adult specimen of small size (Callison and Quimby, 1984; Weishampel *et al.* 1993). The height estimate of 1.5 m, suggests an animal of about 3 or 4 m long. This means that the track-maker has a length that corresponds to 20%-25% of that of the titanosaurids. An example of small sauropods has been reported from the Hateg Basin, Romania, where isolated material of a small size sauropod occurs (Jianu and Weishampel, 1999; Mussell and Weishampel, 2000). Romania is well known for being an archipelago during the Cretaceous, a very similar situation to the central part of Mexico at that time. The small size of the dinosaur in Romania was probably due to geographic barriers such as islands (Rivera-Sylva, 2003).

Another late Cretaceous track site from central Mexico is known as El Aguaje,

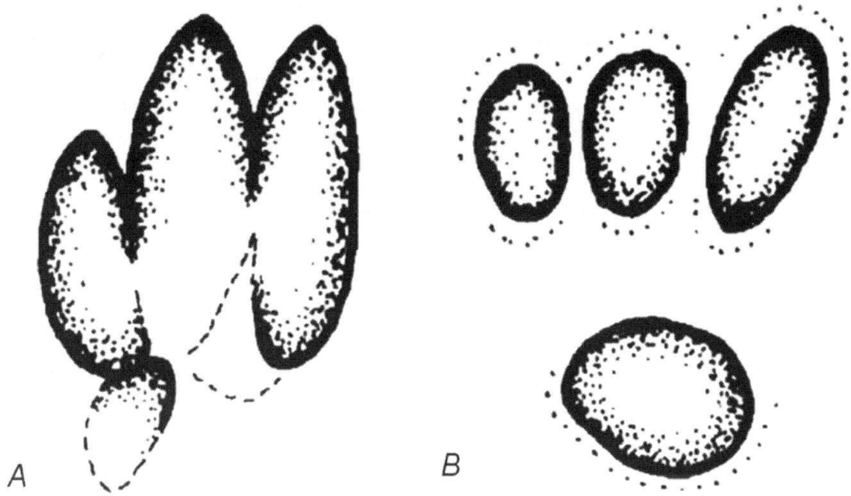

FIGURE 2. The dinosaur footprints from El Aguaje, which are similar to footprints from Mitepec, represent the most austral fauna for North America. a) Theropod footprint where the allux can be confidentally attributed to the Family Tyrannosauridae. b) An unusually big sized Ornithopod impression attributed to the Family Hadrosauridae.

from the Aguilillas Sandstone of Michoacán (Ortíz-Mendieta *et al.* 2000; Ortíz-Mendieta, 2001). The El Aguaje paleoichnofauna includes morphotypes referred to Tyrannosauridae, Dromaeosauridae, Ornithomimidae and Hadrosauridae (Figure 2). However, it has been commented that the diversity of these tracks may represent extramorphological variation of a lesser number of morphotypes (Rodríguez-de la Rosa *et al.* 2004). The El Aguaje locality represents the most austral Cretaceous theropod footprint site found in North America. From osteological evidence the El Aguaje locality is similar to the one found in Coahuila. In addition, the El Aguaje footprints were found within continental sediments, which adds detail to the paloegeographic interpretation of Mexico during the Late Cretaceous. Several vertebrate track sites are currently known from the Late Campanian Cerro del Pueblo Formation in the southeastern part of Coahuila. One of these track sites known as Las Águilas covers approximately 5,000 m2 and is currently the largest documented dinosaur track site in Mexico (Rodríguez-de la Rosa *et al.* 2002; 2004).

Tracks and trackways of small to large theropods (e.g., ornithomimids and tyrannosaurids), and large ornithopods in bipedal and quadrupedal gaits (e.g., the Las Águilas Tracksite) are known from several localities in southeastern Coahuila. Other Coahuila vertebrate tracks include those of turtles, crocodiles, pterosaurs (cf. *Pteraichnus*), and birds (Aguillón-Martínez *et al.* 1998; Brinkman *et al.* 2002; Eberth *et al.* 2003; Rodríguez-de la Rosa *et al.* 2002; 2003; 2004). The El Pelillal Track site within the Late Campanian Muerto Formation, preserves tracks of small theropod dinosaurs in association with those of turtles, crocodiles, pterosaurs (*Pteraichnus*), birds and possibly a mammal (Rodríguez-de la Rosa, 2001; 2003; Rodríguez-de la Rosa *et al.* 2003; 2004). Within the stratigraphic sequence of the Olmos Formation located near Sabinas, in northcentral Coahuila another Late Cretaceous (Early Maastrichtian) track site has yielded deep tracks of medium-sized theropod dinosaurs, as well as those of small ornithopods and associated tracks including those of turtles, crocodiles and small birds (Rodríguez-de la Rosa *et al.* 2004).

4. Final Comments

The study of dinosaurs from Mexico has a history of nearly 80 years with a significant increase in scientific dinosaurian research in the last few years. However the knowledge of dinosaurs from this area of southern North America is still, for the most part, unexplored when compared to other parts of the world with extensive Mesozoic, particularly Cretaceous, outcrops. Therefore the following points could be adopted as final comments about the record of dinosaurs from Mexico (Figure 3).

The Middle Jurassic to Late Cretaceous dinosaur record of Mexico comprises a moderate diversity of taxa including several track sites. Although the osteological remains of some Late Cretaceous dinosaurs in Mexico are restricted to southeastern Coahuila the presence of ichnofaunas of the same age in Central Mexico establishes, with confidence, the presence of dinosaurs to lower latitudes.

Within the state of Coahuila it is possible to find abundant Late Cretaceous vertebrate remains, including dinosaurs, in association with track sites. The association of dinosaur bones and dinosaur tracks provides a unique opportunity to compare dinosaur tracks with possible track-makers. The finding of hadrosaurian tracks in direct association to at least three lamberosaurine hadrosaur partial skeletons at the Las Águilas track site, in southeastern Coahuila, exemplifies this point (Rodríguez-de la Rosa *et al.* 2003).

As mentioned earlier, the Las Aguilas track site is currently the largest dinosaur track site in Mexico. Recent work has revealed this site to be a mega-track site (Lockley, 1991) similar to those found in other areas of North America.

Additionally Mexico has the only known Middle Jurassic dinosaurian fauna from North America represented by two localities; an ichnofauna from Oaxaca and saurischian and ornithischian remains from Tamaulipas (Clark *et al.* 1994; Ferrusquía-Villafranca *et al.* 1996).

The study of dinosaurs from Mexico was sporadic in the past and has only recently been the focus of renewed scientific research. The authors believe that ongoing and continued research will undoubtedly result in new taxa and valuable and varied information about dinosaur diversity, as well as paleobiological, paleoecological and paleobiogeographical aspects of dinosaurs from this portion of southern North America.

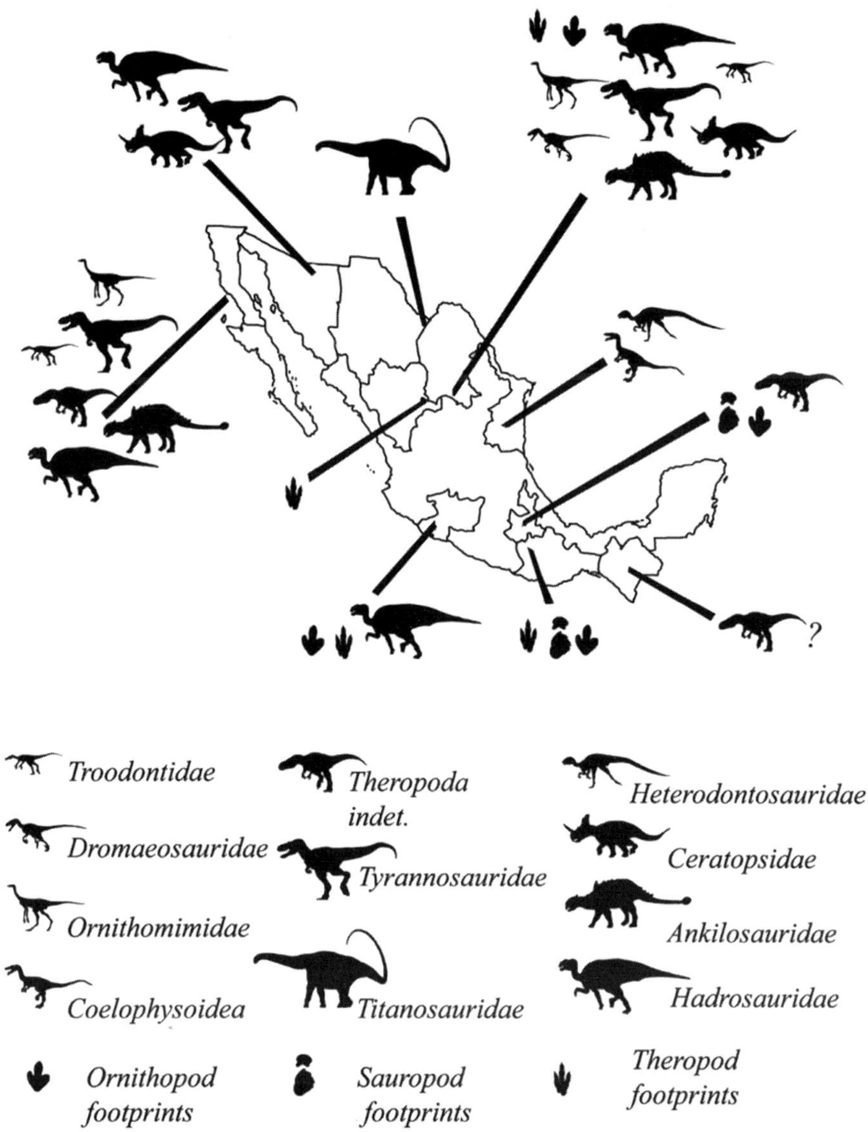

FIGURE 3. Map showing the dinosaur localities from Mexico.

Appendix: Summary of the Dinosaur-Bearing Localities of Mexico

Jurassic localities

1. Tamaulipas

   a) Huizachal Formation (Early Jurassic)
      Theropoda
         Coelurosauridae
            Syntarsus
      Ornithopoda
         Heterodontosauridae
            Heterodontosaurus

2. Puebla

   a) Oltaltepec, Unknown Formation (Late Jurassic – Early Cretaceous)
      ?Theropoda indet.

3. Oaxaca

   a) Xochixtlapilco, Tecocoyunca Group (Middle Jurassic)
      Theropod footprints
      Sauropod footprints

4. Michoacán

   a) Chuta (Late Jurassic)
      Theropod footprints
      Ornithopod footprints
         Iguanodontidae

5. Durango

   a) Unknown Formation (?Jurassic)
      Theropod footprint

Cretaceous localities

1. Puebla

   a) Mitepec, Mexcala Formation (Late Cretaceous)
      Theropod footprints
      Sauropod footprints

Titanosauridae
Ornithopod footprints
Hadrosauridae

b) San Martin Atexcal, San Juan Raya Formation (Early Cretaceous)
Theropod footprints
Sauropod footprints
Ornithopod footprints

2. Baja California

a) El Rosario, La Bocana Roja Formation (Late Cretaceous)
Theropoda
Labocania anomala

b) El Rosario, El Gallo Formation (Late Cretaceous)
Theropoda
Tyrannosauridae
cf. *Albertosaurus*
Aublysodon
Troodontidae
Ornithomimidae
Ornithopoda
Hadrosauridae
Lambeosaurus laticaudus
Thyreophora
Ankylosauridae
cf. *Euoplocephalus*

3. Chihuahua

a) Snake Ride Formation (Late Cretaceous)
Theropoda
Ornithopoda
Hadrosauridae
Ceratopsia
Ceratopsidae

b) Altares, Aguja Formation (Late Cretaceous)
Sauropoda
Titanosauridae indet.

4. Sonora

a) Cabullona Group (Late Cretaceous)

Theropoda
Tyrannosauridae
Ornithopoda
Hadrosauridae
Ceratopsia
Ceratopsidae

5. Coahuila

a) Soledad Beds (Late Cretaceous)
Ceratopsia
Ceratopsidae
Centrosaurus

b) Cerro del Pueblo Formation (Late Cretaceous)
Theropoda
Ornithomimidae
Dromiceiomimus
Tyrannosauridae
Albertosaurus
Dromaeosauridae
Saurornitholestes
Ornithopoda
Hadrosauridae
Kritosaurus
Lambeosaurus
Thyreophora
Ankylosauridae
cf. *Euoplocephalus*

c) Sabinas, Olmos Formation (Late Cretaceous)
Ornithopoda
Hadrosauridae
Kritosaurus
Theropod footprints
Ornithopod footprints

d) Las Águilas, Cerro del Pueblo Formation (Late Cretaceous)
Theropod footprints
Ornithopod footprints

e) El Pelillal, Muerto Formation (Late Cretaceous)
Theropod footprints

6. Michoacán

a) El Aguaje, Aguililla Sandstone (Late Cretaceous)
Theropod footprints
Tyrannosauridae
Ornithomomimidae
Dromeosauridae
Ornithopod footprints
Hadrosauridae

b) (Late Cretaceous)
Ornithopod
Hadrosauridae

7. Chiapas

a) Ocozocoautla Formation (Late Cretaceous)
Theropoda indet.

Acknowledgments

We thank S. Clabby, B. G. Grace and A. S. Smith for their careful reviews.

References

Aguillón-Martínez, M. C., Vallejo-González, I., Hernández, R., and Kirkland, J. I., 1998, Dinosaur trackway from the Cerro del Pueblo Formation, Difunta Group (Latest Campanian, Cretaceous), Coahuila, Mexico, *J. Vert. Paleont.*, 18: 23A.

Andrade-Ramos, P., Montellano-Ballesteros, M., Cevallos-Ferriz, S.R.S., and Lozano-García, S., 2002, A Late Cretaceous association from Altares, Chihuahua, Mexico, *J. Vert. Paleont.*, 22: 32A.

Brinkman, D. B., Eberth, D. A., Sampson, S. D., Aguillón, M. C., Delgado-de Jesús, C. R., and Rodríguez-de la Rosa, R.A., 2002, Paleontology and stratigraphy of the dinosaur-bearing Cerro del Pueblo Formation, southern Coahuila, Mexico, *J. Vert. Paleont.*, 22: 38A.

Callison, G., and Quimby, H. M., 1984, Tiny dinosaurs: are they fully grown. *J. Vert. Paleont.*, 3: 200-209.

Carrillo-Montiel, E., Ortiz-Ubilla A., Bravo-Cuevas, V. M., and Rodríguez-de la Rosa, R. A., (In press), Huellas de dinosaurios en el Cretácico Inferior de San Martín Atexcal, Puebla. *Paleos-Antiguo.*

Clark, J. M., Montellano M., Hopson, J. A., Hernández, R., and Fastovsky, D. E., 1994, An Early or Middle Jurassic tetrapod assemblage from the La Boca Formation, northeastern Mexico, in: *The shadow of the dinosaurs (Early Mesozoic tetrapods)* (N. C. Fraser and H.-D. Sues, eds.), Cambridge University Press, pp. 295-302.

Eberth, D. A., Sampson, S. D., Rodríguez-de la Rosa, R. A., Aguillón-Martínez, M. C., Brinkman, D. B., and López-Espinoza, J., 2003, Las Águilas: An unusuallly rich Campanian-age vertebrate locality in southern Coahuila, Mexico. *J. Vert. Paleont.*, 23 (Suppl. to No. 3): 47A.

Eberth, D. A., Delgado-de Jesús, C. R., Lerbekmo, J. F., Brinkman, D. B. Rodríguez-de la Rosa, R. A., and Sampson, S. D., 2005, Cerro del Pueblo Fm. (Difunta Group, Upper Cretaceous), Parras Basin, southern Coahuila, Mexico: Reference Sections, Age, and Correlation. *Instituto de Geología, Rev.*

Mex. Ciencias Geol.

Fastovsky, D. E., Clark, J. M., Strater, N. H., Montellano, M., Hernández, R., and Hopson, J. A., 1995, Depositional environments of a Middle Jurassic terrestrial vertebrate assemblage, Huizachal Canyon, Mexico, *J. Vert. Paleont.*, 15(3): 561-575.

Ferrusquía-Villafranca, I., Applegate, S. P., and Espinosa-Arrubarrena, L., 1978a, Las huellas más australes de dinosaurios en Norte América y su significación geobiológica. Actas del II Congreso de Paleontología y Bioestratigrafía y I Congreso Latinoamericano de Paleontología, Buenos Aires, Argentina: 249-263.

Ferrusquía-Villafranca, I., Applegate, S. P., and Espinosa-Arrubarrena, L., 1978b, Rocas volcanosedimentarias mesozoicas y huellas de dinosaurios en la región suroccidental pacífica de México, *Univ. Nal. Autón. México, Inst. Geol., Rev.* 2: 150-162.

Ferrusquía-Villafranca, I., Tilton, T. L., Lang, H. R., Pittman, J. G., and Lockley, M. G., 1993, Dinosauricnitas tardicretácicas en Puebla suroccidental y su significación geológico-paleontológica, Memorias del IV Congreso Nacional de Paleontología: 33-34.

Ferrusquía-Villafranca, I., Jiménez-Hidalgo, E., and Bravo-Cuevas, V. M., 1995, Jurassic and Cretaceous dinosaur footprints from México: additions and revisions, *J. Vert. Paleont.,* 15 (Suppl. to No. 3): 28A.

Ferrusquía-Villafranca, I., Jiménez-Hidalgo, E., and Bravo-Cuevas, V. M., 1996, Footprints of small sauropods from the Middle Jurassic of Oaxaca, southeastern México, in: *The Continental Jurassic* (M. Morales,ed.), *Mus. North. Ariz. Bull.* 60. *Flagstaf, Arizona*, pp. 119-126.

Ford, T. L. and Chure, D. J., 2001, "Aublysodon" teeth from the El Gallo Formation (Late Campanian) of Baja California: the southernmost record of tyrannosaurid theropods. McCord, R. D. and Boaz, D. (eds.), *Western Association of Vertebrate Paleontologists and Southwest Paleontological Symposium Proceedings 2001, Mesa Southwest Museum Bulletin No. 8,* pp 75-89.

Gillette, D. D. and Lockley, M. G., 1989, Dinosaur tracks and traces, Cambridge University Press, Cambridge, 454 p.

Hernández, R., 1992, New dinosaur finds in the Cerro del Pueblo Formation (Upper Cretaceus, Campanian) from Coahuila, Mexico. *J. Vert. Paleont.*, 12 (Suppl. to No. 3): 32A.

Hernández, R., 1996, Los dinosaurios y otros vertebrados fósiles del Cretácico Superior en Coahuila, *Geo UNAM,* 3(1): 39-44.

Hernández, R., 1997, Mexican dinosaurs, in: *Encyclopedia of Dinosaurs* (Currie, P. J. and K. Padian, eds.), Academic Press, San Diego, California, pp. 433-437.

Hernández, R. and Espinosa-Arrubarrena, L., 1990, Dinosaur localities in Mexico, *J. Vert. Paleont.,* 10 (Suppl. to No. 3): 27A.

Hernández, R. and Kirkland, J. I., 1993, The rediscovery of a rich uppermost Campanian dinosaur locality in the Cerro del Pueblo Fm, Coahuila, Mexico, *J. Vert. Paleont.*, 13: 41A.

Hernández, R. and Delgado-de Jesús, C. R., 2000, Hadrosaur skin impression and associated skeletal remains from Cerro del Pueblo Fm. (Uppermost Campanian) southeastern Coahuila, Mexico, *J. Vert. Paleont.,* 20: 48A.

Hunt, A. P., 1992, Late Pennsylvanian coprolites from the Kinney Brick Quarry, central New Mexico, with notes on the classification and utility of coprolites. *New Mexico Bur. Min. Miner. Resch. Bull.*, 138: 221-229.

Janensch, V. W., 1926, Dinosaurier-Reste aus Mexiko. *Centr. Min., Geol. Pal.*, 1926B: 192-197.

Jianu, C. M., and Weishampel, D. B., 1999, The smallest of the largest: A new look at possible dwarfing in sauropod dinosaurs, *Geol. Mijn.*, 78: 335-343.

Kirkland, J. I., Hernández, R., Aguillón-Martínez, M. C., Delgado-de Jesús, C. R., Gómez-Núñez, R., and Vallejo-González, I., 2000, The Late Cretaceous Difunta Group of the Parras Basin, Coahuila, Mexico, and its vertebrate fauna, *Field Trip Guide Book, Society of Vertebrate Paleontology Annual Meeting,* pp. 133-172.

Langston, Jr. W. and Oakes, M. H., 1954, Hadrosaurs in Baja California. *Geol. Soc. Amer.* Abstracts (65): 1344.

Lim, S. K., Yang, Y. S., and Lockley, M. G., 1989, Large Dinosaurs footprints assemblages from the Cretaceous Jindong Formation of Korea, in: *Dinosaurs tracks and traces* (Gillete, D. D., and M. G.,

Lockley, eds.), Cambridge Univ. Press, pp. 333-336.

Lockley, M. G., 1991, Tracking *Dinosaurs: A New Look at an Ancient World,* Cambridge University Press.

Lucas, S. G., and González-León, C. M., 1990, Reporte preliminar sobre los dinosaurios del Cretácico Tardío de la Cuenca de Cabullona. *Bol. Dept. Geol. Univ. Sonora,* 7(1-2): 1-6.

Lucas, S. G., and González-León, C. M., 1993, Fossil vertebrates from the Upper Cretaceous Cabullona Group, north-eastern Sonora, Mexico, *J. Vert. Paleont.*, 13: 47A.

Lucas, S. G., and González-León, C. M., 1996, Dinosaurios del Cretácico Tardío del Grupo Cabullona, Sonora, *Geol. Noroeste*, 1: 20-25.

Lull, R. S., and Wright, N. E., 1942, Hadrosaurian dinosaurs of North America. *Geol. Soc. Amer., Spec. Pap.*, 40:1-242.

Molnar, R. E., 1974, A distinctive theropod dinosaur from the Upper Cretaceous of Baja California (Mexico), *J. Paleont.,* 48: 1009-1017.

Montellano-Ballesteros, M., 2003, A titanosaurid sauropod from the Upper Cretaceous of Chihuahua, Mexico, *Rev. Mex. Cien. Geol.*, 20: 160-164.

Montellano-Ballesteros, M., Hernández, R., Álvarez-Reyes, G., Andrade-Ramos, P. and Martín-Medrano, L., 2000, Discovery of Late Cretaceous vertebrate local faunas in northern Mexico, *J. Vert. Paleont.*, 20: 58A-59A.

Morris, W. J., 1967, Baja California: Late Cretaceous dinosaurs, *Science*, 155: 1539-1541.

Morris, W. J., 1971, Mesozoic and Tertiary Vertebrates in Baja California, *Natl. Geogr. Soc. Resch. Repts.*, 1965: 195-198.

Morris, W. J., 1972, A giant hadrosaurian dinosaur from Baja California, *J. Paleont.*, 46: 777-779.

Morris, W. J., 1973a, Mesozoic and Tertiary Vertebrates in Baja California, *Natl. Geogr. Soc. Resch. Repts.*, 1973: 197-209.

Morris, W. J., 1973b, A review of Pacific Coast hadrosaurs, J. Paleont., 47: 551-561.

Morris, W. J., 1981, A new species of hadrosaurian dinosaur from the Upper Cretaceous of Baja California – ?*Lambeosaurus laticaudus. J. Paleont.*, 55: 453-462.

Morris, W. J., 1982, California dinosaurs, in: *Late Cretaceous Depositional Environments and Paleogeography, Santa Ana Mountains, Southern California* (Bottjer, D. J., Colburn, I. P. and Cooper, J. D., eds.), Pacific Section, SEPM, *Field Trip Volume and Guidebook*, pp. 89-90.

Morris, W. J., 1983, A Paleontologic reconnaissance of Baja California, Mexico, 1974, *Natl. Geogr. Soc. Resch. Repts.*, 15: 157-174.

Munter, R. C., 1999, Two theropod dinosaur specimens from Huizachal Canyon, Mexico, *J. Vert. Paleont.,* 19: 65A.

Murray, G. E., Boyd, D. R., Wolleben, J. A. and Wilson, J. A., 1960, Late Cretaceous fossil locality, Eastern Parras Basin, Coahuila, Mexico, *J. Paleont.*, 34: 368-373.

Mussell, J. and Weishampel, D., 2000, *Magyarosaurus* and the possible role of progenesis in sauropod development. *J. Vert. Paleont.* 20: 59A.

Ortega-Guerrero, B. 1989. Paleomagnetismo y geología de las Unidades Clásticas Mesozoicas del Área Totoltepec-Ixcaquixtla, Estado de Puebla y Oaxaca. UNAM, Fac. Ciencias, Tesis de Maestría. 122 pp.

Ortíz-Mendieta, J. A., 2001, Dinosauricnitas Cretácico-tardías de El Aguaje, Michoacán, Región Suroccidental de México y sus implicaciones geológico-paleontológicas, Tesis de Licenciatura, Facultad de Ciencias, Universidad Nacional Autónoma de México, 75 pp.

Ortíz-Mendieta, J. A., García-Orozco, J., Ferrusquía-Villafranca, I., Jiménez-Hidalgo, E. and Bravo-Cuevas, V. M., 2000, Primer hallazgo de huellas de dinosaurio en Michoacán suroccidental: Reporte preliminar, *VII Congreso Nacional de Paleontología y I Simposio Geológico en el Noreste de México, Memorias,* 109-110.

Paul, G.S. 1988. *Predatory Dinosaurs of the World: A Complete Illustrated Guide*. Simon & Schuster. New York.

Rivera-Sylva, H. E., 2003, *Dinosaurs from the Mixteca Region in Central Mexico,* Master Degree Thesis, Department of Earth Sciences, University of Bristol, 56 pp.

Rodríguez-de la Rosa, R. A., 1998a, Cáscaras de huevo avianas (Neognathae) y de Ornithopoda (Dinosauria) del Cretácico Tardío de Baja California, México. *VI Congreso Nacional de Paleontología, Memorias*, 59-60.

Rodríguez-de la Rosa, R. A., 1998b, ¿Aves o dinosaurios en Durango, México? *VI Congreso Nacional de Paleontología, Memorias*, 58-59.

Rodríguez-de la Rosa, R. A., 2001, Pterosaur tracks from the Late Cretaceous of northern México: Paleoecological and anatomical implications. *Two hundred years of pterosaurs*, Toulouse, France, *Strata*, série 1 (11): 85-86.

Rodríguez-de la Rosa, R. A., 2003, Pterosaur tracks from the Latest Campanian Cerro del Pueblo Formation of south-eastern Coahuila, Mexico, in: *Evolution and Palaeobiology of Pterosaurs* (Buffetaut, E., and J. M. Mazin, eds.), Geological Society, London, Special Publications, pp. 275-282

Rodríguez-de la Rosa, R. A. and Aranda-Manteca, F. J., 1999, Theropod teeth from the Late Cretaceous El Gallo Formation, Baja California, Mexico. *VII International Symposium on Mesozoic Terrestrial Ecosystems*, Buenos Aires, Abstracts: 56.

Rodríguez-de la Rosa, R. A. and Aranda-Manteca, F. J., 2000a, Were there venomous theropods?, *J. Vert. Paleont.*, 20: 64A.

Rodríguez-de la Rosa, R. A. and Aranda-Manteca, F. J., 2000b, Nuevos hallazgos de vertebrados de la Formación El Gallo (Campaniano Tardío) del Estado de Baja California, *VI Congreso Nacional de Paleontología y I Simposio Geológico en el noreste de México, Memorias:* 65-66.

Rodríguez-de la Rosa, R. A. and Cevallos-Ferriz, S. R. S. 1998. Vertebrates of the El Pelillal Locality (Campanian, Cerro del Pueblo Formation), south-eastern Coahuila, Mexico, *J. Vert. Paleont.*, 18:751-764.

Rodríguez-de la Rosa, R. A., Cevallos-Ferriz, S. R. S. and Silva-Pineda, F. J., 1998, Paleobiological implications of Campanian coprolites, *Palaeogeogr., Palaeoclimat., Palaeoecol.,* 142: 231-254.

Rodríguez-de la Rosa, R. A., López-Espinoza, J., Vallejo-González, J. I., Eberth, D. A. and Simith, J. A., 2002, Huellas de vertebrados cretácicos (Campaniano Tardío, Formación Cerro del Pueblo) del sureste de Coahuila, México, *VIII Congreso Nacional de Paleontología, Memorias,* pp. 75-76..

Rodríguez-de la Rosa, R. A., Eberth, D. A., Brinkman, D. B., Sampson, S. D. and López-Espinoza, J., 2003, Dinosaur tracks from the Late Campanian Las Aguilas locality south-eastern Coahuila, Mexico. *J. Vert. Paleont.*, 23: 90A.

Rodríguez-de la Rosa, R. A., Aguillón-Martínez, M. C., López-Espinoza, J. and Eberth, D. A., 2004, The fossil record of vertebrate tracks in Mexico, *Ichnos* 1: 1-11.

Taliaferro, N. L., 1933, An occurrence of Upper Cretaceous sediments in northern Sonora, Mexico, *J. Geol.*, 41:12-37.

Weishampel, D. B., 1990, Dinosaurian distribution, in: *The Dinosauria* (Weishampel, D.B., Dodson, P., and H., Osmólska, eds.) , California University Press, Berkeley, pp. 63-139

Weishampel, D. B., Norman, D. B. and Grigorescu, D., 1993, *Telmatosaurus transsylvanicus* from the Late Cretaceous of Romania: The most basal hadrosaurid dinosaur, *Palaeont.*, 36 :361-368.

Westgate, J. W., Brown, R. B., Cope, D. and J. Pittman, 2000, A Late Cretaceous dinosaur-bearing community from coastal deposits in Chihuahua, Mexico, *J. Vert. Paleont.*, 20: 78A.

Chapter 12

Mexican Fossil Mammals, Who, Where and When?

MARISOL MONTELLANO-BALLESTEROS[1] and EDUARDO JIMÉNEZ-HIDALGO[2]

[1]Instituto de Geología, UNAM. Circuito Exterior, Ciudad Universitaria, Delegación Coyoacán, 04510, D. F., México. marmont@servidor.unam.mx.
[2]Universidad del Mar - Campus Puerto Escondido, Km 3.5 Carretera Puerto Escondido-Oaxaca, Puerto Escondido, 71980 Oaxaca, México. eduardojh@zicatela.umar.mx.

1. Abstract

Although the earliest report of a fossil mammal from Mexico dates from 1799, our knowledge of the group is still poor. The Mexican mammalian fossil record is biased towards the large-sized taxa and younger ages.

The mammalian record in Mexico ranges from the Jurassic to the Quaternary.

F. Vega et al. (eds.), Studies on Mexican Paleontology, 249–273.

Most of the Cenozoic epochs, except for the Paleocene, have mammal bearing deposits. There are exists gaps, such as late Eocene early Oligocene, early Miocene, and early-late Miocene, where no fossils have been found. The Mesozoic mammalian record is poorly represented by three localities in the northern part of Mexico. The early Tertiary is also represented by only a few localities widespread throughout Mexico. The late Tertiary (late Miocene-early Pliocene) is represented from well documented exposures in the central part of Mexico. Although the Quaternary record is very abundant and is present in all the states of Mexico, there are problems with detailed dating of these localities and little is known about small size mammals.

Due to its geological history and geographic position, Mexico bears a great diversity of environmental conditions, which is reflected in its high extant mammal biodiversity. Mexico played an important role in the evolution of several mammal groups during the Cenozoic. During the late Mesozoic and early Cenozoic, Mexico was the southernmost landmass of North America, where several mammal groups evolved, adapting to more tropical conditions. Later, during the Great American Faunal Interchange, Mexico was the path for the South American mammal immigrants to northern lands and it was the region where some immigrants remained. The role that Mexico played during the Ice Ages, whether if it was a refuge for northern forms or not has been questioned.

Additional work is needed in order to have a more comprehensive understanding of Mexican paleomammals.

2. Introduction

To our knowledge, the earliest published record of fossil mammals from Mexico is an anonymous paper, written in 1799, which mentioned the remains of a proboscidean. By the late eighteenth and early nineteenth centuries, natural historians and geographers explorers began to arrive, producing the first scientific observations and descriptions of Mexico. During the nineteenth century native and foreign investigators wrote many papers describing fossil material of Mexico, however these early reports did not provide accurate identification of the fossil localities or stratigraphic position. It is important to mention that by middle nineteenth century a significant improvement in the collecting, describing, and mapping of Mexican minerals and fossils occurred (Montellano-Ballesteros, 1999).

The nineteenth century was also a time when natural history societies and national museums were founded, such as the Sociedad Mexicana de Historia Natural (1868), with its official journal, “La Naturaleza”. In 1884, the Sociedad Científica Manuel Alzate was founded, which later became the Academia Nacional de Ciencias. The Museo Nacional de México was founded in 1825 and included a department of Natural History. At the end of the nineteenth and early twentieth centuries, German contributions were important to the study of ancient life and geology of Mexico.

During the 1950´s, 1960´s and early 1970´s fossil mammal research focused less on the detailed descriptions and more on correlation. Especially during the 1950´s Arellano, a geologist from the Instituto de Geología, brought vertebrate fossils to the attention of paleontologists through his geologic studies of Central Mexico.

Since then, the field of the vertebrate paleontology has been growing and is now developed at several academic institutions, where there are interdisciplinary projects where foreign paleontologists collaborate.

Several attempts to demonstrate the knowledge of the Mexican vertebrate paleontology field have been made, such as Maldonado-Koerdell (1948), Álvarez (1965), Silva-Bárcena (1969), and Barrios-Rivera (1985). Recently, Montellano-Ballesteros and Arroyo-Cabrales (2002) compiled a state of the art monograph on Mexican mammalian paleofaunas.

The main purpose of this contribution is to resume and comment the present situation of the fossil mammalian record in Mexico, appointing the main localities, their ages and some relevant taxa.

3. What Do We know?

3.1. Mammalian Distribution According to Age

Mesozoic

Although continental Mesozoic outcrops are widespread in the northern part of Mexico, there are only two localities that have yielded mammal remains. These are: El Rosario in Baja California of Campanian age, Late Cretaceous; and Cañón del Huizachal, in Tamaulipas of late early Jurassic age (Figure 1). Recently, a new find of a mammal has been reported from the Cerro del Pueblo in Coahuila, a well known area known for its dinosaurs richness.

Tertiary

Faunas and single occurrences of Tertiary age come from 17 of the 32 Mexican states, covering almost all the country´s territory (Figures 1-4), with the exception of some states in the western (Sinaloa, Nayarit, Durango, and Guerrero), central (Estado de México and Distrito Federal), northeastern (Coahuila, Nuevo León, and Tamaulipas) eastern (San Luis Potosí, Veracruz, and Tabasco) and the Yucatán peninsula (Campeche, Yucatán and Quintana Roo).

The absence of the terrestrial fossil mammals from several states can be explained taking into account their geological history, lithic record and the poor prospecting in those areas.

Quaternary

The situation is completely different for the Quaternary record, where the 32 states of the country have yielded mammalian remains. Arroyo-Cabrales *et al.* (2002) showed in their analysis that the Estado de México has the highest percentage of the 776 reported localities for the Quaternary. To the contrary, the states with the poorest record of Quaternary fossil mammals are Colima, Tabasco, Nayarit, and Campeche.

4. Who, Where and When?

4.1. Mesozoic

The oldest Mesozoic mammalian record comes from the late early Jurassic La Boca Formation, at Cañón del Huizachal in Tamaulipas in northeastern Mexico. It includes the tritylodontid *Bocatherium mexicanum*, and a suite of what is best described as mammaliaforms. The La Boca fauna represent new taxa of "triconodont" mammals.

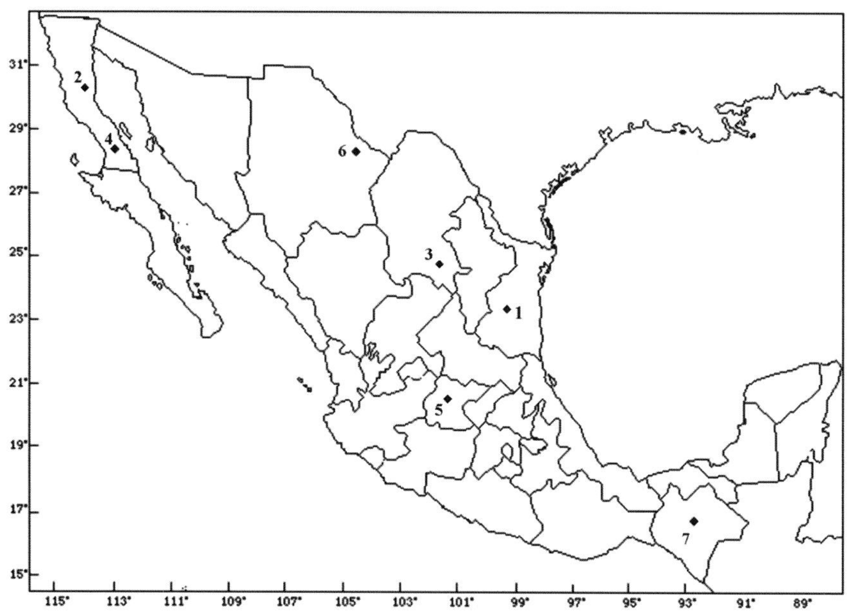

FIGURE 1. Mesozoic and Eocene-Oligocene local faunas bearing mammalian remains and single occurrences. 1. Cañón del Huizachal, late early Jurassic, Tamaulipas; 2. El Rosario, Late Cretaceous, Baja California; 3. Cerro del Pueblo Fm., Late Cretaceous of Coahuila; 4. Lomas Las Tetas de Cabra, Wasatchian, Baja California; 5. El Marfil, Bridgerian-Uintan, Guanajuato; 6. Rancho Gaitán, Chadronian, Chihuahua; 7. Simojovel, Arikareean, Chiapas.

The La Boca fauna is not closely related to Jurassic mammals described elsewhere, suggesting the presence of a radiation of early mammals that show variations on the primitive "triconodont" molar patterns and differing widely in lower jaw morphology (Montellano *et al*, 1995).

Mesozoic mammals of Campanian age have been described by Lillegraven (1972, 1976) and Clemens (1980) from Campanian layers of the El Gallo Formation in Baja California; these include the multituberculates *Mesodma* cf. *M. formosa*, ?*Stygimys* sp., a pediomyd marsupial *Pediomys* sp. and an eutherian of uncertain affinities, *Gallolestes pachymandibularis.* The mammalian assemblage has also been identified at the Judithian Terlingua local fauna in Texas (Rowe *et al.* 1992).

Recently, Aguillón-Martínez *et al.* (2004), reported a mutltituberculate tooth from the Late Cretaceous Cerro del Pueblo Formation, Coahuila, increasing the Mesozoic mammalian record for Mexico.

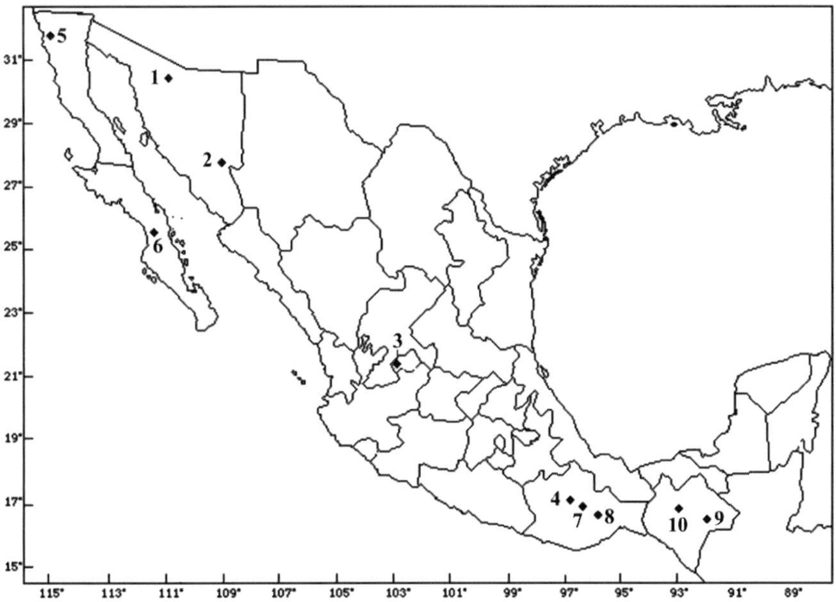

FIGURE 2. Early-Middle Miocene local faunas bearing mammalian remains and single occurrences 1. Tubutama and 2. Yécora, Hemingfordian, Sonora; 3. El Zoyatal, Hemingfordian, Aguascalientes; 4. Suchilquitongo, Hemingfordian, Oaxaca; 5. La Misión, Barstovian, Baja California; 6. La Purísima, Barstovian Baja California Sur; 7. Matatlan and 8. Nejapa, Barstovian, Oaxaca; 9. Ixtapa, Barstovian, Chiapas; 10. Simojovel, Early Miocene, Chiapas.

4.2. Tertiary

4.2.1. The Eocene-Oligocene Record

The oldest Tertiary fauna is the Lomas Las Tetas de Cabra, from the Early Eocene of Baja California (Figures. 1, 4). The mammalian fauna is around 55 million years old, and corresponds to the Wasatchian North American Land Mammal Age (Novaceck *et al.* 1991). It is integrated by 11 orders, 15 families and 16 monotypic genera, of which 45.4 % of the orders, 53.3 % of the families and 56.2 % of the genera are archaic mammals; the most diverse are the condylarths with four species. Also, there are some endemic forms, such as the marsupial *Estelestes ensis,* the condylarth *Ectocion ignotum* and an unnamed genus of pantodont (Novaceck *et al.* 1991;

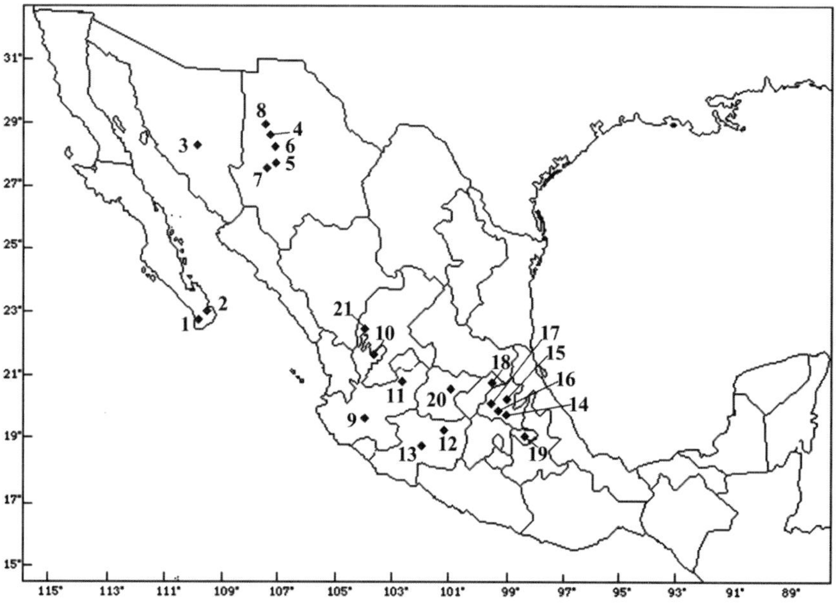

FIGURE 3. Late Miocene-Pliocene local faunas bearing mammalian remains and single occurrences. 1. Las Tunas and 2. Miraflores, early Blancan, Baja California Sur; 3. San José de Pimas, Blancan, Sonora; 4. Yepómera, 5. Basuchil and, 6. Matachic, late Hemphillian, Chihuahua; 7. Miñaca and, 8. Concha, early Blancan.Chihuahua; 9. Tecolotlán, late Hemphillian and ?late Blancan, Jalisco; 10. Colotlán, Hemphillian, Jalisco; 11. Teocaltiche, late Hemphillian, Jalisco; 12. Cinca, Hemphillian, Michoacán; 13. La Goleta, early Blancan, Michoacan; 14. Tula, Hemphillian or Blancan, Hidalgo; 15. Zietla-Tehuichila, early Hemphillian-Blancan, Hidalgo; 16. La Plegaria, late Hemphillian, Hidalgo; 17. Las Arcinas, Blancan, Hidalgo; 18. Landa de Matamoros, Hemphillian, Querétaro; 19. Tlaxcala single ocurrence, Blancan, Tlaxcala; 20. San Miguel de Allende, early Hemphillian-early Blancan Guanajuato; 21. Los Salazar, Hemphillian, Zacatecas.

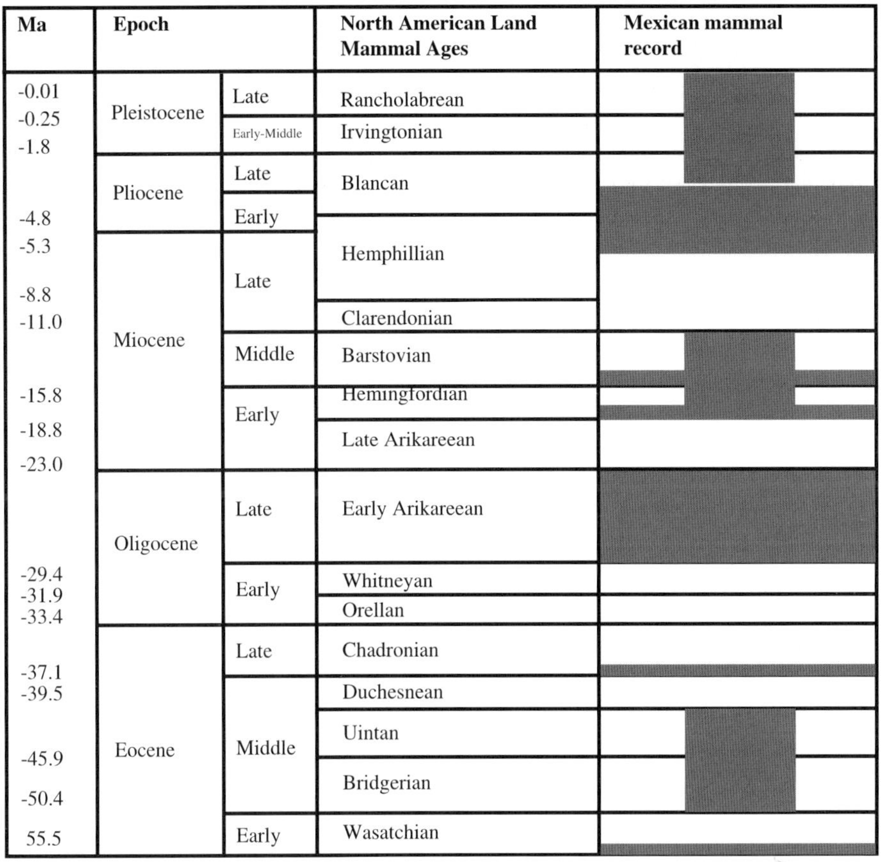

FIGURE 4. The North American Land Mammal Ages and the Mexican mammalian record. The filled rectangle means a more precise age assignment of some faunas.

Ferrusquía-Villafranca *et al.* 2002). Its faunal composition shows a combination of archaic paleogene taxa such as Condylarthra, Mesonychia, Pantodonta and Tillodontia, and modern taxa such as Carnivora, Perissodactyla, Artiodactyla and Rodentia.

The Middle Eocene El Marfil local fauna is located in Guanajuato, central Mexico (Figs. 1, 4). It is considered of Bridgerian-Uintan age based upon the evolutionary stage of some of their endemic forms (Ferrusquía-Villafranca, 1989; Ferrusquía-Villafranca *et al.* 2002). This fauna consists of five orders – one archaic-seven families and seven monotypic genera that displays a large degree of endemism. The most diverse and most studied group is the Rodentia with three species. Almost all the mammals that have been collected from the El Marfil are small to very small forms. There is only one medium-sized organism, a tapiroid (Ferrusquía-Villafranca *et al.*, 2002).

The third Eocene fauna of Mexico is the Rancho Gaitán local fauna, from the Late Eocene of Chihuahua, northern Mexico (Figs. 1, 4). It has been assigned to the Chadronian land mammal age based on faunal composition (Ferrusquía-Villafranca, 1969; Ferrusquia-Villafranca *et al.* 1997; 2002). Four orders, 11 families and 15 genera integrate this fauna, and the only archaic mammal is one species of condylarth, considered a new genus of hyaenodontid. The artiodactyls are the most diverse group with five oreodont species (agriocherids plus merycoidodontids) and two small ruminant forms (Ferrusquía Villafranca *et al.* 2002). The perissodactyls include the equid *Mesohippus texanus*, the rhino *Hyracodon nebraskensis* and the titanothere *Brontops* cf. *B. brachycephalus*. The artiodactyls are represented by four families and seven genera; also the following rodents are present: the paramyid, *Mitonomys gaitania,* and the cylindrodontids *Jaywilsonomys ojinagaensis, J. pintoensis* and *Pseudocylindrodon* cf. *P. medius* (Burke, 1936; Ferrusquía-Villafranca and Wood, 1969; Korth, 1981).

The most recent addition to the Mexican Paleogene mammal record is a single occurrence from the Late Oligocene of Simojovel, Chiapas, southeastern Mexico (Figures 1, 4). It is a new genus and species of a helohyid bunodont artiodactyl of early Arikareean age (Ferrusquía-Villafranca, 2003a).

4.2.2. The Early-Middle Miocene Record

The early-middle Miocene record is represented by two single occurrences and two faunas from the Early Miocene of northwestern, central and southeastern Mexico and five faunas from the Middle Miocene of the northwestern and southeastern parts of Mexico (Figure 2). The Early Miocene faunas are of Hemingfordian age and the Middle Miocene ones fall within the Barstovian land mammal age. Although the faunal record doubles that from the Eocene-Oligocene, it is still meager and includes much less micro-mammals.

The Hemingfordian single occurrences come from Tubutama and Yécora in Sonora (Figure 2), where an endemic species of a stenomyline camel and of a leporid have been described (Jiménez-Hidalgo *et al.* 2002; Ferrusquía-Villafranca, 2003b).

The El Zoyatal Fauna is located in Aguascalientes, central Mexico (Fig. 2). Two orders, four families and four monotypic genera of medium to large size have been described (Dalquest and Mooser, 1974; Stevens, 1977; Jiménez-Hidalgo *et al.* 2002). The faunal composition shows a strong bias against carnivores and micro-mammals, since up to not one of these have been identified. The most diverse group is the artiodactyls with three species, one of them being an endemic genus and species of floridatraguline camel (Jiménez-Hidalgo *et al.* 2000).

The last Hemingfordian fauna is from Suchilquitongo, Oaxaca, southeastern Mexico (Figures 2, 4). The fossil mammal bearing beds overlie a 19.3-20.6 million years volcanic tuff. Two orders, four families and five monotypic genera are present. The artiodactyls are the most diverse taxa with three species, including an endemic genus and species of kyptoceratine protoceratid closely related to a younger (Hemphillian) species from the Gulf Coastal Plain (Jiménez-Hidalgo, 2000; Jiménez-Hidalgo *et al.* 2002; Ferrusquía-Villafranca, 2003b).

On the other hand, two mainly marine mammal faunas, La Misión, Baja California and La Purísima, Baja California Sur, integrate the Barstovian record of northwestern Mexico (Figure 2). At La Misión the only recognized terrestrial mammal is an unidentified camel and at La Purísima the only mammal is one species of a medium-sized canid (Ferrusquía-Villafranca, 2003b).

The other two Barstovian faunas are in Oaxaca (Figures 2, 4). The Matatlán local fauna includes four orders, eight identified families (plus one or two unidentified), nine identified genera (plus one or two unidentified), and 10 species (again, plus one or two unidentified) of medium to large size mammals, such as a mustelid, camel or protoceratid. The only evidence of small-sized organisms are the rodent gnawing marks observed on some elements (Bravo-Cuevas, 2000; Jiménez-Hidalgo, 2000).

The other Barstovian fauna is Nejapa, where five orders, 10 families (plus one or two unidentified), 15 genera (plus one or two unidentified) and 17 species (plus one or two unidentified) are present (Bravo-Cuevas, 2000; Jiménez-Hidalgo, 2000).

The southernmost Barstovian fauna of the country is at Ixtapa in Chiapas (Figure 2). It includes two orders, three families, genera and species. There is a strong bias towards medium to very large size mammals, such as a horse, a rhino and a gomphothere (Ferrusquía-Villafranca, 1990; Jiménez-Hidalgo *et al.*, 2002).

Recently, a new species of protoceratine protoceratid from the Early Miocene of Simojovel, Chiapas was described, which apparently was an isolated occurrence (Webb *et al.* 2003).

4.2.3. The Late Miocene-Pliocene Record

Without doubt, the mammals from this time span are the best studied of all the Mexican Tertiary. The record consists of seven faunas and a single occurrence from the northwestern Mexico and nine faunas and three single occurrences from central Mexico (Figure 3).

These faunas have been dated as Late Miocene-earliest Pliocene (Hemphillian mammal age), Early Pliocene (Blancan mammal age) or both based on a combination of faunal composition, magnetostratigraphy and radiometric dates.

Northern part of the country. Two early Blancan faunas are known from Baja California Sur, northwestern Mexico (Fig. 3). Las Tunas local fauna comprises six orders, nine families and 11 monotypic genera. Only 27% of the recognized species are rodents and lagomorphs. The canids and camels are the most diverse groups with two species each (Miller, 1980; Miller and Carranza-Castañeda, 1984).

The Miraflores local fauna includes seven orders, eight families and eight monotypic genera. There are only two species of small size, a lagomorph and a rodent.

This fauna shares several taxa with the Las Tunas local fauna. A remarkable find is a ground sloth, which represents the first record of this South American mammal in the Baja California Peninsula (Carranza-Castañeda and Miller, 1999).

In Sonora, near San José de Pimas (Figure 3) there is a single occurrence of a gomphothere species of Blancan age (Lindsay, 1984; Miller and Carranza-Castañeda, 2002).

In Chihuahua there are three late Hemphillian and two early Blancan localities (Figure 3), which for a long time represented the best-known late Tertiary faunas of Mexico. None of them have any record of South American mammals.

The late Hemphillian Yepómera local fauna includes eight orders, 17 families, 30 genera and 32 species; of them, 46.8 % are small species represented by an insectivore, a bat, rodents and lagomorphs. The most diverse taxa are the equids and two rodent families (Heteromyidae and Cricetidae), with four species each (Lindsay, 1984; Lindsay and Jacobs, 1985).

The other two late Hemphillian faunas are Basuchil and Matachic (Figure 3). At Basuchil there are five orders and seven families, genera and species, of which only one is a small mammal. At Matachic there are four orders and families and seven monotypic genera; only two species are micro-mammals. In Matachic the most diverse taxa are the equids with three species (Lindsay, 1984). Matachic, Yepómera and Basuchil are geographically close; the first two share all the identified mammal taxa, and Basuchil shares some mammal taxa; possibly these faunas constitute three samples of a larger mammal assemblage.

The early Blancan Miñaca Fauna (Figure 3) includes six orders, 10 families and 12 monotypic genera; 16.6 % of species are of small size (Lindsay, 1984; Lindsay and Jacobs, 1985). The most diverse group is the Equidae with the two typical Blancan species: *Nannipus peninsulatus* and *Equus* (*Dolichohippus*) *simplicidens*. It is important to note the presence of a cervid in this fauna, because it is the only record of this group reported for the Tertiary of Mexico.

The other early Blancan local fauna of Chihuahua is Concha, where four orders, six families, nine genera and 11 species occur. Only one species is a large mammal, all others are rodents, insectivores and lagomorphs (Lindsay, 1984).

Central Mexico. Regarding the fossil record from central Mexico, there is a single occurrence of a Hemphillian equid species from Zacatecas (Figure 3) (Carranza-Castañeda and Miller, 1998).

In the state of Jalisco, three faunas are reported (Figure 3). Tecolotlán includes mammalian faunas of late Hemphillian and ?late Blancan age. The diversity of the Hemphillian taxa is represented by seven orders, 12 families and 16 monotypic genera; here are only two species of micro-mammals. The most diverse group is the Equidae with three species. Additionally, and there is a record of a ground sloth recovered from a dated volcanic ash of 4.89 million years (Miller and Carranza-Castañeda, 1998; Carranza-Castañeda and Miller, 2000, 2002a). The Tecolotlán Blancan fauna comprise four orders, five families and monotypic genera; two of which are South American immigrants (Carranza-Castañeda and Miller, 2002a).

In Colotlán (Figure 3) three Hemphillian equid species have been recorded (Carranza-Castañeda and Miller, 1998; Carranza-Castañeda *et al.* 1998). The late Hemphillian Teocaltiche local fauna includes four orders and families and five monotypic genera (Montellano-Ballesteros, 1997) of medium to large size. Of them, two are equids. This fauna share their taxa with Tecolotlán, therefore, the habitat for both probably was similar.

In the state of Michoacán two faunas are known, Cinqua of Hemphillian age and the early Blancan La Goleta (Figure 3). At Cinqua peccary species, two equid species and one antilocaprid species are recorded (Carranza-Castañeda *et al.,* 1998). The La Goleta local fauna consists of six orders, nine families and 10 monotypic genera, of which 22.2% are of small size (Miller and Carranza-Castañeda, 1984; Carranza-Castañeda and Miller, 1998). A form of a hyena is present in this fauna and along with the one from Miñaca (Chihuahua) represents the only record of a hyena taxon in Mexico.

In the state of Hidalgo three faunas and one single occurrence are known (Figure 3). In Tula one Hemphillian or Blancan equid species has been reported (Carranza-Castañeda *et al.* 1998). The Tehuichila local fauna has been known since the 1800's, being one of the first Tertiary faunas studied in Mexico. Recently, close to Tehuichila, in the same basin of Zacualtipán, another locality, named Zietla, of early Hemphillian age was discovered (Carranza-Castañeda, 1994). The Zietla-Tehuichila local fauna is early Hemphillian to Blancan in age. As a whole, it includes six orders, nine families and 13 monotypic genera, of which five are early Hemphillian, one is Blancan, and at least one is of late Hemphillian (Miller and Carranza-Castañeda, 1984; Carranza-Castañeda and Espinosa-Arrubarrena, 1994; Castillo-Cerón, 2000). All the recovered mammals are of medium to large size, the equids are the most diverse and there is a record of ground sloth that probably represents one of the oldest records for Mexico (Carranza-Castañeda and Miller, 2000).

Another fauna is La Plegaria, of late Hemphillian age (Figure 3), where six orders, 11 families and 16 monotypic genera are present; there are three small-sized species and a record of a ground sloth (Carranza-Castañeda *et al.* 1998; Padilla-Gutiérrez, 2004). Camels and equids are the most diverse taxa with three species each.

The last known locality from the state of Hidalgo, Las Arcinas, is of Blancan age; two equid, one tayassuid, one camelid and one gomphothere species are present (Carranza-Castañeda *et al.* 1998).

In the state of Querétaro the Hemphillian local fauna of Landa de Matamoros (Fig. 3) includes an equid, a camel and an antilocaprid species (Carranza-Castañeda *et al.* 1998).

In the state of Tlaxcala, there is the record of a gomphothere of Blancan age, which represents the first fossil vertebrate, reported for the late Tertiary of Mexico (Miller and Carranza-Castañeda, 2002).

The best-known fauna of late Tertiary age from central Mexico and probably for all the country, at least for macromammals, is the San Miguel de Allende local fauna in the state of Guanajuato. It includes a superimposed succession of three Tertiary North American land mammal ages: early Hemphillian, late Hemphillian and early Blancan (Carranza-Castañeda, 1989; Carranza-Castañeda and Walton, 1992; Carranza-Castañeda and Miller, 1998; Carranza-Castañeda and Miller, 2000; Miller and Carranza-Castañeda, 2002; Montellano-Ballesteros, 1989, among others).

The early Hemphillian taxa identified are two equids, one camel, and one antilocaprid species (Carranza-Castañeda and Miller, 1998; 2000; Jiménez-Hidalgo and Carranza-Castañeda, 2002; Jiménez-Hidalgo, 2004). The late Hemphillian fauna includes seven orders, 16 families, 29 genera and 34 species, of which 35.2 % are small-sized. The most diverse groups are the cricetid rodents with 10 species, and the camels and equids with four species each (Carranza-Castañeda, 1989; Carranza-Castañeda and Walton, 1992; Carranza-Castañeda and Miller, 1998; Carranza-Castañeda and Miller, 2000; Miller and Carranza-Castañeda, 2002; Montellano-Ballesteros, 1989; Jiménez-Hidalgo, 2004).

The early Blancan fauna from San Miguel de Allende consists of seven orders, 13 families, 22 genera and 24 species, of which rabbits and squirrels represent 25% of the fauna . The most diverse groups are camels with five species and rabbits with four (Miller and Carranza-Castañeda, 1982; Carranza-Castañeda and Miller, 2000; Jiménez-Hidalgo, 2004). Size range of the mammals varies from a squirrel to a gomphothere.

In San Miguel de Allende there are some Hemphillian and early Blancan taxa, such as felids, canids, camels and antilocaprids, that show up earlier or lasted longer than in the United States faunas (Carranza-Castañeda and Miller, 1996; Miller and Carranza-Castañeda, 1998; Jiménez-Hidalgo and Carranza-Castañeda, 2002; Jiménez-Hidalgo, 2004; Jiménez-Hidalgo *et al.* 2004).

Future regional work should consider these new temporal ranges in order to construct a detailed picture of the biochronology of North America.

4.3. Quaternary

In Mexico, although the pleistocenic sediments are widely distributed and the remains of fossil megafauna are usually found during agriculture activities or civil constructions, the taxonomic diversity, geographical and temporal fossil mammal distributions are not well understood. The Pleistocene is divided into two North American land mammal ages: Irvingtonian and Rancholabrean. Most Pleistocene Mexican localities are assigned to the later. There are only two local faunas that had been assigned to the Irvingtonian: Arroyo El Cedazo in Aguascalientes (Mooser, 1955; Mooser and Dalquest, 1975; Montellano-Ballesteros, 1992) and El Golfo in Sonora (Shaw, 1981). The reason of this situation might be the lack of radiometric dating methods, lack of prospecting and field work, or maybe there are no faunas of this age (Montellano-Ballesteros, 1991; Arroyo-Cabrales *et al.* 2002). The last alternative is the least probable. Therefore there is an important temporal gap in our understanding of the whole Pleistocene picture.

Arroyo-Cabrales *et al.* (2002), presented a database where information of localities, taxa, known geographic locations and age are included; it should be mentioned that the study is restricted to the last 120,000 years.

In their analysis, 13 orders, 44 families, 146 genera and 286 species are recorded. Of the total number of genera 27 are extinct, and eight are not present now in Mexico. Also, of the total number of species 85 are extinct and 12 are not present now in Mexico. The number of recent species doubles the Quaternary´s record. The difference strives in the number of small-sized mammals, in the recent fauna mammals are much more abundant; although this can be explained by the bias in the method of collecting. Another important difference is that in the recent faunas, large megaherbivores such as Xenarthra, Perissodactyla, Artiodactyla, and Proboscidea are not present.

5. Some Interesting Topics

Mexico is located between the temperate and tropical belts, and the contact or interfingering of the Neotropical and Nearctic biogeographic regions lies in its territory (Fa and Morales, 1998). Due to its geological history and geographic position, Mexico bears a great diversity of environmental conditions, that is reflected in the richness of its terrestrial vertebrates.

5.1. Mexico and the History of Some Taxa

The presence of certain species that yield important information for understanding the evolution of specific groups should be mentioned. Some examples are given below.

At the El Marfil local fauna the rodent *Marfilomys aewoodi* is very similar to the Eocene *Platypitamys* (Wood and Patterson, 1959; Patterson and Wood, 1982) from Patagonia, which is considered an ancestral form of the South American caviomorph rodents (Patterson and Wood, 1982). *Marfilomys* is then considered a survivor of the early Cenozoic protogomorph rodents that is morphologically close to the ancestral form of the caviomophs.

The presence of a new kyptoceratine protoceratid in the Hemingfordian of southeastern Mexico is very interesting, since at present it could be considered the sister taxon of *Kyptoceras*, the youngest-known protoceratid. Its presence in a tropical region agrees with Webb's assertion that probably a large part of the evolutionary history of the Kyptoceratini occurred in the subtropical savannas of Mesoamerica (Webb, 1981).

In the early Blancan faunas of central Mexico, some specimens show intermediate characters between the equids *Dinohippus mexicanus* (the ancestor of the genus *Equus*) and *Equus simplicidens* (the first true *Equus*), reflecting that probably the gradual or phyletic transition between the two species took place in this region of the country (Carranza-Castañeda and Miller, 1998; Miller and Carranza-Castañeda, 2002; MacFadden and Carranza-Castañeda, 2002).

5.2. A Paleosynecological Topic

This is a preliminary analysis of the feeding strategies of the mammals through the Cenozoic. The goal is to detect changes in the structure of the communities and appearances of new ways of feeding. One word of caution, this is a first attempt, using the information from some chosen local faunas which are considered moderately well studied, we only use macrofauna, and we are aware that according to several authors some of the species used are considered invalid species.

In the oldest Tertiary local fauna Lomas Tetas de Cabra of Wasatchian age, the faunal association suggests a deciduous forest (Ferrusquía-Villafranca *et al.* 2002), the herbivores are more diverse than the carnivores as expected, and no grazers are identified (Figure 5).

The paleoenvironment of the only Mexican Chadronian local fauna, Rancho Gaitán, has been interpreted as a subtropical forest (Ferrusquía-Villafranca *et al.* 2002). Here the carnivores are extremely scarce, also no grazers are identified and the artiodactyls, perissodactyls and, micromammals were browsers (Figure 6).

Matatlán and Nejapa are located in the southern part of the country and are the two best known Barstovian local faunas. The environment has been interpreted as a tropical woodland savanna (Jiménez-Hidalgo *et al.* 2002). The division of the feeding strategies are quite similar in both of them (Figures 7, 8). In addition, grazers are present. In these local faunas merychippines, hipparionines and equines are present, which suggests the presence of different equid lineages. The co-occurrence of all these species of equids together does not occur in Barstovian U.S. faunas. Carnivores are poorly represented.

The Late Miocene and Pliocene is the only time where it is possible to compare faunas from different latitudes in Mexico; here the local faunas from Chihuahua and central Mexico are comparable. In the northern local fauna of Yepómera (late Hemphillian), the carnivores are well represented, and grazing is the dominant feeding strategy among the herbivores (Figure 9). In comparison the fauna from central Mexico (here represented by the San Miguel local fauna) the carnivores are not as abundant as in Yepómera and the browsers are better represented (Figure 10). This difference might reflect a true difference in the structure of the savanna´s vegetation between northern and central Mexico.

It is interesting to note the rarity or absence of some browsing forms, such as cervids or dromomerycids. These taxa had a low diversity and were not as abundant as other artiodactyls in the late Hemphillian and early Blancan in the U. S. local faunas (Janis and Manning, 1998; Webb, 1998). Possible future paleontological work will disclose their presence in Mexican faunas.

Comparing the early Blancan local faunas Miñaca (Chihuahua) and San Miguel de Allende (central Mexico), the only difference is that in the lack of browser-like mixed feeders at Miñaca (Figures 11, 12).

Concerning the possible feeding strategies of the Pleistocene macromammals, carnivores constitute almost a quarter of the total species recorded from Mexico country (Figure 13). The herbivores main feeding preference is grazing, followed by browsing and mixed feeders. The proportions of the herbivore categories suggest that trees and shrubs continued being an important element of Mexican habitats during the Pleistocene.

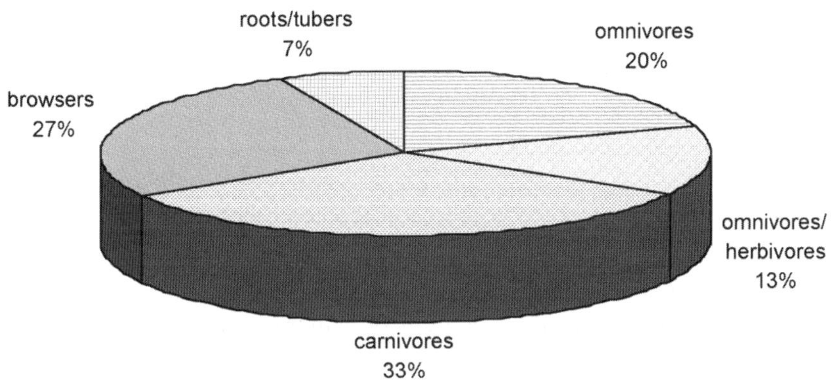

FIGURE 5. Probable feeding strategies of mammals from the Lomas Las Tetas de Cabra local fauna, Wasatchian, Baja California.

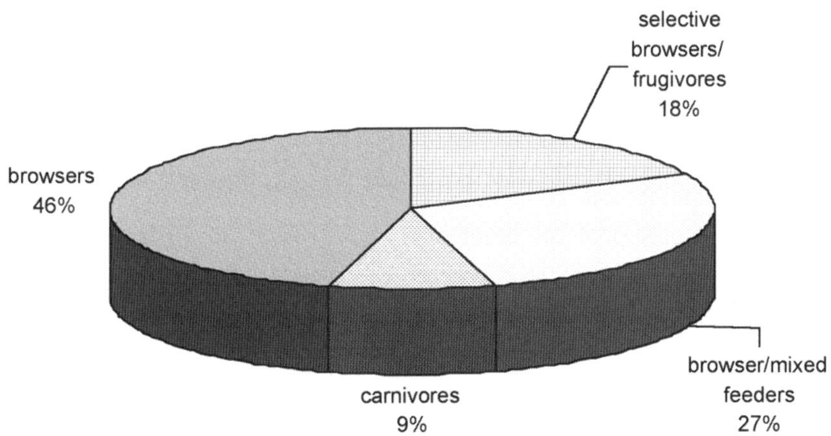

FIGURE 6. Probable feeding strategies of the mammals from the Rancho Gaitan local fauna, Chadronian, Chihuahua.

5.3. Mexico and the Great American Faunal Interchange

Because of its geographic position Mexico played an important role in the event of what has been called the "Great American Faunal Interchange" that took place during the late Tertiary. The study of the late Tertiary faunas in Mexico indicates the presence of several South American lineages at least since Hemphillian times.

Species of ground sloths, pampatheres, glyptodonts and capybaras, among others, have been identified from several localities. Strata of 4.8 million years (late Hemphillian) in Guanajuato has yielded two kinds of ground sloths; a capybara, and a

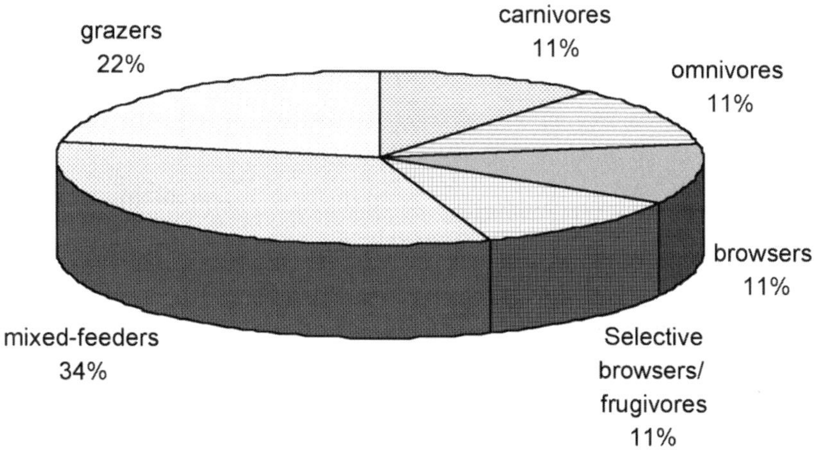

FIGURE 7. Probable feeding strategies of the mammals from the Matatlán local fauna, Barstovian, Oaxaca.

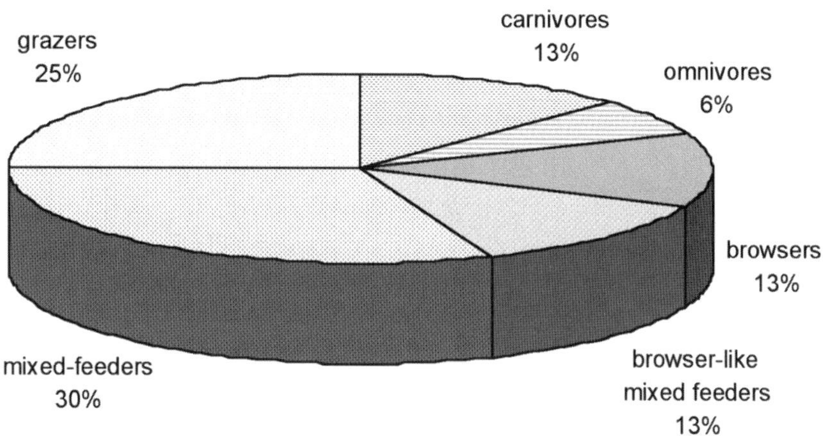

FIGURE 8. Probable feeding strategies of the mammals from the Nejapa local fauna, Barstovian, Oaxaca.

glyptodont have been recorded in beds of 3.9-3.3 million years (early Blancan). This is the oldest record for capybaras in North America. The presence of ground sloths in other Hemphillian localities (Zietla, and La Plegaria, in Hidalgo and Tecolotlán in Jalisco) suggests that these South American immigrants were quite successful, quickly reaching a wide distribution.

The Great American Faunal Interchange has been postulated to have occurred between 2.5 and 3.0 million years ago (Marshall, 1985). Discoveries in central Mexico strongly suggests that the migration of at least the South American forms towards the northern lands started earlier than previously considered (Miller and Carranza-Castañeda, 1999; Carranza-Castañeda and Miller, 2002a, 2002b).

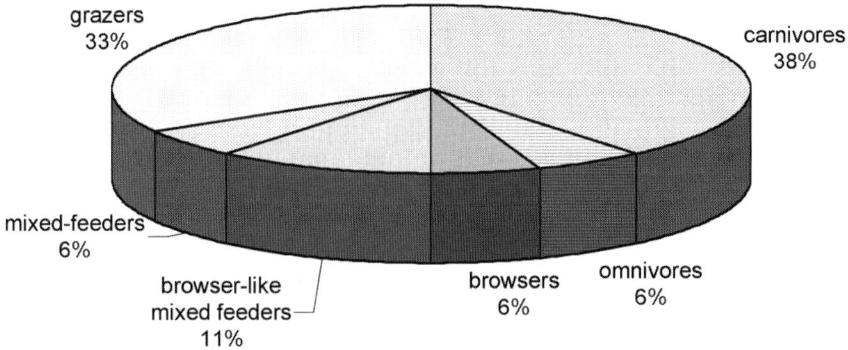

FIGURE 9. Probable feeding strategies of the mammals from the Yepómera local fauna, late Hemphillian, Chihuahua.

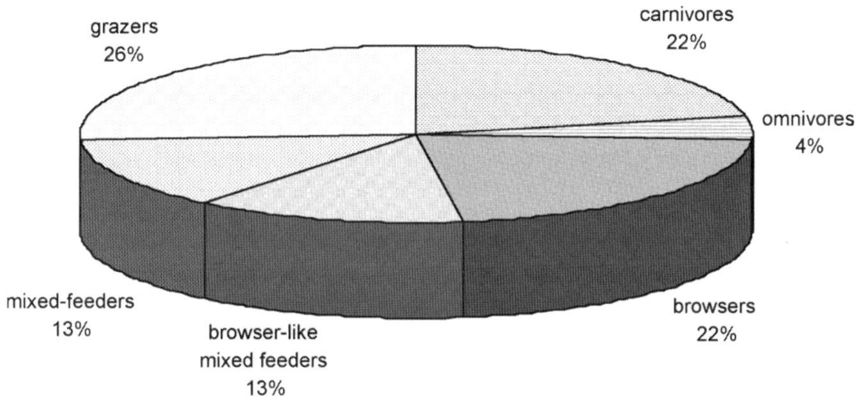

FIGURE 10. Probable feeding strategies of mammals from the late Hemphillian local faunas, San Miguel de Allende, Guanajuato.

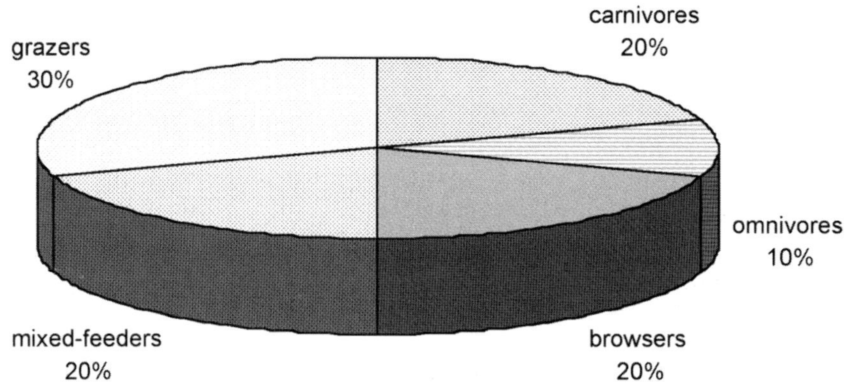

FIGURE 11. Probable feeding strategies of early Blancan mammals from the Miñaca Fauna of Chihuahua.

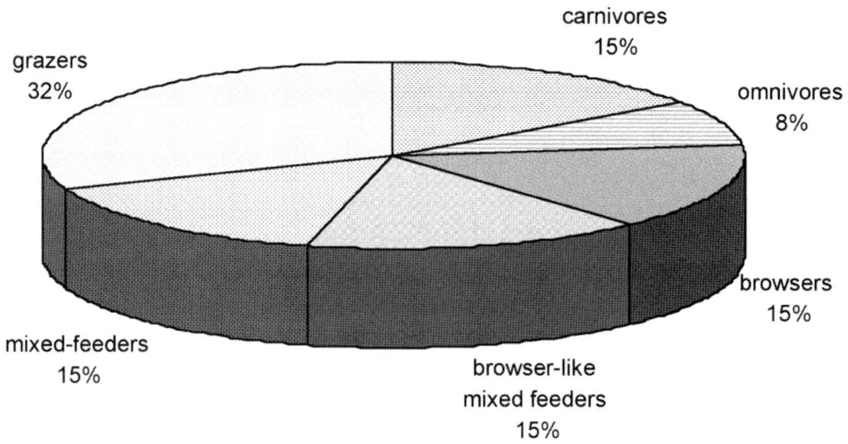

FIGURE 12. Probable feeding strategies of mammals from Las Golondrinas, late Hemphillian of Hidalgo.

With increasing field work in late Tertiary and Pleistocene deposits, additional remains of the South American immigrants: pampatheres, glyptodonts, ground sloths and capybaras are being discovered suggesting that the South American taxa were quite a successful group in Mexico.

5.4. Isotopic Studies

Stable carbon isotopes have proven been very useful in reconstructing dietary preferences of herbivore fossil mammals, thus they have been used in several

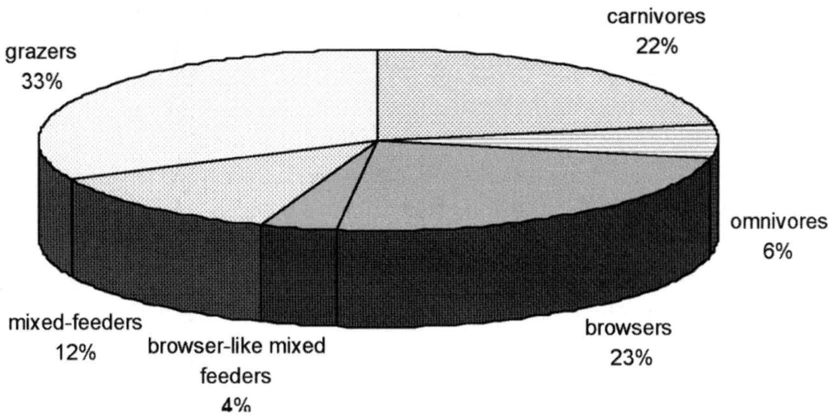

FIGURE 13. Probable feeding strategies of mammals from the San Miguel de Allende local fauna, late Hemphillian, Guanajuato.

paleoecologic studies of some Tertiary and Pleistocene faunas and specific taxa of North America (MacFadden and Cerling, 1996; Koch, 1998; Koch *et al.* 1998; Feranec, 2003; Feranec and MacFadden, 2000). The basic principle is that different photosynthetic pathways (C3, C4 and CAM) give different ratios of $^{13}C/^{12}C$ in different kinds of plants, which are reflected in the enamel of the teeth of the animals that ate them. Tropical warm-season sedges and grasses (C4) are enriched with ^{13}C; C3 plants including most browse and high-latitude grasses are enriched with ^{12}C and CAM plants incorporate intermediate ratios of both (O'Leary, 1988; Ehleringer *et al.* 1991; Ehleringer and Monson, 1993).

North American studies have been done with fossils that proceed from high and mid-latitudes, but to our knowledge none have been done with mammals from southern North America. So, what will be the results in applying the carbon isotopes techniques to the low-latitude Mexican taxa? Will the species maintain the same inferred dietary preference (browser, mixed-feeder or grazer) independently of the place where they lived or it will change according to latitude? Will dietary patterns of dietary change across time such as in some taxa of horses (browsing to grazing), and will the same timing be observed all along North America or will they could be decoupled? What can the isotopic signals tell us about the kinds of vegetation present in different geographic regions of Mexico? Future work will help answering these and other interesting questions.

5.5. Mexico During the Pleistocene

The Pleistocene epoch is characterized by rapid and repeated climatic fluctuations. Environmental changes have involved the increasing of ice sheets, the sizes and courses of rivers, as well as the development and drying of extensive lakes, as well as fluctuations in sea level were observed. All these climate changes affected the distribution, diversity and structure of North American mammal communities.

Finally, the end of the Pleistocene is marked by a significant extinction event that decimated the large mammal herbivores and carnivores of several continents (Graham, 1990).

Two models have been proposed to explain the way that the terrestrial biota of the non-glaciated areas of North America responded to the fluctuating environmental conditions. The Clementsian model proposes that large groups of species shifted as tightly linked and coevolved assemblages. The Gleasonian model suggests that individual species responded to these changes in accordance with their own tolerance limits. Graham, *et al.* (1996) suggested that the mammal communities in United States have responded in a Gleasonian manner.

The mammals response to Late Quaternary environmental fluctuations in Mexico is an enigma. Based upon endemic taxa, tree species richness, topography, climate, and geographic distribution of some animal groups such as butterflies, Toledo (1982) suggested the presence of five primary refuges located in the Lacandona and Soconusco regions within the southernmost part of Mexico and three secondary refuges in the state of Veracruz. This has been questioned by some studies on vertebrates (Flores-Villela, 1998).

The role that Mexico played during the Pleistocene is not well understood. It has been suggested that during the maximum extension of the ice sheet, the Sierras Madres and the Transmexican Volcanic Belt behaved as a route of dispersal of temperate species. Some northern species (mainly rodents) expanded their distribution towards the south. The recent relict distributions of *Glaucomys, Microtus* and *Sorex* are considered as proof of this migration during the Wisconsinian (Fa and Morales, 1998).

As far as we know no studies had been made analyzing how the distribution and structure of the communities of mammals changed during the Pleistocene.

The study of the Pleistocene faunas and understanding the response of terrestrial biotic communities to environmental changes is fundamental to modeling and planning for future climate change.

6. Final Remarks

Our knowledge of the mammalian fossil record has improved greatly during the twentieth century. Álvarez (1965) recorded 133 species included in 88 genera and belonging to 12 orders. All of his records are of Pliocene or Pleistocene in age. Twenty years later, Barrios-Rivera (1985) reported 267 species included in 178 genera, representing 19 orders, and 71 families, and ranges from the Late Cretaceous until Pleistocene. In the monograph compiled by Montellano-Ballesteros and Arroyo-Cabrales (2002) the mammalian record ranges from the Jurassic until the Pleistocene. The Paleogene is represented by 38 species and 37 genera included in

27 families and 12 orders. The Miocene record includes 50 species, 35 genera, 18 families and 7 orders; the Late Tertiary record includes 138 species, 59 genera, 18 families and 9 orders. And the Pleistocene record includes 286 species, 146 genera, 44 families and 13 orders. It should be noted that the diversity of the micromammals is underestimated, considering that in the extant mammalian faunas the most abundant groups are the rodents and the bats, which represent the 79% of the total extant fauna (Ramírez-Pulido and Müdespacher, 1987). This situation is mainly due to the method of collecting.

Although our knowledge of fossil mammals has improved, information on Mesozoic mammals is still poorly and is reduced to only two known localities. The Late Cretaceous taxa are also present in the northern local faunas such as in Montana and Wyoming, the exception is the new genus and species *Gallolestes pachymandibularis*, which recently was reported from Texas as noted above. With further fieldwork and prospecting the diversity will be increased and the relationships with the northern faunas would be clearer.

The Mexican Tertiary and Quaternary local faunas show strong similarity to those described in United States, suggesting that as a whole, the North American land mammal faunas constituted a large community with some local or endemic forms. Unfortunately, little is known from the southernmost tropical region of Mexico, where the environmental conditions should have been different from the northern latitudes and it is here where the majority of the endemic forms should be located.

Based on Figure 4 it is clear that most of the Cenozoic epochs are represented in the Mexican mammalian record, only the Paleocene is not recorded. The fossil mammalian record is filled with gaps.

Much work needs to be done in order to develop a detailed and deep understanding of the Mexican fossil mammals in all its different disciplines. It should be noted that extensive field and laboratory work in ongoing. Results promise to add useful information to fill the gaps of our knowledge.

References

Aguillón-Martínez, M. Zelenitsky, D., Brinkman, D., and Eberth, D., 2004, Eggshell fragments from the Uppermost Cerro del Pueblo Formation (Upper Cretaceous, México, *J. Vert. Paleontol.*, 24:33A.

Álvarez, T., 1965, Catálogo paleomastozoológico mexicano, *Ints. Nal. Antrop. His., Depto. Prehis. Publ.*, 17, 70 p.

Arroyo-Cabrales, J., Polaco, O., and Johnson, E., 2002, La mastofauna del cuaternario de México, in: *Avances en los estudios paleomastozoológicos* (M. Montellano-Ballesteros, and J. Arroyo-Cabrales coord.), *Inst. Nal. Antrop. Hist., Col. Cient.*, p. 103-123.

Barrios-Rivera, H., 1985, Estudio analítico del registro paleovertebradológico de México, Unpublished Professional thesis, Universidad Nacional Autónoma de México.

Bravo-Cuevas, V. M., 2000, Los équidos mesomiocénicos del Estado de Oaxaca: taxonomía e implicaciones paleobiológicas. Unpublished M.S. dissertation, Universidad Nacional Autónoma de México.

Burke, J. J., 1936, *Ardynomys* and *Desmatolagus* in the North American Oligocene, *Ann. Carnegie Mus.*, 25:134-54.

Carranza-Castañeda, O., 1989, Rinocerontes de la fauna del Rancho El Ocote, Mioceno tardío (Henfiliano tardío) de Guanajuato, México, *Univ. Nal. Autón. Méx., Rev. Inst. Geol.*, 8:88-99.

Carranza-Castañeda, O., 1994, Mastofauna del Mioceno Tardío de la Cuenca Carbonífera de Zacualtipán, Hidalgo, México, *Univ. Autón. Edo. Hidalgo, Rev. Inv. Pancromo,* 23:40-49.

Carranza-Castañeda, O., and Espinosa-Arrubarrena, L., 1994, Late Tertiary equids from the State of Hidalgo, México, *Univ. Nal. Autón. Méx., Inst. Geol., Rev.*, 11:182-192.

Carranza-Castañeda, O., and Miller, W. E., 1996, Hemphillian and Blancan age felids from central Mexico, *J. Paleont.*, 70:509-518.

Carranza-Castañeda, O., and Miller, W. E., 1998, Paleofaunas de vertebrados de las cuencas sedimentarias del Terciario Tardío de la Faja Volcánica Transmexicana, in: *Avances en Investigación, Paleontología de Vertebrados* (O. Carranza-Castañeda, and D. Córdoba-Méndez, eds.), *Univ. Autón. Edo. Hidalgo, Publ. Espec.*, p. 85-95.

Carranza-Castañeda, O., and Miller, W. E., 1999, A new Pliocene terrestrial vertebrate locality from southern Baja California, *J. Vert. Paleont.*, 19:36A

Caranza-Castañeda, O., and Miller, W. E., 2000, Selected late Cenozoic vertebrate localities in the states of Hidalgo and Guanajuato, Mexico, in: *Guide Book of the Field Trips, Society of Vertebrate Paleontology, 60th Annual Meeting, Mexico City,* (O. Carranza-Castañeda, ed.), *Avances en Investigación, Univ. Autón. Edo. Hidalgo, Publ. Esp.*, p. 1-48.

Carranza-Castañeda, O., and Miller, W. E., 2002a, Paleontology and stratigraphy of the Tecolotlán Basin, Jalisco, Mexico, *J. Vert. Paleont.*, 22:41A.

Carranza-Castañeda, O., and Miller, W. E., 2002b, Inmigrantes sudamericanos en las faunas del Terciario tardío del centro de México, in: *Avances en los estudios paleomastozoológicos* (M. Montellano-Ballesteros and J. Arroyo-Cabrales, coord.), *Instit. Nal. Antropol. Hist., Col. Cient.*, p. 69-81.

Carranza-Castañeda, O, Miller, W. E., and Kowallis, B. J., 1998, New vertebrate faunas from the Transmexican Volcanic Belt, Central Mexico, *J. Vert. Paleont.*, 18:31A.

Carranza-Castañeda, O., and Walton, A., 1992, Cricetid rodents from the Rancho el Ocote Fauna, late Hemphillian (Miocene), Guanajuato, México, *Univ. Nal. Autón. Méx., Rev. Inst. Geol.*, 10:71-93.

Castillo-Cerón, J., 2000, Fossil vertebrates from the Miocene of Hidalgo, Mexico, *J. Vert. Paleont.*, 20:34A.

Clemens, W. A., 1980, *Gallolestes pachymandibularis* (Theria: *incertae sedis*; Mammalia) from the Late Cretaceous deposits in Baja California del Norte, Mexico, *PaleoBios*, 33:1-10.

Dalquest, W. W., and Mooser, O. B., 1974, Miocene vertebrates from Aguascalientes, central Mexico, *Texas Mem. Mus., Pearce-Sellards Series,* 21:1-10.

Ehleringer, J. R., and Monson, R. K., 1993, Evolutionary and ecological aspects of photosynthetic path variation, *Ann. Rev. Ecol. Syst.,* 24:411-439.

Ehleringer, J. R, Sage, R. F., Flanagan. L. B., and Pearcy, R. W., 1991, Climate change and evolution of C4 photosynthesis, *Trends Ecol. Evol.*, 6:95-99.

Fa, J. E., and Morales, L. M., 1998, Patrones de diversidad de mamíferos de México, in: *Diversidad biológica de México: orígenes y distribución* (T. P. Ramamoorthy, R. Bye, A. Lot, and J. Fa, comp.), Univ. Nal. Autón. Méx., Inst. Biol., p. 315-352.

Feranec, R. S., 2003, Stable isotopes, hypsodonty, and the paleodiet of *Hemiauchenia* (Mammalia: Camelidae): a morphological specialization creating ecological generalization, *Paleobiology*, 29:230-242.

Feranec, R., and MacFadden, B. J., 2000, Evolution of the grazing niche in the Pleistocene mammals from Florida: evidence from stable isotopes, *Palaeogeogr. Palaeoclimatol. Palaeoecol.*, 162:155-169.

Ferrusquía-Villafranca, I., 1969, Rancho Gaitán local fauna, Early Chadronian, northeastern Chihuahua, *Soc. Geol. Mex., Bol.*, 30:99-138.

Ferrusquía-Villafranca, I., 1989, A new rodent genus from Central Mexico and its bearing on the origin of the Caviomorpha, in: *Papers on fossil rodents in Honor of Albert Elmer Wood* (C.C., Black and M. R. Dawson eds.), *Nat. Hist. Mus., Los Angeles County, Science Ser.* 3, p. 91-117.

Ferrusquía-Villafranca, I., 1990, Biostratigraphy of the Mexican continental Miocene: Part I, Introduction and the Northwestern and central faunas; Part II, The Southeastern (Oaxacan) faunas; Part III, The southernmost (Chiapas) fauna and concluding remarks on the discussed vertebrate record, *Univ. Nal. Autón. México, Inst. Geol., Paleont. Mex.*, 56, 149 p.

Ferrusquía-Villafranca, I., 2003a, The first Paleogene mammal record of middle-America: Helohyidae new gen. and sp., *J. Vert. Paleont.*, 23:49A.

Ferrusquía-Villafranca, I., 2003b, Mexico's Middle Miocene mammalian assemblages: an overview, *Am. Mus. Nat. Hist., Bull.*, 13:321-347.

Ferrusquía-Villafranca, I., Galindo-Hernández, C., and Barrios-Rivera, H., 1997, Los mamíferos oligocénicos de México; revisión y adición a la Fauna Local Rancho Gaitán, Formación Prietos, Chadroniano de Chihuahua nororiental, in: *Homenaje al Profesor Ticul Álvarez* (J., Arroyo-Cabrales, and O. J. Polaco, coord.), *Inst. Nal. Antropol. Hist., Col. Cient.*, p. 97-134.

Ferrusquía-Villafranca, I., Jiménez-Hidalgo, E., Ortiz-Mendieta, J. A., and Bravo-Cuevas, V. M., 2002, El registro paleogénico de mamíferos de México y su significación geológica-paleontológica, in: *Avances en los estudios paleomastozoológicos* (M., Montellano-Ballesteros and J. Arroyo-Cabrales, coord.), *Inst. Nal. Antropol. Hist., Col. Cient.*, p. 25-45.

Ferrusquía-Villafranca, I., and Wood A. E., 1969, New fossil rodents from the Early Oligocene Rancho Gaitán local fauna, northeastern Chihuahua, Mexico, *Texas Mem. Mus., Pearce-Sellards Ser.*, 16:1-13.

Flores-Villela, O., 1998, Herpetofauna de México:distribución y endemismo, in: *Diversidad biológica de México: orígenes y distribución* (T. P. Ramamoorthy, R. Bye, A. Lot, and J. Fa, comp.), *Univ. Nal. Autón. Méx., Inst. Biol.*, p. 251-278.

Graham, R. W., 1990, Evolution of new ecosystems at the end of the Pleistocene, in: The Mammoth site of Hot Springs, South Dakota (L. Agendbroad, J. I. Mead, and L. W. Nelson, eds.), Inc. Sci. Pap., 1:54-60.

Graham, R. W., Lundelius, E. L. Jr,, Graham, M. A., Schroeder, E. K., Toomey, R. S. III, Anderson, E., Barnosky, A. D., Burns, J. A., Churcher, Ch., Grayson, D. K., Guthire, R. D., Harrington, C. R., Jefferson, G. T., Martin, L. D., McDonald, H. G., Morlan, R. E., Semken, H. A. Jr., Webb, D. S., Werdelin, L., and Wilson, M. C., 1996, Spatial response of mammals to Late Quaternary environmental fluctuations, *Science*, 272:1601-1606.

Janis Ch. M., and Manning, E., 1998, Dromomerycidae, in: *Evolution of Tertiary Mammals of North America. Volume. 1, Carnivores, ungulates and ungulate-like mammals* (Ch. M. Janis, K. M. Scott and L. L. Jacobs, eds.), Cambridge University Press, New York. p. 477-490.

Jiménez-Hidalgo, E., 2000, Las mastofaunas mesomiocénicas del Sureste de México y su significación paleobiológic, Unpublished M.S. dissertation, Universidad Nacional Autónoma de México.

Jiménez-Hidalgo, E., 2004, Bioestratigrafía de los artiodáctilos (Tylopoda y Ruminantia) de las faunas del Terciario tardío de la Faja Volcánica Transmexicana, Unpublished Ph.D. dissertation, Universidad Nacional Autónoma de México.

Jiménez-Hidalgo, E., and Carranza-Castañeda, O., 2002, Bioestratigrafía de los artiodáctilos de las faunas del Terciario tardío del área de San Miguel de Allende, Guanajuato, *XVI Convención Geológica Nacional, Soc. Geol. Mex., Abstracts*, p. 307.

Jiménez-Hidalgo, E., Carranza-Castañeda, O., and Montellano-Ballesteros, M., 2004, A Pliocene record of *Capromeryx* (Mammalia: Antilocapridae) in Mexico, *J. Paleont.*, 78:1179-1186.

Jiménez-Hidalgo, E., Ferrusquía-Villafranca, I., and Bravo-Cuevas, V. M., 2002, El registro mastofaunístico miocénico en México y sus implicaciones geológico-paleontológicas, in: *Avances en los estudios paleomastozoológicos* (M. Montellano-Ballesteros and J. Arroyo-Cabrales, coord.), *Inst. Nal. Antropol. Hist., Col. Cient.*, p.47-68.

Koch, P. L., 1998, Isotopic reconstruction of past continental environments, *Ann. Rev. Earth Plan. Sci.*, 26:573-613.

Koch, P. L., Hope, K. A., and Webb, S. D., 1998, The isotopic ecology of late Pleistocene mammals in North America, Part 1, *Florida Chem., Geol.*, 152:119-138.

Korth, W. W., 1981, New Oligocene rodents from western North America, *Carnegie Mus. Nat. Hist., Ann.*, 50:289-318.

Lillegraven, J. A., 1972, Preliminary Report on Late Cretaceous mammals from the El Gallo formation, Baja California Norte, Mexico, *Contrib. Sci. Nat. Hist. Mus., Los Angeles County,* 232:1-11.

Lillegraven, J. A., 1976, A new genus of therian mammals from the Late Cretaceous, El Gallo Formation Baja California, Mexico, *J. Paleont.*, 50:437-444.

Lindsay, E. H., 1984, Late Cenozoic mammals from Northwestern Mexico, *J. Vert. Paleontol.*, 4:208-215.

Lindsay, E. H., and Jacobs, L. L., 1985, Pliocene small mammal fossils from Chihuahua, Mexico, *Univ. Nal. Autón. Méx., Inst. Geol., Paleont. Mex.*, 51:1-53.

MacFadden, B., and Carranza-Castañeda, O., 2002, Cranium of *Dinohippus mexicanus* (Mammalia: Equidae) from the early Pliocene of central Mexico, and the origin of *Equus*, *Bull. Florida Mus. Nat. Hist.*, 43:163-185.

MacFadden, B. J., and Cerling, T., 1996, Mammalian herbivore communities, ancient feeding ecology, and carbon isotopes: a 10 million-year sequence from the Neogene of Florida, *J. Vert. Paleont.*, 16:103-115.

Maldonado-Koerdell, M., 1948, Los vertebrados fósiles del Cuaternario de México, *Rev. Soc. Mex. Hist. Nat.*, 9:1-35.

Marshall, L. G., 1985, Geochronology and Land Mammal biochronology of the Transamerican faunal interchange, in: *The Great American Biotic Interchange* (F. G. Stehli, and S. D. Webb, eds.), Plenum Press, N.Y., p. 49-85.

Miller, W. E., 1980, The late Pliocene Las Tunas Local Fauna from southernmost Baja California, Mexico, *J. Paleont.*, 54:762-805.

Miller, W. E., and Carranza-Castañeda, O., 1982, New lagomorphs from the Pliocene of central Mexico, *J. Vert. Paleont.*, 2:95-107.

Miller, W. E., and Carranza-Castañeda, O., 1984, Late Cenozoic mammals from central Mexico, *J. Vert. Paleont.*, 4:216-236.

Miller, W. E., and Carranza-Castañeda, O., 1998, Late Tertiary canids from Central Mexico, *J. Paleont.*, 72:546-556.

Miller, W. E., and Carranza-Castañeda, O., 1999, Early South American immigrants in central Mexico, and times of their appearances, *J. Vert. Paleont.*, 19:64A.

Miller, W. E., and Carranza-Castañeda, O., 2002, Importante of Mexico´s late Tertiary mammalian faunas, in: *Avances en los estudios paleomastozoológicos* (M. Montellano-Ballesteros and J. Arroyo-Cabrales, coord.), *Inst. Nal. Antropol. Hist., Col. Cient.*, p. 83-101.

Montellano-Ballesteros, M. 1989 Pliocene Camelidae of Rancho El Ocote, central México, *J. Mamm.*, 70:359-369.

Montellano, M., 1991, Panorama de las faunas pleistocénicas de México, *Convención sobre la Evolución Geológica de México, Soc. Geol. Mex.*, Memorias, p. 123-125.

Montellano-Ballesteros, M., 1992, Una edad del irvingtoniano al rancholabreano para la fauna Cedazo del estado de Aguascalientes, *Univ. Nal. Autón. Méx., Inst. Geol., Rev.*, 9:195-203.

Montellano-Ballesteros, M., 1997, New vertebrate locality of late Hemphillian age in Teocaltiche, Jalisco. Mexico, *Rev. Mex. Cienc. Geol.*, 14:84-90.

Montellano-Ballesteros, M., 1999, Mexico and Central America, in: *Enclyclopedia of Paleontology. Vol. 2.* Ed. Ronald Singer. Fitzroy Dearborn Publishers, p. 728-734.

Montellano-Ballesteros, M., Hopson, J. A., Clark, J. M., Fastovsky, D. E., and Hernández-Rivera, H., 1995, Mammals from the middle Jurassic of Huizachal Canyon, Tamaulipas, Mexico, *J. Vert. Paleont.*, 15:45A.

Montellano-Ballesteros, M., and J. Arroyo-Cabrales, 2002, *Avances en los estudios paleomastozoológicos, Inst. Nal. Antropol. Hist., Col. Cient.*, 248 p.

Mooser, O., 1955, La fauna "Cedazo" del Pleistoceno en Aguascalientes, *Univ. Nal. Autón. Méx., An. Inst. Biol.*, 29:409-452.

Mooser, O., and Dalquest, W. W., 1975, Pleistocene mammals from Aguascalientes, Central México, *J. Mamm.*, 56:781-820.

Novaceck, M. J., Ferrusquía-Villafranca, I., Flynn, J. J., Wyss, A. R., and Norell, M., 1991, Wasatchian (early Eocene) mammals and other vertebrates from Baja California, Mexico: the Lomas Las Tetas de Cabra Fauna, *Bull. Am. Mus. Nat. Hist.*, 208:1-88.

O'Leary, M. H., 1988, Carbon isotopes in photosynthesis, *Bio-Science*, 38:328-336.

Padilla-Gutiérrez, J. M., 2004, Bioestratigrafía del área de las Golondrinas, Terciario Tardío de Tepeji del Río, Estado de Hidalgo, México. Unpublished M.S. dissertation, Universidad Nacional Autónoma de México.

Patterson, B., and Wood, E. A., 1982, Rodents from the Deseadan Oligocene of Bolivia and the relationships of the Caviomorpha, *Univ. Harvard, Mus. Comp. Zool., Bull.*, 149:371-543.

Ramírez-Pulido, J., and Müdespacher, C., 1987, Estado actual y perspectivas del conocimiento de los mamíferos de México, *Ciencia*, 38:49-67.

Rowe, T., Cifelli, R., Lehman, Th., and Weil, A., 1992, The Campanian Terlingua local fauna, with a summary of other vertebrates from the Aguja Formation, Trans Pecos Texas, *J. Vert. Paleont.*, 12:472-493.

Shaw, A. Ch., 1981, The middle Pleistocene El Golfo local fauna from northwestern Sonora, Mexico, Unpublished M.S. thesis, California State University.

Silva-Bárcena, A., 1969, Localidades de vertebrados fósiles en la República Mexicana. *UNAM, Instituto de Geología, Univ. Nal. Autón. Méx., Inst. Geol., Paleont. Mex.*, 28:1-34.

Stevens, S. M., 1977, Further study of Castolon local fauna (Early Miocene), Big Bend National Park, Texas, *Texas Mem. Mus., Pearce-Sellards Series,* 28:1-69.

Toledo, V. M., 1982, Pleistocene changes of vegetation in Tropical Mexico, in: *Biological diversification* in the tropics (G. T. Prance, ed.), Columbia University Press, p. 93-112

Webb, S. D., 1981, *Kyptoceras amatorum,* new genus and species from the Pliocene of Florida, the last protoceratid artiodactyls, *J. Vert. Paleont.*, 1:357-365.

Webb, S. D., 1998, Cervidae and Bovidae, in: *Evolution of Tertiary Mammals of North America. Volume. 1, Carnivores, ungulates and ungulate-like mammals* (Ch. M. Janis, K. M. Scott and L. L. Jacobs, eds.), Cambridge University Press, New York, p. 508-510.

Webb, S. D., Beatty, B. L., and Poinar, G. Jr., 2003, New evidence of Miocene Protoceratidae including a new species from Chiapas, Mexico, *Bull. Am. Mus. Nat. Hist.*, 279:348-367.

Wood, A. E., and Patterson B., 1959, The rodents of the Deseadan Oligocene of Patagonia and the beginnings of South American rodent evolution, *Univ. Harvard, Mus. Comp. Zool., Bull.*, 120:279-428.

Chapter 13

Revision of Recent Investigations in the Tlayúa Quarry

SHELTON P. APPLEGATE[1], LUIS ESPINOSA-ARRUBARRENA[1], JESÚS ALVARADO-ORTEGA[1], and MOULOUD BENAMMI[2].

[1] • Instituto de Geología, Universidad Nacional Autónoma de México, Ciudad Universitaria, 04510, D. F., México.
[2] • Instituto de Geofísica, Universidad Nacional Autónoma de México, Ciudad Universitaria, 04510, D. F., México.

F. Vega et al. (eds.), Studies on Mexican Paleontology, 275–304.

1. Abstract

The Tlayúa Quarry in Tepexi de Rodríguez, Puebla, México, is one of the most important localities in the New World because of its uniquely rich biota. For the last twenty years, its exceptionally well-preserved fossil content and its taphonomical and paleoecological implications have caught the attention of many geologists and paleontologists since this Lagerstätte represents an important new asset in Mexican Paleontology. In this paper, we review all the recent investigations related to a development of the paleoecological models to explain the extraordinary conservation of the rich and diverse fossil assemblage of Tlayúa.

2. Introduction

The Tlayúa project has yielded so far more than 6000 specimens many considered world class fossils, including the preservation of soft parts in fishes (Alvarado-Ortega and Applegate, 1999; Alvarado-Ortega and Espinosa-Arrubarrena, 2004). Field work, conservation, and the investigation of the fossil materials have been largely carried out by the Geological Institute of the Universidad Nacional Autónoma de México (UNAM). The Tlayúa Collection is housed at the Paleontological Museum of UNAM (Colección Nacional de Paleontología) with very few (and controlled) exceptions.

The main effort of the Tlayúa investigations has been focused on the characterization of this Albian biota. Many of the Tlayúa specimens represent new taxa, nearly 95% of these fossils are identified at the supraspecific level. To date, among Masters and Ph.D. theses, and through collaboration with foreign Paleontologists only 5% of this unique assemblage has been described. The taxonomical groups present in Tlayúa have been reviewed by Espinosa-Arrubarrena and Applegate (1996) and more recently by Kashiyama *et al.* (2004), including information derived from new species of fishes, reptiles, arthropods, ammonites, belemnites, and several known species of pollen and benthonic foraminifera. Nevertheless, we feel that the current literature does not give a sufficient picture of what has been collected, including the yet undescribed biota. As a partial remedy, herein we have assembled the whole 6000+ paleobiota (see Appendix).

Besides the taxonomical aspects of Tlayúa that still are in a beginning stage, it has also been an important goal for the Geological Institute to elucidate the possible paleobiological conditions the Quarry represents, and through the study of the complementary geological, sedimentological and paleontological information, explain the paleoambiental and post-depositional conditions that produced this extraordinary fossil deposit. There are several paleoecological models to explain the deposit of the laminated limestones of Tlayúa and its extraordinary fossils (Applegate, 1987; Pantoja-Alor, 1992; Espinosa-Arrubarrena and Applegate, 1996; and Kashiyama *et al.* 2004). Nevertheless, these investigations assess some contrasting ideas that include: 1) tidal-influenced shallow water environment with dissecation periods; 2) a comparison with the Upper Jurassic Solnhofen Lagerstätte; and 3) more recently, Tlayúa has been considered as an open marine basin with storms dominated sedimentation and bottom waters with restricted circulation.

In this paper, we attempt to update the results of our investigations at the Tlayúa Quarry, following the results of two areas:

a) New magnetostratigraphical data to refine the previous Albian age assignment that recently has been challenged, placing the Tlayúa Quarry 10 million years older within the Aptian.

b) Paleontological evidence that will add in the understanding of the nature of the pre-and post depositional conditions (e. g. the presence of algal mats and the nature of the hematitic layers present in the bedding planes of limestones).

With all this new and some already published elements we attempt to re-enforce our shallow water back-reef lagoonal interpretation, as opposed to an open marine, deep-water basin.

3. Geological Aspects

3.1. Location of the Tlayúa Quarry

The Tlayúa Quarry is located in the southern part of the Mexican State of Puebla, approximately 200 km southeast of Mexico City (Figure 1). The deposit consists of a series of commercially active limestone quarries in the Tlayúa ravine (Barranca Tlayúa), near the town of Tepexi de Rodríguez. This is a site that for the last 50 years has been exploited for building facings and flagstones. The large number (over 6000) of macrofossils collected and the superb preservation of the paleontological materials make this locality one of the most important areas for research in the New World.

3.2. Regional Geology

The Tlayúa Formation was described by Pantoja-Alor (1992), and Applegate (1992) discussed additional geological settings. The Tepexi de Rodríguez area lies within a region of Paleozoic (Cambrian-Ordovician) metamorphic rocks of the Acatlán Complex. During Jurassic times, this complex represented a positive land surface (Ortega-Guerrero, 1989), a condition that could have persisted until the Early Cretaceous (encompassing the time for the deposition for the Tlayúa beds). Since Pantoja-Alor (1992) described the Tlayúa Formation (the fossil bearing unit of the Tlayúa Quarry) we have detected, in the surrounding region a number of pre-Tlayúa units that warrant investigation and characterization.

The top of the Acatlán Complex is eroded. Topping this unit there are sandstone beds without fossils, and above these layers, just under the Tlayúa Formation, there is a kilometer thick sequence of Upper Jurassic or Lower Cretaceous limestones.

Resting above the Tlayúa beds there is an unknown thickness of Tertiary rocks

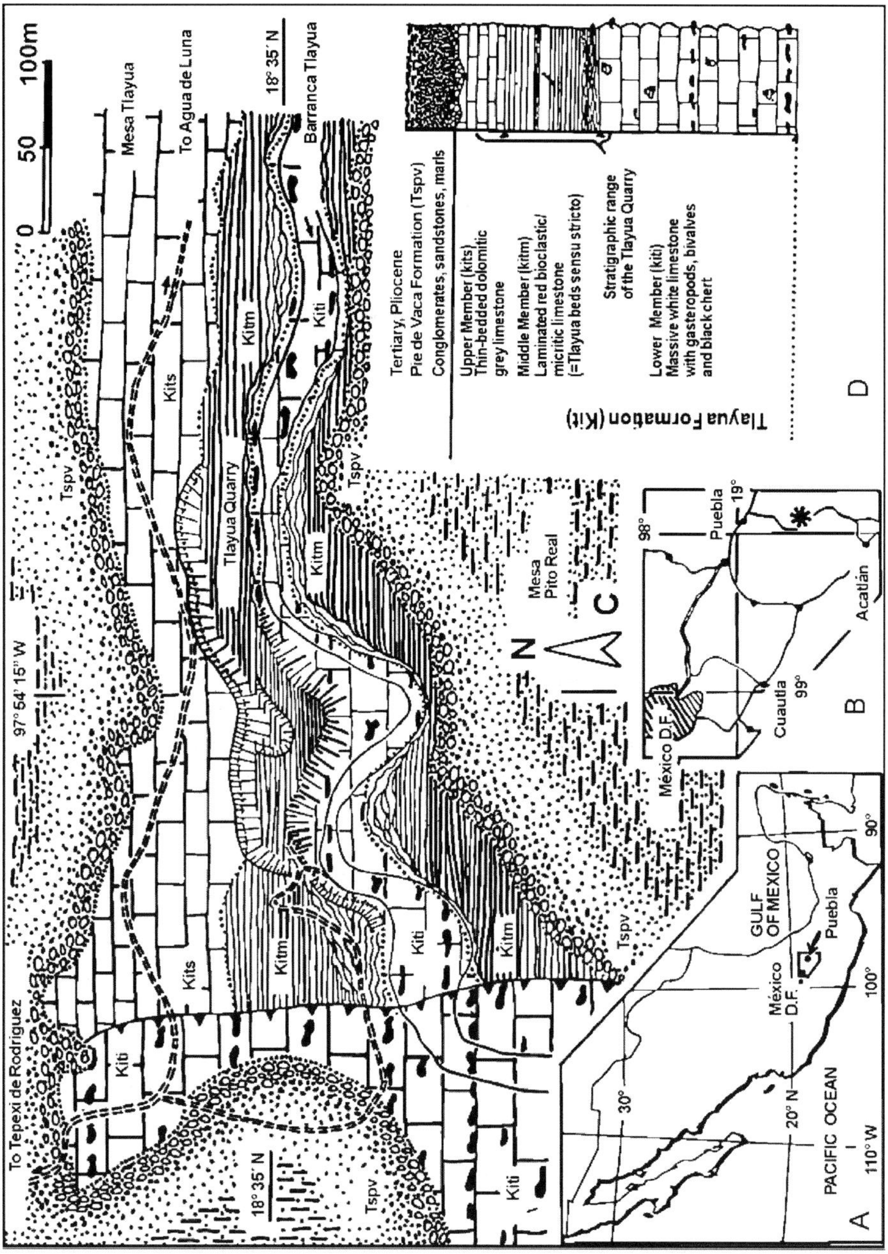

FIGURE 1. A) Location and geology of Tlayúa Quarry (after Seibertz and Spaeth, 2002). B). Tlayúa Quarry indicated by asterisk between the cities of Puebla and Acatlán, Puebla. C) Geologic sketch of the immediate surrounding of the Tlayúa Quarry in the upper Barranca Tlayúa. D) Generalized lithostratigraphic section of the Barranca Tlayúa.

of the Pie de Vaca and Agua de Luna Formations. From this last unit come the large travertine blocks that represent one of the most important economical sources of the Tepexi Region.

Pantoja-Alor (1992), proposed the name Tlayúa Formation for the Albian carbonates that appear in the Tlayúa ravine, and subdivided them into three members (Figure 1). The base of the Tlayúa Formation is not exposed and the top is truncated and covered by the Pie de Vaca (Pliocene) and Agua de Luna (Pleistocene) formations.

3.3. The Tlayúa Quarry (Local Geology)

According to Pantoja-Alor (1992), the Lower Member consists of at least 50 m. of bioturbated, tightly folded bluish gray micritic limestones. Throughout this unit there are numerous beds with miliolids and it is caped by an undescribed mollusk fauna of very small bivalves and gastropods (Espinosa-Arrubarrena and Applegate, 1996). The presence of the rudist *Toucasia polygyra* and the bivalve *Chondrodonta* characterize this member. The biostratigraphy of these mollusks sets the Lower Member of the Tlayúa Formation within the Lower Albian (Alencáster, 1973).

The Middle Member of the Tlayúa Formation corresponds to what in this chapter will be called "the Tlayúa Quarry". It is the most important unit due to its numerous and diverse fossil assemblage. The sequence is about 35 m. thick and consists of multiple laminae of extremely fine-grained yellow-brown micritic litographic limestones. The bedding planes show very distinctive reddish-yellow to purplish-red hematitic layers, which bear a large array of exceptionally well-preserved and articulated fossils. Horizontally, this limestone exposure measures 700 m. and it is bounded to the west, by an over thrust and to the east, it is covered by vegetation. Although the outcrop seems too small, our recent investigations have yield new promising areas with outcrops of Tlayúa limestones, which are in the process of being mapped and published.

Finally, the Upper Member of this formation, consist of an incomplete sequence of gray dolomites, and its age is considered as Cenomanian due to the presence of the miliolid species *Dicyclina schlumbergeri* (Fernández-Becerra in Pantoja-Alor, 1992).

3.4. Age Assignments (Biostratigraphy and Magnetostratigraphy)

Benammi *et al.* (2003, 2004) reported a magnetostratigraphycal analysis to determine the paleomagnetic signal in a vertical section of the Tlayúa Quarry, including 95 samples from 31 horizons (Figure 2).

In this locality, the limestones are weakly magnetized and exhibit in general two component magnetizations. The magnetic properties and characteristic remanence are dominated by soft and hard coercivity magnetic minerals, most of which probably formed early in the depositional and diagenetic history. The section displays a mean

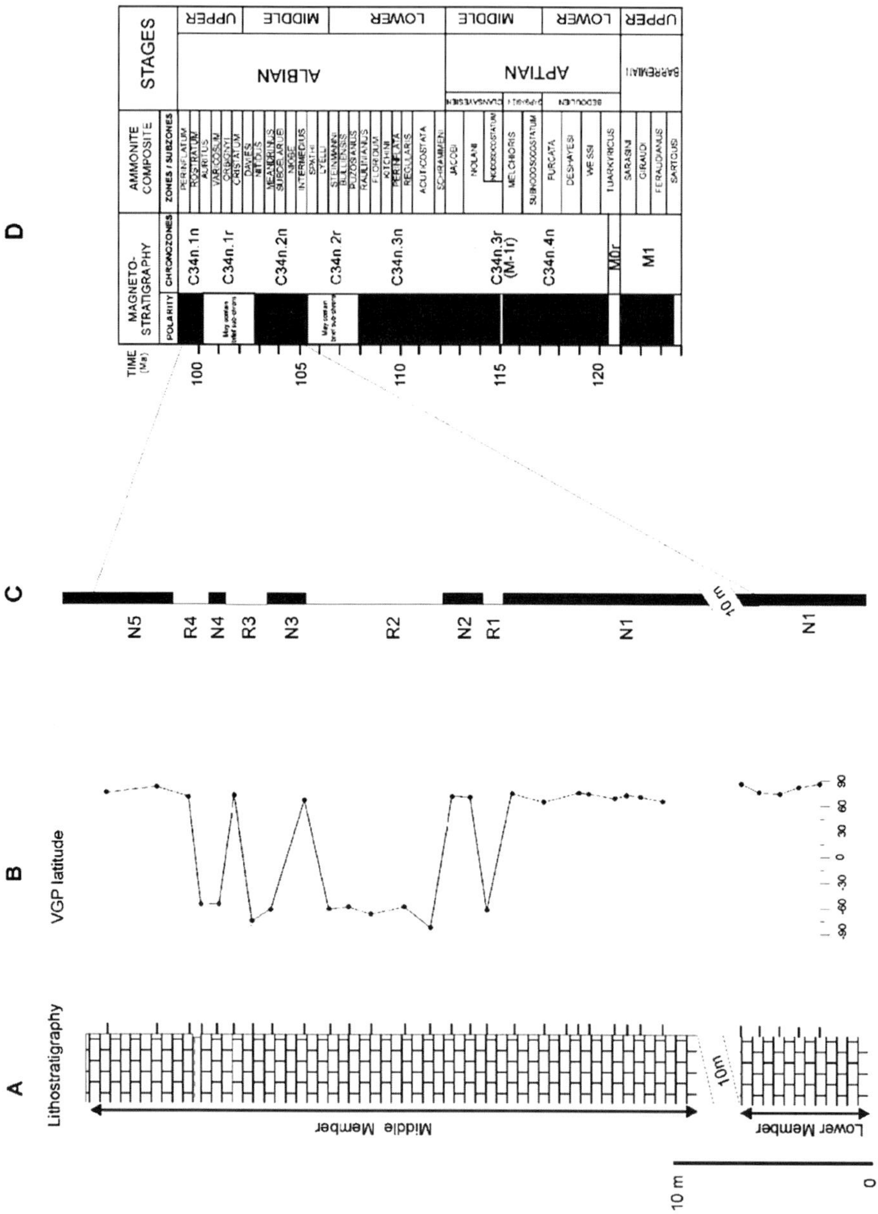

FIGURE 2. Magnetostratigraphy of the Tlayúa Quarry. Variation of the geomagnetic paleomagnetic signal (B) trough out the Middle Member and part of the Lower Member of the Tlayúa Formation (A). Correlation of the magneto zones -traced from B- of the Tlayúa Quarry (C) within the geomagnetic polarity time scale (GPTS) of the Lower Cretaceous (D). GPTS after Gradstein et al. (1995).

directions: D = 344.9°, I = 32.4°, k= 21, _95= 4.2°; and D = 149.4°, I = -36.6°, k= 17, _95= 8.7°.

Our analyzed section of the Tlayúa Quarry provides reverse and normal polarities respectively; supporting a primary origin of the ChRM, following the reversal test (McFadden and McElhinny, 1990) the result was positive. However, a fold test was not possible as bedding is uniform at the local section. The two directions are not antipodal probably due to the strong overprint of the present-day field, which may not have been removed completely. To limit the biasing effect due to the present-day overprint, we referred all directions to common (lower) hemisphere and calculated an overall mean of Dec = 341°; Inc = 35.4°; k= 35; _95 = 4.3°. In addition, we compared our directions with the reference directions for stable North America derived from the APW path of Besse and Courtillot (2002).

Finally, we estimated the magnetic direction for the Tlayúa Formation, 110 Ma ago, as really close to that expected in a stable North America (Dec= 342.4°, Inc= 40.8°), with a slightly lower inclination. The inclination value is lower than expected at the site latitude for the middle Cretaceous magnetic field. The depositional geometry of the Tlayúa beds may be produced by the difference between the found and expected inclinations. This is typical of detrital remnant (DRM). Paleomagnetic data therefore suggests no significant relative movements of the Tlayúa Formation in respect to stable North America during the middle Cretaceous. Urrutia-Fucugauchi (1988), based on paleomagnetic data from the Morelos Formation, also suggested this tectonic stability between southern Mexico and the North America.

Since the discovery of the Tlayúa Quarry, paleontologists have attempted to assign a date to this locality. Kashiyama *et al.* (2004:155) recognized the following benthic foraminifera: *Spiroloculina cretacea, Trochamminoides coronus, Derventina filipescui, Sabaudia minuta, Rumanoloculina robusta, Valvulineria* sp., *Earlandia*? *conradi*, and probably *Praechrysalidina infracretacea* and *Nautiloculina* sp.. These authors proposed a lower Aptian age for this association, which is typical of lower Aptian inner shelf sediments in the Apennines as well as the West Alps-Vercors, and Chartreuse regions. However, one sees that some species have large stratigraphic ranges that exceed the Aptian. For example, *Praechrysalidina infracretacea* is known from the lower Aptian to the end of the Albian, *Sabaudia minuta* stretches as far as middle Albian, while *Spiroloculina cretacea* is known only from lower Cenomanian (Chiocchini *et al.*, 1984). From these remarks, the age of this association could be younger than the age proposed by these authors.

In North America, ammonite ranges of high resolution are available for the Cenomanian through lower Maastrichtian interval (Cobban *et al.* 1994). The lower Cretaceous is subdivided into 66 ammonite zones and subzones (Gradstein *et al.* 1995), from the upper Valanginian to the Albian. Cantu-Chapa (1987) recognized the genus *Hysteroceras, Mortoniceras* and cf. *Anisoceras*, which represent a typical late Albian ammonite assemblage of the European zone, being similar to the *Mortoniceras inflatum* zone at Mont Risou south-east France, proposed as the stratotype for the

Albian-Cenomanian boundary (Gale *et al.*, 1996). In conclusion, a Late Albian age can be assigned undoubtedly to the Tlayúa Quarry.

In the late Albian, the subzones of *Hysteroceras orbignyi* and *Hysteroceras varicosum* are capable of geographically very widespread recognition outside of the European province (Wiedmann and Owen, 2001). According to our polarity chron interpretation (Figure 2), *Mortoniceras* and *Hysteroceras* occur at the polarity chron C34n.1r (magnetozones R1-R4 correlated with chron C34n.1r); in fact, Gradstein *et al.* (1995) noted the presence of a short interval of normal polarity in this chron. This interpretation appears consistent with a proposal by Gradstein *et al.* (1995), who placed the *Mortoniceras inflatum* Zone of the European province in the late Albian. This polarity magnetozone sequence correlates with C34n.1n–C34n.2n chrons, with an age of 100 to 105 Ma for the Tlayúa Formation, having a sedimentation rate of about 0.6 cm/ka.

4. Sedimentological and Taphonomical Aspects

4.1. The Nature of the Sediments of Tlayúa

According to Pantoja-Alor (1992), the Middle Member of the Tlayúa Formation corresponds to a sequence of thinly laminated limestones (mudstones), formed by micrite and to a lesser degree microesparite (Malpica-Cruz *et al.*, 1989). Other interpretations (Applegate, 1987 and Espinosa-Arrubarrena and Applegate, 1996) regard a close comparison of these sediments and those described from Solnhofen: a) slabs of almost pure micrite (a tan or honey-colored aspect) related to Barthel *et al.* (1990) "*flinz*"; and b) the fissile, shaly layers that form the stratification planes that were analogized with the Solnhofen´s "*fäule*" (Barthel *et al.* (1990).

More recently, Kashiyama *et al.* (2004) through microfacies analysis by optical petrography and statistical facies-transition analysis, report six primary textures, termed microfacies, that consist of:

1) Bioclastic grainstone, consisting of abundant, poorly sorted foraminifera and micritic peloids.
2) Peloidal grainstone, consisting of very well sorted micritic peloids.
3) Wackstone/packstone microfacies, consisting of well-sorted micritic peloids in a micritic matrix; co-occurs with rare foraminifera tests.
4) Textured mudstone.
5) Massive mudstone
6) Finely laminated mudstone. The reconstructed microfacies succession indicated a recurring upward-fining sequences.

As the previous authors, Kashiyama *et al.* (2004), recognized the nature of Tlayúa sediments as largely composed of laminated micrites; nevertheless, they furthermore, describe a more complex amalgam of primary and diagenetically altered sedimentological fabrics. More important, they discuss the rusty colored bedding planes (that can vary to a reddish yellow, or even to purplish red tones), stained by hematite as "largely secondary and mobilized through textural discontinuities during diagenesis". These results contrast with the authigenic nature of the hematite that was reported by Espinosa-Arrubarrena and Applegate (1996), regarding the evidence that the crevasses and folding within the fossils are filled with what at the time of deposition could have been a "hematitic soup" that, when settled in-filled any holes or depressions where some decay may have occurred. Also, within the sequence of laminations studied in Tlayúa by Kashiyama *et al.* (2004), they interpreted the presence of "thin reddish clayey laminae as numerous serrate stylolites", structures previously regarded by Seibertz and Buitrón (1987) as cyanobacterial mats (Figure 3). Before and since the publication of Kashiyama *et al.* (2004) we have observed little evidence of stylolites in these strata, which do not replace those superficial structures that we and other authors (i.e. Seibertz and Spaeth, 2002) have considered as cyanobacterial mats.

After the challenge made by Kashiyama *et al.* (2004) of the bimodal Solnhofen´s *flinz* and *fäule* paradigm (Espinosa-Arrubarrena and Applegate, 1996), our present observations tend to focus in at least four types of events (or phenomena) that have to be considered to understand the sediments at Tlayúa. Two of them are derived and strongly controlled by storms that could correspond to the double-monsoon influences reported by Kashiyama *et al.* (2004). These sediments involve: a) the authigenic hematite (*sensu* Espinosa-Arrubarrena and Applegate, 1996), present in all the bedding planes of the Middle Member section and that at all times cover the mega fossils, which, as stated above are profusely infiltrated with the hematite grains and other clay minerals; and b) the other type of "storm controlled" sediments related to cyanobacterial mats. In this case the mats acted as sedimentary traps, consisting nearly of 40% hematite (and other clay minerals) and 60% CaC03. It is interesting to note the strong relationship found between many megafossils (mostly fishes) and the cyanobacterial fabric in which they are imbedded.

The other two elements to be considered correspond to sediments that were not related to storms but are: a) honey colored layers that bare a diverse planktonic biota; and b) thicker layers of almost pure (over 95%) CaC03 with no apparent fossil evidence (maybe due to recrystallization) dubbed "lithographic limestones". Interestingly, since 1998 the Mexican Art Community, in producing high-quality lithographs has used these slabs successfully (Vila, 1999). The characteristics of the lithographs from the Tlayúa slabs have been compared with the ones obtained from the limestones at Solnhofen by German artists and scientists, with favorable results (Gunther Viohl, former director of the Jura Museum at Eichtätt, Germany, per. com. 2003). It is important to note the similitude between these two lithologies, because for the last 200 years, high quality lithographs have been only obtained from the Solenshofen´s limestones.

Present investigations will reveal to what extent these four sedimentological

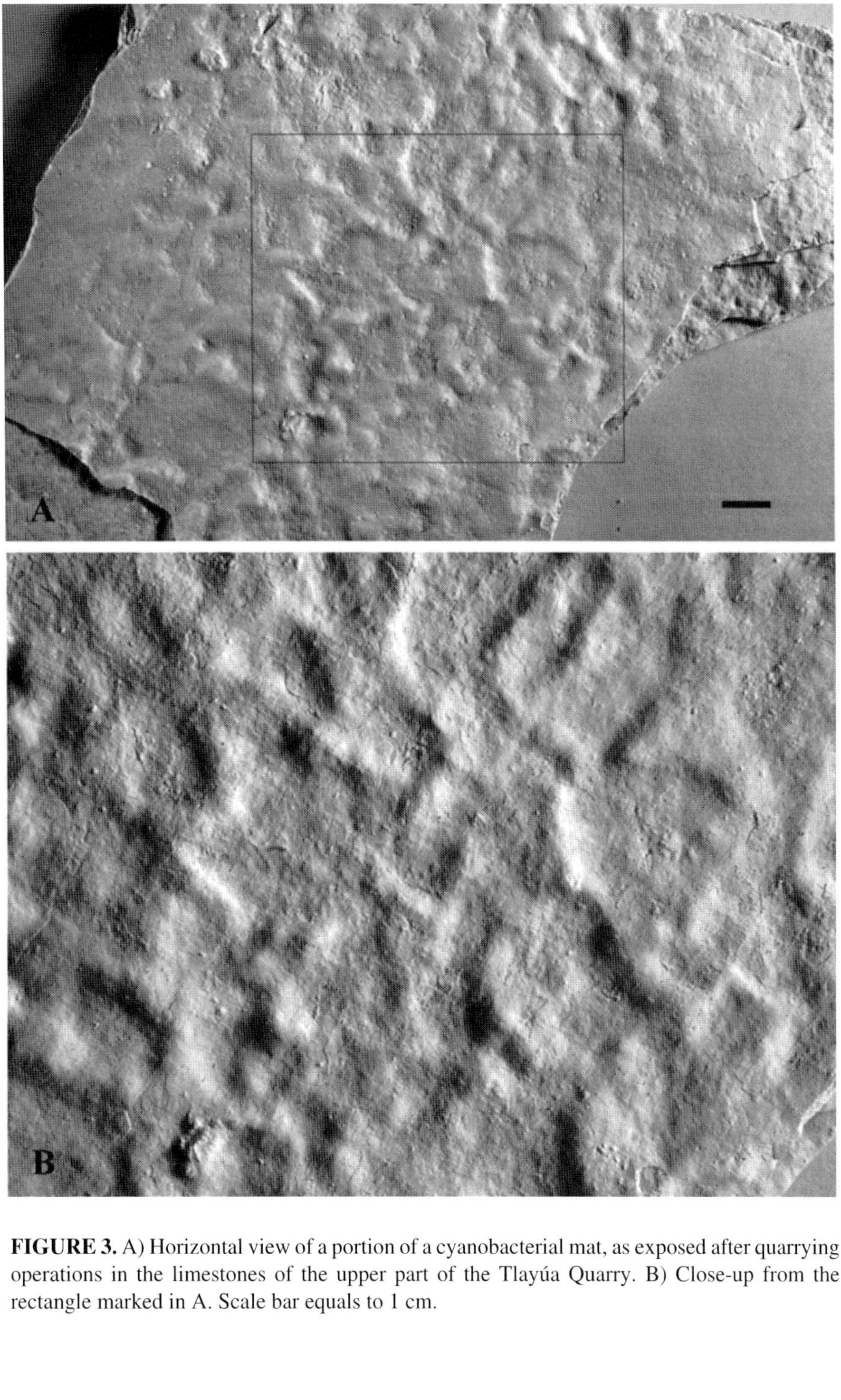

FIGURE 3. A) Horizontal view of a portion of a cyanobacterial mat, as exposed after quarrying operations in the limestones of the upper part of the Tlayúa Quarry. B) Close-up from the rectangle marked in A. Scale bar equals to 1 cm.

types correspond to Kashiyama *et al.* (2004) six microfacies and sedimentological fabrics.

4.2. Cyanobacterial Mats, Hypersalinity and Anoxicity

Within the middle and upper sections of the Tlayúa beds (Middle Member of the Tlayúa Formation), numerous undulating beds are present. These structures with an averaged thickness ranging between 0.5 and 2 cm. have been identified as cyanobacterial mats (Seibertz and Buitron, 1987; Espinosa-Arrubarrena and Applegate, 1996; Seibertz and Spaeth, 2002). Recently, the presence of these microbial mats has received a widely differing interpretation. Kashiyama and colaborators (2004) regarded these structures as hematite-stained serrate stylolites, precluding all previous interpretations of cyanobacterial mats.

The controversy between the occurrence of these biogenic structures or their absolutely absence in the Tlayúa beds becomes of major importance when the depth of the water, in which deposit was formed, is considered. According to our recent observations, the middle and upper Tlayúa beds bear a number of examples that support the idea that such structures are actually present.

In Tlayúa, cyanobacterial mats are expressed in patches, covering areas no larger than 4 square meters (Figure 3). Many of the Tlayúa fossils have been found imbedded within the undulations of these algal-mat structures. Figure 4 shows a *Tepexichthys aranguthyorum* specimen in which the caudal fin is firmly adhered to the mat, while the rest of the body has moved. This cohesive texture, produced by the limy nature of the microbial mats, has also been documented in fossils from Solnhofen (Viohl, 1990). In fact, this enveloping of the fishes and other organisms, encapsulating them in a film of calcium carbonate, could be considered one of the major factors in understanding the exceptional preservation of this site

The most convincing evidence of the presence of cyanobacteria in Tlayúa comes from samples taken in cross sections and plain views from the upper part of the Middle Member. Figure 5A shows the typical undulations of the mats. Whereas Figure 5B, taken in a plane view at the same level, shows some microbial filaments found within the Tlayúa layers. Bernier and collaborators (1991, fig. 2) have documented similar evidence in the micritic Kimmeridgian limestones of Cerin, France.

The importance of detecting the presence of cyanobacterial mats, at least in the middle and upper part of the Tlayúa sections, rests on the known fact that these organisms show very set limits when considering the water depth in which they live. Cyanobacteria reflect the blue-green wavelengths of the light and use the red for photosynthesis. As the red is a long-wave end of the visible light spectrum, this color is the first to be absorbed when passing through the water column. Accordingly, in clear tropical water the intensity of red light would be too low for efficient photosynthesis below 60 meters (Barthel *et al.* 1990). Therefore, the presence of cyanobacterial mats represents a strong argument in favor Tlayúa being a shallow water deposit.

One significant element to consider in Tlayúa is the lack of bioturbation by macrobenthonic organisms, one of the reasons why the deposit is so finely laminated, as is in the case of Solnhofen (Barthel *et al.* 1990). The best explanation for this is an excessive concentration of salt in the water, known as hypersalinity. Although so far only demonstrated indirectly, the expression of this toxic condition is present in the *post-mortem* contortions found in many of the fossils recovered; in a reaction due to the osmotic withdraw of the water from their tissues, in what is known as "the brine effect" (Barthel *et al.* 1990; Viohl, 1990). Among these *post-mortem* features in Tlayúa we can mention: 1) The unnatural curling of the arms of serpent stars (see the chapter "Fossil Ophiuroids from Mexico" in this book); and 2) The conspicuous contraction of the back muscles in many fish larvae and in some of the slender-bodied adult fishes (i.e. elongated ichthyodectids, *Belenostomus*, clupeomorph larvae, etc.). Viohl (1990), reported this necrolitic feature in the fossil bearing beds of Solnhofen, illustrating a *Tharsis dobius* specimen strongly bent by dehydration from a brine type environment.

Although, the instances discovered at Tlayúa strongly support the occurrence

FIGURE 4. *Tepexichthys aranguthyorum* Applegate, 1992, imbedded in a cyanobacterial mat.

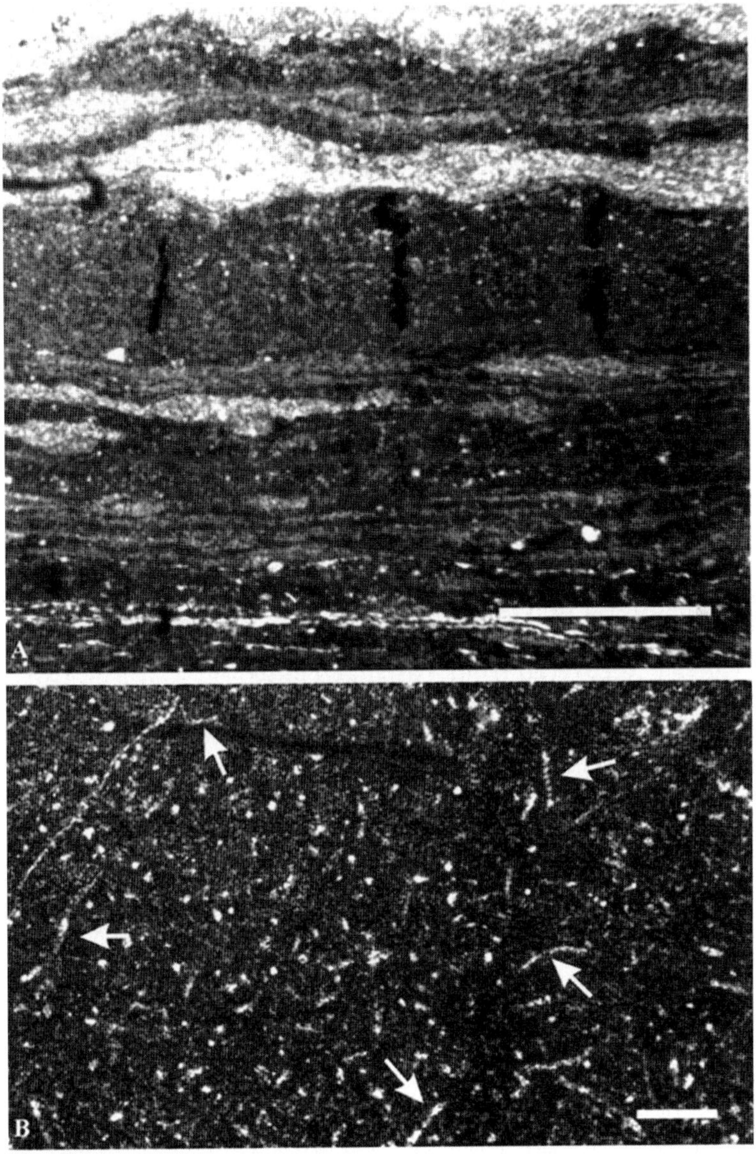

FIGURE 5. Microscopic view of cyanobacterial mats from the upper part of the Tlayúa Quarry. A) Cross section of a series of cyanobaterial mats sowing the typical undulations in the two upper layers, as well as the large amount of trapped materials within the calcareous matrix (scale bar equals to 1 cm.). B) Horizontal (plane view) section of the same sample shown in A, in which numerous cyanobaterial filaments (pointed by arrows) can be seen (scale bar equals to 20 mm.).

of a hypersaline environment along the section, this scenario should be analyzed in the context that no evaporitic minerals, such as gypsum and or pseudomorph salts, have been found so far. Interestingly, this has been the same condition reported from Solnhofen by Barthel and collaborators (1990). According to this point of view, the waters at Solnhofen (as in Tlayúa) never reached the high salinity needed to start the precipitation of evaporitic minerals, due to a constant influx of normal marine water. In Tlayúa, besides the influx of less saline conditions, there was probably an extra input of fresh or brackish water affecting the depositional environment.

The idea regarding the presence of anoxic conditions in the Middle Member of the Tlayúa Formations must be considered. This interpretation is based on the occurrence of little or no sedimentary structures attributable to traction transport (mainly wave action) in the Tlayúa beds. This condition is obtained whether in an open marine basin (*sensu* Kashiyama *et al.* 2004) or in a closed and protected (with no strong wave action, except during monsoonal storms) shallow basin (*sensu* Espinosa-Arrubarrena and Applegate, 1996). In this quiet "stagnant" environment oxygen concentration was depleted to a point where the bottom waters turned completely anoxic. Particularly, as evaporation increased, salinity significantly rose to a level in which the water column was not able to hold any more oxygen in solution.

5. The Paleobiota of Tlayúa

5.1. Evidence Recovered from Vascular Plants

The fossil plants at Tlayúa are quite rare and still in need of study. The first report of a terrestrial (vascular) plant was made by Martill (1989), where he mentions the presence of *Frenolepsis* that for the time being will be considered as cf. *Frenolepsis* (or a very close taxon). Other conifers identified in Tlayúa are cf. *Araucaria* and *Brachyphyllum*, as well as the Bennettitales *Zamites*, and many small fragments of fossil wood.

Contrary to other gymnosperms, *Brachyphyllum* lacked a well-developed woody "skeleton" and therefore is regarded as a weak-stemmed, shrubby plant, similar to what today is found on the salty soils of deserts or extremely arid areas, where halophytic vegetation is prevalent (Barthel *et al.* 1990). The fact that *Brachyphyllum* only lignified the central part of its branches and stem may explain why although fossil wood has been collected in Tlayúa, it always appears as small brittle pieces. The other genera of vascular plants identified, do not seem to contradict the salty-arid or (semi arid) paleoenvironment that has been described for *Brachyphyllum*, particularly *Frenolepsis*, which is also commonly reported from "stressed" environments. It is important to note that to our knowledge, no mesophytic vegetation has yet been correctly identified in Tlayúa (e.g. ferns, horsetails, club mosses, etc), and therefore the land adjacent or near the Tlayúa deposit could be compared to what today are the coasts of Arabia.

5.2. Sessile Organisms (Invertebrates and Algae) Transported by Storms

An important wealth of organisms that normally are found firmly attached to the substrate in which they live, by means of a "root" system or holdfast, are often found in the Tlayúa Quarry lacking their anchoring mechanisms. These include: at least two different forms of sponges, housed in the fossil collection by more than 10 individuals; more than 60 gorgonian corals (representing at least to two forms); and several stem crinoids that, as the groups above mentioned, lack their holdfast. This condition occurs when these organisms were ripped off from their substrate by a strong storm. It should be noted that some thalloid algae (possibly brown), have been discovered in the paleobiota. These organisms also seem to have lost their holdfasts and therefore we include them in this "taphonomical" category.

5.3. Mobile organisms "Dragged" to the Deposit by Storms, or that Entered Floating after Death

This is probably the category that includes the largest (and most diverse) array of invertebrates that have been found in Tlayúa so far. Within this category are: annelid worms, bryozoa?, starfishes, sea urchins, serpent stars, crabs, and isopods (there is some unexplored evidence that the isopods could be living in the deposit as fish scavengers). Evidently, due to their small size and strong morphological structure, all of these organisms could be easily dragged by strong storms without losing significant parts of their anatomy, or causing any conspicuous marks on the bottom of the deposit. Nevertheless, we also have delicate organisms that filled with gas could have floated (not dragged) into the fossil beds. Among these, we find comparatively poorly preserved ammonites, in all cases missing the aptychi. In something that contrasts with what has been reported in Solnhofen, where a large number of ammonites are preserved with the aptychi inside the body chamber (Barthel *et al.* 1990). The guards of belemnites represent another good example of cephalopods that could have entered the deposit lifted by decomposition gases (see Seibertz and Spaeth, 2002). Some echinoderms such as the holothurians (seven or eight forms recognized and in the process of being described) possibly were filled with gases as they were carried into Tlayúa by waves and water turbulences caused by strong storms. The paleobiota also includes multiple examples (mostly unidentified) of small-sized clams, oysters and gastropods. These mollusks may have arrived attached to gorgonians or algae that have been found in Tlayúa. One algae type is similar to *Sargassum* and the other could resemble "kelp". Although the "algal" remains are rare (in most of the cases they were not preserved), the mentioned mollusks are commonly found in clusters, formed by six or seven specimens, that were buried following that position after the algae decayed.

It is important to note the presence of an odonate dragonfly nymph, suggesting the "fresh water" nature of at least some of the beds of the Tlayúa Quarry. And finally, the presence of a spider and a small diptera (Tupulidae) "may fly" is interpreted as "wind blown".

5.4. Evidence recovered from the Fish Fauna

As has been frequently reported the Tlayúa Quarry represents a fossil fish locality (Martill, 1989; Pantoja-Alor, 1992; Applegate, 1996; Applegate *et al.* 2000; and Alvarado-Ortega *et al.* in this book; among others), therefore, the paleoecological evidence derived from this group deserves special attention.

With more than 5000 fossil fishes collected to date, there are examples of five "biotic realms" that have been categorized as follows:

Cosmopolitan-Coastal

Fishes commonly reported along all the coastal shallow waters of the Tethyan realm (in the old and new world). Within this category we have recovered *Lepidotes*, *Vinctifer*, and *Belenostomus*. Also, several elopomorphs from Tlayúa can be included in this coastal category. One *Brannerion*-like form, *Megalops* sp., *Notelops*, and *Rhacolephis*-like form are species that although reported from coastal shallow water deposits from many places of the world (similar to present day *Elops* as a typical coastal form), their strong body structure does not discard the possibility of finding them in deeper oceanic fossil fish localities.

Reef Dwellers

Reef dwellers are an important group of near-shore fishes associated with coral reefs. Within this category pycnodonts have received more scientific attention than any other of the fossils present in Tlayúa. After the clupeomorphs, the pycnodonts are in Tlayúa by far the most diverse group in the paleobiota. In 1992, Applegate described *Tepexichthys aranguthyorum* and since, seven genera with at least ten new species have been recognized and today are in the process of being described. Although there are reports of freshwater pycnodonts, in Tlayúa there is evidence of their close association with coral reefs. The first author, described several anatomical elements found in pycnodonts that relate them ecologically with the present-day brilliantly colored parrotfish, a common dweller of coral reefs (e.g. body shape and structure of the teeth). In addition, Applegate (1992) analyzed pycnodontid gut contents and coprolite masses, finding white powder that, when viewed under a microscope, resulted in pure white angular calcite grains that were thought to represent the crushed coral that was passing through the gut. In Tlayúa, besides the pycnodonts, macrosemiid fishes have also been recognized. This group is well represented in the fish fauna, and recently *Macrosemiocotzus americanus* was described by González-Rodríguez *et al.* (2004). This is one species of at least four other new forms that are in process of being studied. Macrosemiids are usually found associated with reefs within the Tethyan realm and they also were reported from low-energy lagoonal environments (Bartram, 1977).

Finally, according to Applegate *et al.* (2000) and Alvarado-Ortega and Espinosa-Arrubarrena (2001), two forms of Ionocospiformes have been recognized (one is *Teoichthys kallistos* and the other is a new genus in process of being described), and due to their crushing teeth, these fishes were adapted to feed on invertebrates protected with shells or hardened structures. In addition, they have been interpreted as reef dwellers.

Freshwater Provenience

Even though it is difficult to prove, there are several yet undescribed forms in the Tlayúa fish fauna that are thought, through comparisons with their counterparts in other described fossil localities, to be freshwater forms. This category probably comprises several of the many clupeomorph fishes, the most abundant group in Tlayúa (e.g. species of *Clupea*-like forms, *Ellimmichthys*, and other Clupeiformes). In addition, the presence of the amiid *Pachyamia mexicana* can be regarded as the last marine amid of its lineage, or one of the first species of this taxon to venture into freshwater. Grande and Bemis (1998) described *P. mexicana* from marine waters, based on pycnodont jaw teeth in its stomach region, yet we do not discard a possible freshwater provenience for this species. The same goes for the sarcopterygian *Axelrodichthys* cf. *A. araripensis* reported in Tlayúa by Espinosa-Arrubarrena *et al.* (1996). There are several examples in the literature of well-documented marine and freshwater species of actinistians (coelacanths); therefore we think that there is need for more detailed studies to determinate the correct freshwater or marine origin of the remains of this coelacanth.

The Autochthonous Fishes (Living in the Lagoon)

Besides the toxicity (due to hypersalinity) of the bottom waters of the basin in which the Tlayúa deposit was formed, the upper part of the water column was able to support a distinctive fish fauna. This is evidenced by the presence of numerous fish coprolites and regurgitations, where fish parts and or invertebrates structures can be observed (see Feldmann *et al.* 1998). Another line of evidence that speak for an autochthonous fish fauna, rests on several forms of clupeomorphs and some interesting ichthyodectids that show a large array of sizes, that possibly represent life cycles, in which series of larval to adult stages have been recorded. Also, there are well-documented examples of "fish to fish" predation in the surface waters of the deposit, in which the preyed fish can be still seen in the mouth of the fish predator. In other specimens the preys can be seen intact inside the digestive tracts, suggesting that the predator caught its prey within the lagoon or near to the site of deposition.

There is additional evidence to show that enough "life" could thrive temporally in the area of deposition, which probably could also include several reptiles (e.g. crocodiles, and one form of a semi-aquatic sphenodontid). In addition, it is thought that upon the death of the fish, arthropods such as isopods (Feldmann *et al.* 1998) and a large array of benthic miliolid foraminifera could have scavenged on their corpses (Espinosa-Arrubarrena and Applegate, 1996). Recently, the possibility of life on the bottom of the Tlayúa "lagoon" was reported by Guerrero-Arenas (2004), where she describes five distinctive ethological categories of benthic "invertebrate tracks" from the upper part of the Middle Member of the Tlayúa Formation.

Open Ocean Fishes

It is possible that some large ichthyodectids, *Rhacolepis*-like forms and elopid allies could thrive in open oceanic conditions and that whether "carried" or "swimming", reached the Tlayúa lagoon during high tides or storms.

5.5. The Reptiles of the Tlayúa Quarry (The Insular Concept)

After the fish fauna, fossil reptiles represent the second most important taxonomic group present in the paleobiota of Tlayúa. Reynoso *et al.* (2000) reported the presence of at least five major reptilian groups. Isolated bones similar to *Nychtosaurus* and *Pteranodon* represent pterosaurs. Recently, a complete articulated pterosaur specimen was found. Crocodiles are known (two specimens, one a complete juvenile, and an almost complete disarticulated adult, which belong to the same taxon similar to the Late Jurassic *Crocodileimus*). Turtles are represented by four complete skeletons, three adults and one hatchling (in the process of study). As a late Mesozoic record the sphenodontian fauna consists of: *Pamizinsaunus tlayuaensis* Reynoso, 1997, a terrestrial sphenodontid with its body and tail covered with small rounded osteoderms, as well as a fat tail similar to that of the living Gila monster (*Heloderma suspectum*), an adaptation that relates this specimen to dry open environments; and *Ankylosphenodon pachyostosus* Reynoso 2000, an unusual herbivorous sphenodont with a short and very stout pachyostotic ribs suggesting a semi-aquatic habitat. Two lizards are also present, *Huehuecuetzpalli mixtecus* Reynoso, 1998, a terrestrial lizard, that even retaining primitive features, resembles modern iguanids, and *Tepexisaurus tepexii* Reynoso and Callison, 2000, another terrestrial lizard, modern in appearance.

Evidently, the herpetofauna of Tlayúa bears representatives of several adaptations: flight (pterosaurs); semi-aquatic and amphibian life (one sphenodontian *Ankylosphenodon pachyostosus* and the crocodilians respectively); and strictly terrestrial, possibly adapted to dry open environments (one sphenodontian *Pamizinsaurus tlayuaensis* and the two lizards).

It is important to add this reptilian information to the paleoecological evidence derived from the fish fauna, due to the interesting mixture of organisms from different environments recorded in Tlayúa, mainly to explain the origin and transportation of each of the fossil elements. As the first pterosaur evidence found in the quarry was isolated bones, it was believed that the remains had reached the deposit floating from comparatively long distances. Recently (June 2004), a complete articulated specimen was found indicating that, as in Solnhofen, some pterosaurs may have sunk to the bottom of the lagoon without disarticulation by the strong storms or monsoon winds. Unfortunately, the turtles have not been studied enough to disclose their possible (terrestrial or marine) origin, and therefore the possibilities of their transportation into the Tlayúa beds is controversial. In the case of the crocodiles, their amphibian habits suggest that they could had lived so close to the deposit that their transportation should not been seen as "problematic", in terms of the absence of fraction transport structures within the quarry limestones.

The information derived from the strictly terrestrial lepidosaurs (one sphenodentid and two lizards) represents a strong argument for considering the proximity of land to the depositional environment. Regarding this land mass adjacent to Tlayúa Reynoso *et al.* (2000) also reported morphological, phylogenetic, and biogeographical analyses of these terrestrial sphenodontians and lizards, showing that: a) the taxa bear a considerably different morphology from their ancestral types; and b) they seem to be preserved as relicts in relatively late deposits.

The coexistence of relatively primitive forms with atypical morphologies is probably explained by the evolution of the lineages in an insular system. Where isolation preserves ancient forms and genetic drift plays an important role in fixation of evolutionary novelties (Reynoso *et al.* 2000). This information seems to be consistent with the presence of Albian shallow and/or deep waters in south-central México. Therefore, the influence of the terrestrial elements, present in the mixed assemblage at Tlayúa, seems to correspond to an insular system rather than to a well- connected extension of the southern extreme of "middle" Cretaceous North America. However, the later possibility of a North American peninsula cannot be ruled out.

6. Paleoecological Models of Tlayúa

Since the Tlayúa "Lagerstätte" was first reported by Applegate and Espinosa Arrubarrena (1982) one of the main goals in all subsequent investigations has been focused in the study of the paleoenvironmental and post-depositional conditions that produced the remarkable preservation found at the Tlayúa Quarry. The attempts to interpret the paleoenvironments represented at Tlayúa include a growing list of investigations (Applegate, 1987; Martill, 1989; Malpica-Cruz *et al.* 1989; Pantoja-Alor, 1992; Espinosa-Arrubarrena and Applegate, 1996; Feldmann *et al.* 1998, Kashiyama *et al.* 2004; and Guerrero-Arenas, 2004).

For a complete discussion of the different models suggested for the paleoenvironments of Tlayúa see Espinosa-Arrubarrena and Applegate (1996) and Kashiyama *et al.* (2004). For the purpose of this chapter, we will only discuss two groups of ideas that one way or another encompass the paleoecological and sedimentological information derived from past and recent investigations.

6.1. Shallow Water Back-Reef Lagoonal Interpretation

In this category, the Tlayúa deposit represents a shallow coastal lagoon (intertidal zone) behind a barrier reef that according to Pantoja-Alor (1992) was connected to the open sea. Within the same context (a coastal lagoon, behind a barrier reef), Applegate (1987), as well as Espinosa-Arrubarrena and Applegate (1996), proposed a double-enclosed lagoon in which the sedimentary environment regarded was an extremely shallow closed lagoon with stagnant, anaerobic, and hypersaline water (Figure 6). According to Espinosa-Arrubarrena and Applegate , 1996). Evidently, these ideas call for shallow basins, an interpretation based on the observation of desiccation cracks (Malpica-Cruz *et al.* 1989; Pantoja-Alor, 1992), evidence that has been challenged by Kashiyama *et al.* (2004), regarding the supposed desiccation structures as synaeresis cracks. Because of our recent investigations in Figure 7 we show another example of what we think may represent a real "mud crack" depicting an episode in which the lagoon reached subaereal exposure. Also, these shallow water back-reef lagoon interpretations call for the presence of cyanobacterial mats (see section on cyanobacterial mats in this chapter) bracketing the lower depth of the lagoon in no more than 60 m, an estimated depth that in clear subtropical waters cyanobacteria can still photosynthesize (Barthel *et al.* 1990).

6.2. Open Marine Basin with Storm-Dominated Sedimentation Model

As part of the recent investigations in Tlayúa, Kashiyama *et al.* (2004) combined statistically based microfacies-transition analysis with spectral analysis of depth-series measurements, in an attempt to better understand the lithologies and other taphonomical factors involved in the paleoenvironmental setting. In variance with the shallow water, back-reef lagoonal interpretations, for Kashiyama *et al.* (2004) the Tlayúa deposit represents an open marine basin with storm-dominated sedimentation and bottom waters with restricted circulation. They also found a repetitive pattern in the sedimentary sequence, interpreted as Milankovitch cycles, attributed to double-monsoon influences from both the northern and southern hemispheres (Kashiyama *et al.* 2004). Although, this model discards tidal deposits some of the characteristics of this shallow-water type of deposit, as described by Feldmann *et al.* (1993) are acknowledged by Kashiyama *et al.* (2004), e.g. terrestrial and marine assemblages present in the deposit. Evidently, this open marine basin model calls for the lack of any trace of shallow water deposit features (e.g. cyanobacterial mats, desiccation cracks, etc.), elements that we think are still controversial. In addition, the model calls for a site of deposition located in a distal part of a shelf, an oceanic basin, or an intracratonic sea. This distal-shelf and basin sedimentation conflicts with the presence of small terrestrial fossils (lizards, spiders and insects). As a possible solution to this inconsistency Kashiyama *et al.* (2004), theorize a re-inundated rough Karst topography, in which a basin trapping storm-induced suspension deposit fairly close to

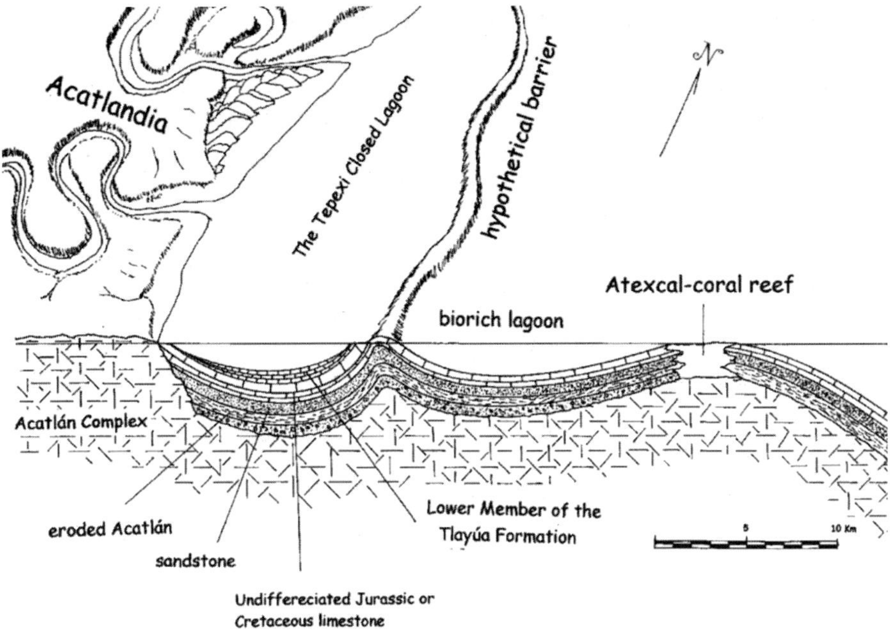

FIGURE 6. Sketch map of the paleoenvironment of Tlayúa.

a terrestrial source could occur. According to this idea, a deep and steep depression in proximity to land would receive only occasional storm-induced suspended sediments and form laminae in the bottom of the depression. Furthermore, if the relict karst topography were extreme enough, then the basin could have been protected from physical disturbance of deposits and from the formation of sedimentary structures due to traction transport.

As in previous models, Tlayúa has been compared with other well-known fossil localities (e.g. Solhhofen). Kashiyama *et al.* (2004) proposed the intracratonic Mowry Shale of Wyoming, as a setting that could explain the paleoenvironmental conditions that prevailed in Tlayúa. We believe that in paleocological terms the biotas of both units (Tlayúa and the Mowry Shale) are widely different. Tlayúa is much richer and bears an almost exclusive "shallow water" fauna (see section on the paleobiota of Tlayúa).

In an attempt to stress the significance of the nature of the paleontological contents in a study of a deposit like Tlayúa, we will discuss the fish fauna of what, to our understanding, represents one of the few well-known Cretaceous oceanic (deep water) fish fauna. The chalk deposits of the Moorevile Formation of Alabama (see Applegate, 1970) are fossiliferous enough to serve as a meaningful contrast with what has been recorded in the "shallow water" Middle Member of the Tlayúa Formation. At the Moorvile Chalk the following fish groups have been recorded:

a) Deep water chimaerids (at least at least three species).
b) Several species of pelagic sharks (measuring over three meters long).
c) A large non-reef pycnodont (e.g. *Hadrodus*).
d) At least three forms of pachyrhizodontids (strong fast swimmers, that can be considered equivalents to present day tuna).
e) Two species of large ichthyodectids (e.g. *Ichthyodectes* and *Xiphactinus*). Here we note that these large ichthyodectids (over two meters long) are very different from the comparatively small forms that occur in Tlayúa (see Alvarado-Ortega, 1998, 2004).
f) Scales of deep-water beryciformes.
g) The very common presence of scales belonging to myctophids (lantern fishes), which can be considered one of the best indicators of the deep-water environment where the Moorvile Chalk was deposited.

It is interesting to mention that none of these groups of fishes are found in Tlayúa in a collection that to date is represented by more that 5000 fishes (see section on the paleobiota of Tlayúa).

We can also mention the presence of abundant large marine reptiles, like mosasaurs and gigantic turtles in the Mooreville, (see Russell, 1970). Such large animals are absent in Tlayúa, that on the other hand, shows a reptilian fauna mostly composed of small terrestrial lizards; several flying (coastal) pterosaurs; one small (possibly amphibian) sphenodontian; two medium to small sized crocodiles; and three small turtles (that possibly can be fresh water or terrestrial).

6.3. The Freshwater Signal

Recently, investigations based on isotopic analysis, focused on the marine or fresh water nature of the limestones in the deposit have been carried on. Guerrero-Arenas (2004) preliminary results show that, at least in the upper layers of the Middle Member of the Tlayúa Formation (upper part of the section), the "lagoon" was highly influenced by fresh water, with a scarce but important benthic community (with five distinctive ethological categories). Part of these results is in variance with some of the previous ideas proposed for the ("anoxic-bottom toxic") Tlayúa beds. Mostly based on in paleontological evidence Espinosa-Arrubarrena and Applegate (1996) and Feldmann *et al.* (1998), recognized the influence of the fresh water in Tlayúa (mainly the presence in the paleobiota of an odonate dragonfly nymph, a small diptera "may fly" and many fresh water fishes). Nevertheless, nobody has yet elucidated if this "fresh water signal" is unique to a single layer, or if it is sequential and if it can be traced (horizontally) along the entire quarry.

As it can be seen, there are still many inconsistencies and contradictions among the different ideas regarding the paleoenvironmental and post-depositional conditions that produced Tlayúa. Every new result confirms the complex nature of this extraordinary fossil deposit, and calls for more interdisciplinary investigations.

FIGURE 7. Cracked slab from the Tlayúa Quarry.

7. Final Remarks

It is evident that the Tlayúa Quarry, in Tepexi de Rodríguez, Puebla State, Mexico, is a unique deposit that shows similarities with other famous Mesozoic fossiliferous localities of the world (e.g. Solnhofen, Cerin, and Nusplinger, among others). It is also clear that the deposit bears particularities that call for more thorough studies aimed in those details that today are forming part of the polemic issues that have arose from the results obtained from the recent investigations discussed herein.

From our perspective, one important point of contention corresponds to the shallow water *versus* oceanic nature of the deposit. This controversy, to be considered a shallow water deposit, requires adequate characterization of the presence of cyanobacterial mats and/or the discovery of undeniable evidence of evaporitic minerals or desiccation cracks. In the case of the deep water oriented model, it is important to study and re-evaluate the paleoecological nature of the biota, to find a significant minimal number of "pelagic" or "deep water dwellers" within the more than 6000 fossils collected to date.

In addition, to support the "open marine, deep water concept" it is very important to locate the setting of the deposit in context with the regional geology of the middle Albian of this part of Mexico, particularly, when the model needs to explain the mixed assemblage (marine, fresh water and terrestrial organisms). These future investigations should take into consideration the tectonic events known for eastern Mexico, associated with the formation of the Gulf of Mexico, and their consequent differential block faulting. This tectonic setting could serve as a better scenario to conceive comparatively deep basins adjacent to land, with borders shallow enough to develop coral reefs (Morán-Zenteno, pers. com., July, 2004).

Another set of research involves sedimentological and geochemical analyses, these include: a) the problem of the fining-upwards laminations in a close lagoonal type of environment (the shallow water-deep water oceanic basin controversy); b) the diagenetic or authigenic origin of features such as the hematite found along the bedding planes of the limestones slabs; and c) the presence of pressure-solution indicators, synaeresis cracks (Figure 7).

In addition, the record of a fresh water isotopic signal within the section requires an extensive (layer by layer) analysis of the Middle Member of the Tlayúa Formation, with the objective to find any cycles of eventual fresh water influence in the deposit. These studies, as well as the sedimentological investigations involve detailed sections across the quarry, to asses all vertical and lateral changes present in the outcrop that measures 700 m. from one extreme of the quarry complex to the other.

Finally, the magnetic polarity stratigraphy, combined with ammonite biostratigraphic age assignments performed by Benammi *et al.* (2004, in prep.), have bracketed the age of Tlayúa between Middle and Upper Albian (100-105 Ma). However, there is still some discrepancy between this age assignment and the Aptian? age assessed by Kashiyama *et al.* (2004), based on inner shelf benthic foraminifera. Furthermore, to date, there are two differing estimates of rock accumulation rates, both obtained through different methods. For the spectral analysis of depth-series measurements of magnetic susceptibility of Kashiyama *et al.* (2004) the rate is 2.0 cm/kyr, as for Benammi *et al.* (in press), based on magnetic polarity stratigraphy, the estimated sedimentation rate is 0.6 cm/kyr. These rates need to be analyzed and re-evaluated due to the importance of this aspect in the understanding of the paleoenvironmental setting and its dynamics.

After finishing this chapter, it is clear that the Tlayúa Quarry is still one of the most important deposits with exceptional fossil preservation in the world. After more that twenty years of paleontological investigations, the abundance and diversity of fossils new to science has not decreased and are in need of taxonomic description. The origin and nature of this important Mexican "Konsevat Fossil-Lagerstätten" is still not well understood and, therefore its sedimentary, post-depositional and geochemical processes represent a wealth of present and future investigations.

Acknowledgments

We thank Francisco Vega and María del Carmen Perrilliat Montoya, who invited us to participate in this book. The Faunal and Floral list of Tlayúa was obtained through the generous collaboration of Francisco Vega, María del Carmen Perrilliat, Blanca Estela Buitrón, Víctor Hugo Reynoso and Enrique Martínez Hernández. We would also like to acknowledge the inspiring discussions with Dr. Dante Morán Zenteno and M. C. Gabriela Guzzy Arredondo and Rosalía Guerrero Arenas, regarding the Paleoenviroments and Geological settings of Tlayúa. Ing. Julio Caballero Corona elaborated on the paleoenviromental sketch map, based on paleogeography of Tlayúa. Antonio Altamira-Gallardo took all the photographs shown throughout this chapter.

References

Alencáster, G., 1973, Una nueva especie de *Toucasia* en el Cretácico medio de los estados de Oaxaca y Puebla, *Univ. Nal. Autón. Méx., Inst. Geol., Paleont. Mex.*, 36:4-20.

Alvarado-Ortega, J., 1998, Ichthyodectiformes mexicanos: Un análisis de su diversidad. Avances en Investigación: Paleontología de Vertebrados, *Univ. Autón. Edo. Hidalgo, Pub. Esp.*, 1:109-119.

Alvarado-Ortega, J., 2004, Description and relationships of a new Ichthyodectiform fish from the Tlayúa Formation (Early Cretaceous: Albian), Puebla, Mexico, *J. Vert. Paleont.*, 24:502-813.

Alvarado-Ortega, J., and Applegate, S. P., 1999, The fossil conservation of the Tlayúa Quarry, in Puebla State, Mexico, *33rd Ann. Meet., South-Central Section, Geol. Soc. Am., Abstracts*, p. A1.

Alvarado-Ortega, J., and Espinosa-Arrubarena, L., 2001, Los ionoscopidos de la Cantera Tlayúa, Tepexi de Rodríguez, Puebla, *Reunión Anual 2001, Un. Geof. Mex., and XII Reunión Nacional SELPER*, Resúmenes, p. A3.

Alvarado-Ortega, J., and Espinosa-Arrubarena, L., 2004, Conservación de Tejidos Blandos en los fósiles de la Cantera Tlayúa, *IX Congreso Nacional de Paleontología, Soc. Mex. Paleont.*, Resúmenes, p. 16.

Alvarado Ortega, J., González-Rodríguez, K. A., Blanco-Piñón, A., Espinosa-Arrubarrena, L., and Ovalles-Damián, E., 2005, Mesozoic Osteichthyans of Mexico, This volume.

Applegate, S. P., 1970, The Vertebrate Fauna of the Selma Formation in Alabama. Pt. 8, The Fishes. *Fieldiana, Geol. Mem.*, 3:385-433.

Applegate, S. P., 1987, A preliminary study of the Tlayúa Quarry near Tepexi de Rodríguez, Puebla, *Soc. Mex. Pal., Rev.*, 1:40-54.

Applegate, S. P., 1992, A new genus and species of pycnodont, *Tepexichtys aranguthyorum* from the Cretaceous (Albian) of central Mexico, Tepexi de Rodríguez, Puebla, *Univ. Nal. Autón. Mex., Inst. Geol., Rev.* 10:164-178.

Applegate, S. P., 1996, An overview of the Cretaceous fishes of the quarries near Tepexi de Rodríguez, Puebla, Mexico, in: Arraita, G., and Viohl, G. (eds.) *Mesozoic Fishes - Systematics and Paleoecology* -, Verlag Dr. Friederich Pfeil. Munich, p. 529-538.

Applegate, S. P., and Espinosa-Arrubarrena, L., 1982, Lithographic limestone-like deposits in Tepexi de Rodríguez, Puebla, Mexico, *Univ. Nal. Autón. Méx., Inst. Geol., 42th Ann. Meet., Soc. Vert. Paleont., Fieldtrip, guidebook*, 39 p.

Applegate, S. P., González-Rodríguez, K., and Alvarado-Ortega, J., 2000, Fish fauna of the Tlayúa quarries, in: *Paleontological highlights of the Mixteca Poblana in Central Mexico. Field Guide* (Espinosa-Arrubarrena, L., Montellano-Ballesteros, and M. and Applegate, S. P., coordinators), *60th Annual Meeting of Society of Vertebrate Paleontology*, Univ. Autón. Edo. Hidalgo, Spec. Pub. p. 95-105

Barthel, K. W., Swinburne, N. H. M., Conway Morris, S., 1990, *Solnhofen – A study in Mesozoic palaeontology-*. Cambridge University Press, 236 pp.

Bartram, A. W. H., 1977, The Macrosemiida, a Mesozoic family of Holostean fishes, *British Mus. Nat. Hist., Bull.*, 29:137-234.

Benammi, M, Urrutia-Fucugauchi, J., and Alvarado-Ortega, J., 2003, Magnetic polarity stratigraphy of the Tlayúa Formation near Tepexi de Rodríguez, State of Puebla, Mexico, *Reunión Anual 2003, Un. Geof. Mex.*, Resúmenes, p. 56.

Benammi, M., Urrutia-Fucugauchi, J., and Alvarado-Ortega, J., 2004, Magnetic polarity stratigraphy of the

"Mexican Solnhofen": The Tlayúa Formation near Tepexi de Rodríguez, State of Puebla, Mexico, *European Geosciences Union 2004, Geoph. Res.*, Abstracts, 6, p. 56.

Bernier, P., Gaillard, C., Gall, J. C., Barale, G., Bourseau, J. P., Buffetaut, E., and Wenz, S., 1991, Morphogenetic impact of microbial mats on surface structures of Kimmeridgian micritic limestones (Cerin, France), *Sedimentology*, 38:127-136.

Besse, J., and Courtillot, V., 2002, Apparent and true polar wander and the geometry of the geomagnetic field over the last 200 Myr., *J. Geophys. Res.*, 107, 2300, doi:10.1029/2000JB000050.

Cantú-Chapa, A., 1987, Las amonitas del Albiano Superior de Tepexi de Rodríguez, Puebla, *Soc. Mex. Paleont., Rev.*, 1:159-160.

Cobban, W. A., Merewether, E. A., Fouch, T. D., and Orbadovich, J. D., 1994. Some Cretaceous shorelines in the Western Interior of the United States, in: *Mesozoic Systems of Rocky Moutain Region* (Caputo, M. V., Peterson, J. A., and Franczyk, K. J., eds.), USA, p. 393-425.

Chiocchini, M., Mancinelli, A., and Romano, A., 1984, Stratigraphic distribution of benthic foraminifera in the Aptian, Albian and Cenomanian carbonate sequences of the Aurunci and Ausoni mountains, southern Lazio, Italy, *BENTHOS'83/2eme Symposium international sur les Foraminiferes benthiques Bulletin des Centres de Recherches Exploration-Production Elf-Aquitaine, Mem.*, 6:167-181.

Espinosa-Arrubarena, L., and Applegate, S. P., 1996, A paleoecological model of the vertebrate bearing beds in the Tlayúa Quarries, near Tepexi de Rodríguez, Puebla, México, in: *Mesozoic Fishes-Systematics and Paleoecology* (Arratia, G., and Viohl, G., eds.), Verlag Dr. Friedrich Pfeil. München, p. 539-550.

Espinosa-Arrubarena, L., Applegate, S. P., and González-Rodríguez, K. A., 1996, The first Mexican record of a Coelacanth (Osteichthyes: Sarcopterygii) from the Tlayúa Quarries near Tepexi de Rodríguez, Puebla, with a discussion on the importance of this fossil, Sixth North American Paleontological Convention, *Paleontol. Soc. Spec. Publ.*, 8, p. 116.

Feldmann, R. M., Vega, F. J., Applegate, S. P. and Bishop, G. A., 1998, Early Cretaceous arthropods from the Tlayúa Formation at Tepexi de Rodríguez, Puebla, Mexico, *J. Paleont.*, 72:79-90.

Gale, A. S., Kennedy W. J., Burnett J. A., Caron M., and Kidd B. E., 1996, The Late Albian to Early Cenomanian succession at the Mont Risou near Rosans (Drome, SE France): an integrated study (ammonites, inoceramids, planktonic foraminifera, nannofossils, oxygen and carbon isotopes). *Cret. Res.*, 17, 515-606.

González-Rodríguez, K. A., S. P. Applegate, and L. Espinosa-Arrubarena, 2004. A new Macrosemiid (Pisces: Neopterygii-Alecostomi) from the Albian of Mexico, *J. Vert. Paleont.*, 24:281-289.

Gradstein, F. M., Agterberg, F. P., Ogg, J. G., Hardenbol, J., van Veen, P., Thierry, J., and Huong, A., 1995, A Triassic, Jurassic, and Cretaceous time scale. In: *Geochronology, Time Scales, and Global Stratigraphic Correlation* (Berggren, W. A., Kent, D. V., Aubry, M. P., Hardenbol, J., eds.), *Soc. Econ. Paleont. Mineral., Spec. Publ.*, 54, 95–126.

Grande, L., and W. E., Bemis, 1998, A Comprehensive Phylogenetic Study of Amiid fishes (Amiidae) based on comparative skeletal anatomy. An Empirical Search for Interconnected patterns on Natural History, *Soc. Vert. Paleont., Mem.* 4, (*Suppl. J. Vert. Paleont.,* 18:690).

Guerrero-Arenas, R., 2004, Icnofósiles de invertebrados de la Formación Tlayúa en Tepexi de Rodríguez, Puebla, México, Unpublished M.S. thesis, Universidad Nacional Autónoma de México.

Kashiyama, Y., Fastovsky, D. E., Rutherford, S., King, J., and Montellano, M., 2004, Genesis of a locality of exceptional fossil preservation: paleoenviroments of Tepexi de Rodríguez (mid-Cretaceous, Puebla, Mexico), *Cret. Res.*, 25:153-177.

Malpica-Cruz, V. M., Pantoja-Alor, J., and Galguera-Rosas, G., 1989, Microfacies de la Cantera Tlayúa, Puebla, México, *Simposio sobre Geología Regional de México, Univ. Nal. Autón. Méx., Inst. Geol., Mem.*, 3: 53-56.

Martill, D., 1989, A new Solnhofen in México, *Geology Today, January-February*, p. 25-28.

McFadden, P. L., and McElhinny, M. W., 1990, Classification of the reversal test in palaeomagnetism, *Geoph. J. Int.*, 103:725-729.

Ortega-Guerrero, B . 1989. Paleomagnetismo y geología de las unidades clásticas mesozoicas del area de Totoltepec-Icaquixtla, estados de Puebla y Oaxaca, Unpublished M.S. thesis, Universidad Nacional Autónoma de México.

Pantoja-Alor, J., 1992, Geología y paleoambientes de la Cantera Tlayúa, Tepexi de Rodríguez, estado de Puebla, *Univ. Nal. Autón. Méx., Inst. Geol., Rev.*, 9:156-176.

Reynoso, V. H., 1997, A "beaded" sphenodontian (Diapsida: Lepidosauria) from the Early Cretaceous of Mexico, *J. Vert. Paleont.*, 17:52-59.

Reynoso, V. H., 1998, Acatlántida y el origen insular de la fauna terrestre de Tlayúa, *Univ. Autón. Edo. Hidalgo, Pub. Esp.*, 1:4-11.

Reynoso, V. H., 2000, An unusual aquatic sphenodontian from the Tlayúa Formation (Albian), Central México, *J. Paleont.,* 74:133-148.

Reynoso, V. H., and Callison, G., 2000, A new scincomorph from the early Cretaceous of Puebla, Mexico, *Zool. J. Linn. Soc.*, 130:183–212.

Reynoso, V. H., Cabral-Perdomo, M. A., and Clark, J., 2000, The Reptiles of the Tlayúa Formation, in: *Paleontological highlights of the Mixteca Poblana in Central Mexico. Field Guide* (Espinosa-Arrubarena, L., Montellano-Ballesteros, M., and Applegate, S. P., coordinators.), *60th Annual Meeting, Soc. Vert. Paleont., Univ. Autón. Edo. Hidalgo, Pub. Esp.*, 1:106-111.

Russell, D., 1970 Vertebrate Fauna of the Selma Formation in Alabama. Pt. 7, The Mosasaurs, *Fieldiana, Geol. Mem.*, 3:365-380.

Seibertz, E., and Buitrón, B. E., 1987, Paleontología y Estratigrafía de los *Neohibolites* del Albiano de Tepexi de Rodríguez, Estado de Puebla (Cretácico Medio, Mexico), *Soc. Mex. Paleont., Rev.*, 1:285-299.

Seibertz, E., and Spaeth, C., 2002, Cretaceous belemnites of México III. The Albian Neo- and Mesohibolites of the "Mexican Solnhofen" Tepexi de Rodríguez (state of Puebla) and their biostratonomy (Lower Cretaceous), *N. Jb. Geol. Paläont. Abh.*, 225:55-74.

Urrutia-Fucugauchi J., 1988, Paleomagnetic study of the Cretaceous Morelos Formation, Guerrero State, southern Mexico, *Tectonophysics*, 147:121-125.

Viohl, G., 1990, Solnhofen Litographic Limestones, in: *Paleobiology- A Synthesis-* (Briggs, D. E. G., and P. R. Crowther, eds.), Blackwell Scientific Publications, p. 285-289.

Vila, C., 1999, Litographic Limestone: a close encounter between paleontology and printmaking. in: *Third International Symposium on Litographic Limestones* (Renesto S., ed.), *Rev. Mus. Civ. Sci. Nat. "Enrico Caffi", suppl.* Vol. 20, p. 133-136.

Wiedmann, J., and Owen H. G., 2001, Late Albian ammonite biostratigraphy of the Kirchrode I borehole, Hannover, Germany, *Palaeogeogr. Palaeoclimatol. Palaeoecol.*, 174:161-180.

Appendix: Faunal and Floral List of the Tlayúa Quarry

EUBACTERIA

Cyanobacteria

PROTOCTISTA

Calcareous nannoplankton indet.

Calcisphers indet.

Foraminiferida (Reported in Kashiyama *et al.*,2004:155)

Derventina filipescui Neagu

Earlandia? conradi Arnaud-Vannean

Nautiloculina sp.

Praechrysalidina infracretacea Luperto-Sinni (probable)

Rumanoloculina robusta Neagu

Sabaudia minuta HofKer

Spiroloculina cretacea Reuss

Trochamminoides coronus Loeblich and Tappan

Valvulineria sp.

PLANTAE

Phaeophyceae, brown algae

Two indet. types. One similar to *Sargassum*, the other is a "thalloid" algae.

Gymnospermophyta

Bennettitales

Zamites

Coniferophyta

cf. *Araucaria*

cf. *Frenolopsis*

cf. *Brachyphyllum*

small pieces of indeterminated wood

POLLEN

Under study by E. Martínez-Hernández and E. Ramírez-Arriaga at Instituto de Geología, UNAM.

ANIMALIA

Porifera

Two different forms indet

Cnidaria

Anthozoa

Gorgonian corals, two kinds, indet.

Annelida

Few specimens, indet.

Mollusca

Bivalvia

Inoceramus.

Pectinids indet.

Ostreids indet.

Gastropoda

Several forms that have not been determinated.

Cephalopoda

Belemnoidea

Neohibolites minimus obtusus Stolley, 1911

N. minimus pinguis Stolley, 1911

N. minimus claviformis Seibertz and Buitrón, 1987

N. praeultimus Spaeth, 1971

Mesohibolites semicanaliculatus (Blainville, 1827)

Ammonoidea

Mortoniceras sp.

Hysteroceras sp.

Anisoceras sp.

Arthropoda.
Arachnida
Atocatle ranulfoi Feldmann, Vega, Applegate and Bishop, 1998
Insecta
Odonata
Ixtahua benjamini Feldmann, Vega, Applegate and Bishop, 1998
Unidentified "insect larva", may fly?
Malacostraca
Isopoda
Archaeoniscus aranguthyorum Feldmann, Vega, Applegate and Bishop, 1998
Decapoda
Protaegla minuscula Feldmann, Vega, Applegate and Bishop, 1998
Brachyura
Tepexicarcinus tlayuaensis Feldmann, Vega, Applegate and Bishop, 1998
Ostracoda?

Echinodermata
Echinoidea
At least two forms that have not been determinated
Holoturoidea
Eight taxa that are in the process of description, under study by S. P. Applegate and B. E. Biutrón at Instituto de Geología, UNAM.
Asteroidea
Two or three forms not yet identified
Ophiuroidea
One species of "serpent star" see Martín-Medrano and García-Barrera (in this book)
Crinoidea
One form of an unidentified free living crinoid

VERTEBRATA

Pisces

Fishes represent around 70 % of the fossils recovered in Tlayúa Quarry. Applegate (1996), Applegate and collaborators (2002), and Alvarado-Ortega and collaborators (in this book) published lists of fishes in Tlayúa.

Reptilia

Chelonia

At least two taxa, that include three complete adult specimens and one hatchling.

Pterosauria

More than 10 isolated bones and one articulated specimen, recently collected, some elements have been assigned tentatively to *Nychtosaurus* and *Pteranodon*.

Crocodylia

Two specimens (one juvenile and the other is an adult individual) similar to *Crocodileimus*, but still are considered indet.

Sphenodonta

Pamizinsaurus tlayuaensis Reynoso, 1997
Ankylosphenodon pachyostosus Reynoso, 2000

Lacertilia

Huehuecuetzpalli mixtecus Reynoso, 1998
Tepexisaurus tepexii Reynoso and Callison, 2000

Index